HEALTHCARE TECHNOLOGIES SERIES 19

Neurotechnology

IET Book Series on e-Health Technologies – Call for Authors

Book Series Editor: Professor Joel P. C. Rodrigues, the National Institute of Telecommunications (Inatel), Brazil and Instituto de Telecomunicações, Portugal

While the demographic shifts in populations display significant socio-economic challenges, they trigger opportunities for innovations in e-Health, m-Health, precision and personalized medicine, robotics, sensing, the Internet of Things, cloud computing, Big Data, Software Defined Networks, and network function virtualization. Their integration is however associated with many technological, ethical, legal, social and security issues. This new Book Series aims to disseminate recent advances for e-Health Technologies to improve healthcare and people's wellbeing.

Topics considered include Intelligent e-Health systems, electronic health records, ICT-enabled personal health systems, mobile and cloud computing for eHealth, health monitoring, precision and personalized health, robotics for e-Health, security and privacy in e-Health, ambient assisted living, telemedicine, Big Data and IoT for e-Health, and more.

Proposals for coherently integrated International multi-authored edited or co-authored handbooks and research monographs will be considered for this Book Series. Each proposal will be reviewed by the Book Series Editor with additional external reviews from independent reviewers. Please email your book proposal for the IET Book Series on e-Health Technologies to: Professor Joel Rodrigues at joeljr@ieee.org or joeljr@inatel.br

Neurotechnology

Methods, advances and applications

Edited by
Victor Hugo C. de Albuquerque,
Alkinoos Athanasiou and Sidarta Ribeiro

The Institution of Engineering and Technology

Published by The Institution of Engineering and Technology, London, United Kingdom

The Institution of Engineering and Technology is registered as a Charity in England & Wales (no. 211014) and Scotland (no. SC038698).

First published 2020

The Institution of Engineering and Technology
Michael Faraday House
Six Hills Way, Stevenage
Herts, SG1 2AY, United Kingdom

www.theiet.org

British Library Cataloguing in Publication Data
A catalogue record for this product is available from the British Library

ISBN 978-1-78561-813-0 (hardback)
ISBN 978-1-78561-814-7 (PDF)

Typeset in India by MPS Limited

Contents

About the editors

Victor Hugo C. de Albuquerque [M'17, SM'19] is a professor and senior researcher at the University of Fortaleza, UNIFOR, Brazil and director of Data Science at the Superintendency for Research and Public Safety Strategy of Ceará State (SUPESP/CE), Brazil. He has a Ph.D. in mechanical engineering from the Federal University of Paraíba, an M.Sc. in teleinformatics engineering from the Federal University of Ceará, and he graduated in mechatronics engineering at the Federal Center of Technological Education of Ceará. He is currently a full professor of the Graduate Program in Applied Informatics of UNIFOR and leader of the Industrial Informatics, Electronics and Health Research Group (CNPq). He is an expert, mainly, in IoT, machine/deep learning, pattern recognition, and robotics.

Alkinoos Athanasiou is a board-certified neurosurgeon and a post-doc researcher at the Medical Physics Lab, School of Medicine, Aristotle University of Thessaloniki, Greece. He is employed as associated scientific staff at the Hellenic Open University, teaching Neuronal Restoration Engineering. He serves as elected Secretary and member of the administrative board of the Hellenic Society for Biomedical Technology. His research interests lie with neurorehabilitation and neurotechnology. He has been awarded more than 15 research grants and awards in total.

Sidarta Ribeiro is a professor of neuroscience and director of the Brain Institute at the Federal University of Rio Grande do Norte, Brazil. He has considerable experience in the areas of neuroethology, molecular neurobiology, and systems neurophysiology. He currently coordinates the Brazilian committee of the Pew Latin American Fellows Program in the Biomedical Sciences, and is a member of the steering committee of the Latin American School of Education, Cognitive and Neural Sciences.

Foreword

I am delighted to write the foreword for this edited book on "Neurotechnology: Methods, Advances and Applications". This book highlights novel neurotechnologies which have advanced the fields of neuroscience and biomedical engineering but can be also applied, often ubiquitously, to medical practice and assist health practitioners, or to everyday life scenarios.

Neurotechnology is present in many forms that range from diagnostic tools, research methods, and treatment modalities to devices and apps in everyday life. Ultimately, neurotechnology is used to benefit human neurological outcomes and mental health states and generally secure and promote objective and subjective well-being. That is sometimes achieved with loud ground-breaking advances that can transform entire fields such as neurorehabilitation. Some other times it is achieved through less obvious but gradual and incremental improvements in existing technologies and medical fields, such as in neurology, neurosurgery, and psychiatry. And then again, other times the effects of neurotechnology are even more subtle and long term such as in educational neuroscience, in affective computing and in mobile apps.

To accomplish all these feats, neurotechnology uses patient-centric and human-centric data and designs. Theory is drawn from neuroscience and tools are drawn from engineering to provide technologies that improve human life. As such, a lot of experience, techniques, data, analyses, and tools for the advancement of neurotechnology are drawn from other fields that include robotics, signal and image processing, virtual and augmented reality, computer science, biomedical sensors, and medical practice; to name but a few fields that contribute to an ever-changing technological crossover.

The range of topics covered in this book is quite extensive and every topic is discussed by experts in their own expertise. For example, contributors to the book address the advances in biomedical signal processing and neuroimaging acquisition and analysis. The book discusses different forms of examination and data analysis regarding nervous system structure and function. Such knowledge is in turn employed in virtual reality environments, futuristic brain–computer interfaces, and neurorobotics and biometric systems to design novel applications, devices, and methodologies. Educational aspects are also touched by neuroscience and neurotechnology with an emphasis on school environments but so are the ever-present mobile apps.

This book provides a window to the research and development in the field of neurotechnology in a comprehensive way and enumerates the evolutions of contributing tools and techniques. The advances and challenges are discussed with a focus on successes, failures and lessons learned, open issues, unmet challenges, and future directions. In this brave new world, it is imperative to provide stakeholders, researchers, practitioners, and students with the knowledge of the state-of-the-art and frontiers science that is neural technology.

I highly recommend this book to a variety of audiences, including *neuroscientists*, *biomedical engineers*, and *researchers* in fields that use neurotechnologies, *healthcare practitioners*, *neural technology specialists*, *academics*, as well as *STEM and medical students*. It is my hope and expectation that this book will provide an effective learning experience, a contemporary update and a practical reference for researchers, professionals and students that are interested in the advances of neurotechnology and its integration to the engineering and medical fields.

Prof. Joel J. P. C. Rodrigues, IEEE Fellow
Federal University of Piauí (UFPI), Teresina – PI, Brazil
Instituto de Telecomunicacções, Portugal

Acknowledgements

We would like to express our sincere thanks to The IET for allowing us and helping us to organize and finalize this book. The editorial office staff have been excellently supporting us and guiding us through this long endeavor and we are especially thankful for their support. Among others, we ought to specifically thank Ms Olivia Wilkins and Ms Valerie Moliere. Above all, we are grateful to all the authors who devoted effort and time out of their difficult schedules to compile high-quality chapters on their respective fields of expertise and, ultimately made this book possible. Finally, we would like to thank the reviewers for their thoughtful contributions, suggestions, and improvements. We really hope that the end result, this book on neurotechnology, will be welcomed by the scientific audience as an interesting, timely, robust, and provocative addition to a field that is on the rise.

Chapter 1

A brief introduction to neurotechnology: old challenges and new battlegrounds

Victor Hugo C. de Albuquerque[1,2], *Alkinoos Athanasiou*[3], *and Sidarta Ribeiro*[4]

Neurotechnology encompasses a myriad of different technological advances, ranging from the effective use of previous methods to ground-breaking novel approaches and applications; such technological advances are meeting older challenges and are shaping new frontiers in neuroscience, medicine, and biomedical engineering. While it can be argued that a few scientific fields have been nowadays left untouched by the technological advance, it has been neural technology that has been transforming its related fields in the recent years. Either tackling improvements in accuracy and reliability of signal analysis methods and techniques or providing patients with paralysis with a bright new future, a brief introduction to current neurotechnology has to focus on the problem-solving possibilities, improvement, and novel applications.

The wide application of neurotechnology implies that an overview cannot be, by any means, an easy task. Current advances in neurotechnology are being continuously introduced in the literature and there cannot always be a comprehensive way to keep track of them. For example, accurate, reliable, and fast approaches and methods capable of aiding medical specialists in their decision-making are always being incorporated in clinical practice and are, at times, indistinguishable from the practice itself. One could not perceive neurology or neurosurgery of today without magnetic resonance imaging (MRI) or other advanced neuroimaging methods; and yet, new approaches, algorithms, sequences, or analyses are increasingly complementing medical skill and knowledge. Furthermore, artificial intelligence and automation even offers the promise of reducing or even eliminating subjective errors, fatigue-related errors, and human limitations. This seamless integration of

[1]Laboratory of Bioinformatics (LABIO), University of Fortaleza (UNIFOR), Ceará, Brazil
[2]Data Science Director at the Superintendency for Research and Public Safety Strategy of Ceará State (SUPESP/CE), Brazil
[3]Medical Physics Laboratory, School of Medicine, Faculty of Health Sciences, Aristotle University of Thessaloniki (AUTH), Thessaloniki, Greece
[4]Brain Institute at Universidade Federal do Rio Grande do Norte, Av. Nascimento de Castro, Natal, Brazil

neural technology and neuroscientific expertise is one of the best yet examples of how new discoveries and advances pertaining to diseases and better diagnoses can greatly improve both the capabilities of the experts and, as a consequence, the quality of life of patients. Advances in neural technology should be considered of utmost importance to bring even more new discoveries and propose new methods and alternative treatments.

A significant variety of different technologies have application in many fields of neuroscience, technologies originating, for example, from automation and control theory, signal and image processing and analysis, virtual and augmented reality, computer graphics, biomedical sensors, multicriterial analysis, and data acquisition devices, among others. For example, in neuroscience, it is possible to observe a wide integration of different computational techniques that go beyond of simply improving neuronal signal analysis—those techniques have opened up new possibilities for the existing fields or created entirely new crossover fields. The first path can be examined at the case of neurorehabilitation—a field once dominated by physical therapy—which can now make use of controlled immersive virtual environments for the treatment of neurotrauma (brain and spinal cord injury), stroke, cerebral palsy, and post-traumatic disorder to name but a few examples. Neurorehabilitation can nowadays aim is to promote neuroplasticity and subsequent recovery of brain function lost due to aforementioned lesions through noninvasive brain activity modulation in a way that can be more targeted, more accurate, more personalized and soon, hopefully, faster than the traditional techniques. Neurorobotics, on the other hand, is a relatively new field that Neurotechnology made wholly possible. It deals with the development of artificial limbs controlled by body–machine interfaces and brain–machine interfaces, among other foci, and has become a fast-growing field, currently touching themes of autonomous exoskeletons, anthropomorphic high-definition movement and gestures, sensorial feedback and other easily recognized tropes from the 1980s and 1990s Sci-Fi genre.

Advances in signal acquisition, classification, and analysis have provided new insight into brain function both in normal processes and many pathological conditions; moreover, they have paved the way to real-time brain–computer interface (BCI) and to dynamic study of brain diseases. Biometrics based on electroencephalography (EEG) study chaotic brain and still manage to recognize neurophysiological individual differences as traditional biometrics do based on physical, physiological or behavioral characteristics of a human body such as fingerprint, gait, voice, iris, and gaze. Educational neuroscience to attempts to investigate neural mechanisms behind learning and attention combining cognitive, developmental and educational psychology and (neuro)technology. On top of all that, neurosciences nowadays are being empowered by technologies such as the Internet of Things (IoT) and mobile apps. Furthermore, while our understanding of how our brain works rapidly grows, new realistic computational algorithms are being considered, making it possible to simulate and model specific brain functions for development of new machine learning techniques.

This book, aptly titled *Neurotechnology: Methods, Advances and Applications*, focuses on selected specific themes to provide an overview of and insight into recent advances and future trends in the methods and applications of neural technologies used in neuroscience, neuroengineering, rehabilitation, pattern recognition, computational intelligence and education, but also having interest in other fields such as industrial engineering, control and automation engineering, and robotics. We have sought out experts to provide theoretical background, methodological work and well-established and validated empirical work dealing with these different topics and we have aimed to compile a book of public interest presenting new techniques for the evaluation, diagnosis, and treatment of neurological diseases, advances and innovations in the neurotechnology field to improve the quality of life of patients, help with a better integration into the society and increased independence. This work is addressed to neuroscience and biomedical engineering experts, as well as researchers and students in related fields, and we have aimed to provide a textbook to neurotechnology of today and a basis for the advances and future directions that are yet to come.

This book is comprised of ten (including this introduction) highly thematic chapters on the various aspects of neurotechnology. In each of the remaining chapters, an expert scientist or a group of experts, focus on a central advanced theme of each field as a vantage point, in order to explore those previously mentioned aspects.

In Chapter 2, *Nday et al.* review the **Current Trends of Biomedical Signal Processing in Neuroscience**. In the last few decades, significant progress has been made in the broad field of neural signal processing, aiming at extracting meaningful information directly from raw physiological data. In particular, the automated classification of this data has shown up as a promising strategy for assisting physicians in diagnosing controversies in neural pathologies. This chapter introduces readers to EEG pre-processing and feature extraction and then proceeds to discuss the infamous "inverse problem" and the approaches to tackling it. The authors discuss then the principles of functional connectivity and introduce the reader to network-based statistics analysis and traditional graph theory alike. Finally the authors delve into the utility of EEG analysis to help predict various disorders focusing on the psychiatric EEG data

In Chapter 3, *Thomas Schultz* reviews **Neuroimaging Acquisition and Analysis** and brings together the current state-of-the-art in various multidisciplinary solutions for neuroimaging. This chapter provides a broad overview of human *in-vivo* neuroimaging techniques, and introduces algorithms for neuroimage processing and analysis. Covered modalities include structural, functional, and diffusion MRI, positron emission tomography (PET), and magnetoencephalography (MEG). The author discusses state-of-the-art algorithms for image registration and segmentation, for statistical analysis, and for predictive modeling.

In Chapter 4, *Antoniou et al.* cover **Virtual and Augmented Reality in Neuroscience**. Neuroscience has as a study objective the interpretive interface of humans with their environment, an interface that can be recognized as the central and peripheral neural system. In that context both normal and abnormal neurophysiological function can be triggered for diagnosis, or altered through treatment by affecting evoked responses from environmental sensory input. With the advent of virtual (VR) and augmented reality (AR) as highly immersive and malleable, computer–user interfaces, a new avenue for previously unfeasible environmental input has been realized. This chapter explores the potential of VR/AR in the three axes where they are currently being deployed in neuroscience: first as sensory input modalities for evoked responses study and research, second as brain computer interfaces for seamless integration of robotic prosthetics or for treatment support (e.g., in rehabilitation,) and, finally, as modeling support tools for visualizing, simulating and exploring neurocognitive processes for research and education. Furthermore, some points about the future of immersive experiential advanced interfaces and their potential impact in neuroscience are explored.

In Chapter 5, *Rodrigues et al.* introduce and thoroughly discuss **Electroencephalogram-based Biometric Systems**. Traditional biometrics aim at recognizing the individual differences based on physical, physiological, or behavioral characteristics of a human body such as fingerprint, gait, voice, iris, and gaze. Currently, the state-of-the art methods for biometric authentication are being incorporated in various access control and personal identity management applications. While the hand-based biometrics (including fingerprint) have been the most often used technology so far, there is growing evidence that EEG signals collected during a perception or mental task can be used for reliable person recognition. However, the domain of EEG-based biometry still faces the problems of improving the accuracy, robustness, security, privacy, and ergonomics of EEG-based biometric systems and substantial efforts are needed toward developing efficient sets of stimuli (visual or auditory) that can be used for person identification in BCI systems, for personal authentication on mobile devices, VR headsets, and the Internet.

In Chapter 6, *Sciaraffa et al.* delve into **The Evolution of Passive Brain–Computer Interfaces**, with a focus on **Enhancing the Human–Machine Interaction**. Brain–computer interfaces (BCIs) translate brain activity into interaction with the physical world through computer applications and electromechanical implementations. As the authors state, in the last decade, a real revolution in the field of BCIs led from the "overt" detection of human intention to the "covert" assessment of the actual human mental states. While the first aspect is the basis of the traditional BCI systems, the latter represents the outcome of the passive BCI applications. The authors of this chapter discuss the major achievement of passive BCIs and also the concerns of passive BCIs—the necessity of monitoring human mental states driven by safety-critical application has been just the boost to the passive BCIs developing: more in general, passive BCI represents the implicit channel of information that enhances the goal-oriented cooperation of humans and machines as a whole, the so-called human–machine interaction.

In Chapter 7, *Dimitrousis et al.* touch upon **Neurorobotics** and offer a **Review of the Underlying Technologies, Current Developments and Future Directions**. Neurorobotics is a multidisciplinary field, at the crossover of computational neuroscience, traditional robotic design, and artificial intelligence. It focuses on embodied neural systems and spans scientific theory, research, development, and clinical medical practice. Autonomous systems with decision-making abilities, robots with cognitive aptitudes, neural interfaces for prosthetics, and rehabilitation are among the many applications of this game-changing field. In this chapter, the authors present neuroscience, electronics and artificial intelligence aspects of neurorobotics. It includes research, development, and experimentation taking place in virtual environment, so long as the latter explicitly aims at the development of robotic control devices to be eventually embodied in physical robotics setups. It will also focus on research activity targeting or involving the human central nervous system, so long as the sensorimotor loop is somehow involved in robotic control.

In Chapter 8, *Plate and Kubben* provide an overview of **Mobile Apps for Neuroscience**. This chapter provides the reader with an overview of the recent developments in mobile device technologies and development opportunities, focusing on iOS and Android. For several years already, smartphone applications are an upcoming wave of new technologies within the neuroscience sector. The popularization of smartphones among the general population and health-care providers has resulted in an increasing amount of available medical smartphone applications. The chapter outlines the scientific evidence regarding mobile clinical decision support systems, and elaborates on the upcoming areas for innovation such as machine learning and augmented reality.

In Chapter 9, *Ribeiro et al. introduce us to* **Educational Neuroscience** and brainstorm **Ideas for a School of the Future**, as they discuss cognitive neuroscience and educational psychology the main constituents of the multidisciplinary field of educational neuroscience. The authors make a case in favor of a necessary transformation of the traditional "school" and discuss some key aspects toward this goal. Based on the physiological aspects of learning, computer games, literacy acquisition, and the generation of individual learning curves for the personalized tracking of performance, this chapter covers the recent advances and trends of basic and applied research on educational neuroscience.

Throughout this book, the authors critically review the themes previously touched upon in order to provide the readers with conclusions, lessons learned, and things yet to come. The advances and challenges met within each field of neurotechnology, as presented by the authors, are discussed with a focus on successes and failures and what lessons we have learned from these cases. Finally, open issues, challenges yet to be met, paths that lie ahead and the bright future of Neurotechnology come into discussion.

Chapter 2

Current trends of biomedical signal processing in neuroscience

Christiane M. Nday[1], Christina E. Plomariti[1], Vasilis D. Nigdelis[1], Giorgos Ntakakis[1], Manousos Klados[2], and Panagiotis D. Bamidis[1]

2.1 Introduction

Modern neuroscientific approaches aim to detect the neurophysiological patterns associated with pathological conditions as early as possible with great sensitivity and specificity. Emerging evidence suggests that quantification of the functional brain interactions through a dynamic temporal manner may contribute to the diagnosis of several diseases such as psychiatric disturbances, neurodegeneration and sleep disorders. So, it is important to provide robust computational frameworks based on interdisciplinary concepts stemming from neuroscience, medicine, mathematics as well as computer science.

The present book chapter aims to serve as an integrative tutorial which focuses on the presentation of contemporary techniques for pre-processing and analysing electroencephalographic (EEG) data, based on sections described as follows.

The first section discusses EEG pre-processing and feature extraction which are the two important steps, based on EEG signal-processing techniques, for diagnostic purposes. Raw EEG data are highly unlikely to be clean, thus a series of pre-processing steps are required. These pre-processing techniques help to remove unwanted artefacts from raw EEG data, by separating the noise from the actual signal, hence improve the signal quality and make it suitable for further processing. The aforementioned steps often include the application of filters to suppress noise that contaminates the data, hence complicating the observation of the signal [1]. Filters that are typically applied in EEG data are (a) the high-pass filter, which blocks low frequencies and allows higher frequencies to pass; (b) the low-pass filter, which blocks high frequencies and allows lower frequencies to pass; (c) the

[1]School of Medicine, Faculty of Health Sciences, Aristotle University of Thessaloniki, Thessaloniki, Greece
[2]Department of Psychology, The University of Sheffield International Faculty, CITY College, Thessaloniki, Greece

band-pass filter, which simultaneously blocks low and high frequencies, permitting the in-between frequencies to pass; and (d) the notch filters, which blocks a specific frequency allowing all others to pass. During EEG recording, in case the subject is keeping their eyes open or exhibiting muscle contraction, we may use signals like vertical and horizontal electrooculography (EOG) and chin electromyography (EMG), as a reference in order to remove the related activity [2]. However, it is obvious that EMG signals are not confined solely in chin but may as well be produced by any other body parts. Similarly, we may obtain recordings from electrocardiography (ECG) and respiratory signals, as well as body movements. Having ensured that EEG signals are clean after pre-processing, they are often cut into sets of continuous epochs of up to a few seconds, to extract specific features of interest out of each one. This partitioning allows the calculation of the synchronization features and graph metrics for each epoch enabling the construction of a feature vector for each given epoch [3]. Following the pre-processing steps, data are fit for the main processing phase, indicated as feature extraction component [3], so as to retrieve the most relevant features from the signal. This component employs alternatively two different methodologies for synchronization analysis, the first one calculates the synchronization likelihood (SL) between electrode pairs [4,5] and the second one calculates the relative wavelet entropy (RWE) between the electrode pairs [6,7].

The second section discusses the inverse problem solution. Extensive research has been done to accurately solve the inverse problem, due to its ill-posed nature [8], which makes the solution non-unique and unstable [9]. The EEG inverse problem, also referred to as the neural source imaging problem, estimates the spatial location of ionic current sources within the brain, giving rise to a scalp potential recording [9] by considering the electrode potentials as its inputs [9–12]. Throughout the years, various techniques have been developed to solve the inverse problem for EEG source localization. These techniques fall mainly in two categories: parametric and non-parametric techniques. Parametric techniques estimate the dipole parameters of an *a priori* determined number of dipoles and non-parametric techniques estimate the dipole magnitude and orientation of a number of dipoles at fixed positions distributed in the brain volume. Since in non-parametric techniques the dipole location is not estimated, optimal solutions may be found using techniques such as low-resolution electromagnetic tomography activity (LORETA), standardized LORETA (sLORETA), minimum norm estimation (MNE), variable-resolution-electrical-tomography activity (VARETA) and weighted MNE [13–15].

The third section discusses the principles of functional connectivity (FC) and network-based statistics (NBS) analysis. FC considers neuronal communication showing frequency-dependent results between brain areas. In other words, FC refers to the statistical relationship between temporally specific physiological signals assessed by EEG, as well as by functional magnetic resonance imaging (fMRI) and magnetoencephalography (MEG) [16]. Additionally, FC may refer to any study examining inter-regional correlations in neuronal variability [5,17,18], applied to both resting-state [19,20] and task-state studies [2,21–23], indicating that synchronization between two brain structures (either at resting-state or task-state) suggests communication between these regions [16]. On the other hand, although anatomical connectivity may exert some degree of constrain upon FC, these two not only are not necessarily correlated, but often strong FC may exist despite the lack of a direct anatomical connection [24,25]. Overall, FC suffers from the general problem of interpreting correlations, i.e. whether two time series are correlated either because (i) x influences y, (ii) y influences x, (iii) both influence each other, or (iv) both are influenced by a third variable [26]. Therefore, disambiguating these possibilities requires a model of the causal influences (i.e. effective connectivity). Chriskos *et al.* [3] discuss two previous methods of FC estimation, i.e. SL and RWE, for sleep stage classification purposes. Then, advances in FC estimation and quantifying the statistical significance in the co-operative activity of cortical brain regions is presented through non-parametric NBS. The NBS toolbox identifies functional or structural connectivity differences in neuroimaging data that is modelled as a network [27]. Thus, NBS can be thought of as a translation of conventional cluster statistics [28,29] to a graph and has been used to identify abnormal brain connectivity circuitry in schizophrenia [30], depression [31], autism [32] and attention deficit hyperactivity disorder (ADHD) [33,34].

The fourth section discusses graph theory analysis of functional brain networks, whereby a comprehensive review of estimating global and local network properties based on graph theory principles is presented. In neuroscience, graph theory uses information generated to inform a predefined model containing both key anatomical regions of interest ('nodes' or 'vertices') and their connections ('edges'), representing connectivity [5,35]. These edges represent physical (e.g. an axon between neurons) or statistical (e.g. a correlation between time-series) relationship [36]. By representing brain regions in graph form as nodes connected by edges, the connections of each node can be tested for relationships with behaviour, sub-networks can be identified, and a range of other metrics can be used to investigate properties of the graph [37]. Using metrics derived from graph theory, it is possible to describe EEG activity during sleep on a higher level, providing a macroscopic and holistic system view [3,5]. A new plugin for EEGLAB [38], called FCLAB [39], is able to work with EEG signals in order to estimate and visualize brain FC networks based on a variety of similarity measures as well as run a complete graph analysis procedure followed by a detailed visualization of the ensuing local and global measures distribution.

The fifth section discusses application on sleep analysis through polysomnographic (PSG) data. Researchers worldwide commonly agree on the importance of achieving good quality and adequate quantity of sleep, as a means of preserving a healthy physical and mental state, both in adults and children. Accordingly, recent studies show a significant association between quantitative and qualitative sleep rhythm disturbances and an increasing prevalence of psychiatric disorders, cardiovascular diseases, obesity, type 2 diabetes, memory impairment, cognitive decline, loss of everyday life quality and increased mortality, among many others [40–42]. Therefore, the understanding and analysis of sleep functions may provide valuable insight to overall personal health. Sleep analysis of EEG signals is optimally originated using a pre-processing and subsequent sleep staging pipeline, applied on the entire PSG data derived from each participant. Confirming the above, Chriskos *et al.* [42] propose the aforementioned process to be based upon two novel methods of FC estimation, i.e. SL and RWE, which are comparatively investigated for automatic sleep staging through manually pre-processed EEG recordings [43].

The sixth section discusses application on the psychiatric EEG data. Clinical signs of EEG in psychiatry demonstrate the utility of EEG to help predict various disorders. It exhibits how EEG may contribute to identifying brain diseases and functional neurological disorders, while evaluating the involvement of EEG on the diagnosis of sleep disorders, during both rapid eye movement (REM) and non-REM sleep. However, there are a number of other reasons that establish the importance of EEG in psychiatric practice. For instance, there is comorbidity between severe psychiatric disorder and epilepsy, as in the case of many patients with epilepsy suffering from depression [44]. Similarly, patients with depression show evidence of having a greater risk of developing epilepsy [45]. Accordingly, correlation has also been documented between schizophrenia and epilepsy [46].

2.2 Main sections

2.2.1 EEG pre-processing and feature extraction

After the EEG signal is recorded it is seldom used as it is, and it is pre-processed in order to remove spectral content which is irrelevant to the electrical brain activity recorded on the surface. This content appears in the form of noise which can be produced from multiple sources such as:

- Powerline interference,
- Minor and major body movements,
- Respiration-related movements,
- Electrode malfunction,
- Incorrect placement of electrodes or sweat concentration,
- Interference from other biological signals such as EOG, EMG and ECG and
- Noise from other unknown sources.

All the above are superimposed into the recorded EEG signal reducing its quality and utility. EEG recordings are subsequently used for feature extraction, hence the extracted features will not provide an accurate description of the EEG content.

Pre-processing techniques help to remove unwanted artefacts from raw EEG data, by separating the noise from the actual signal, hence improving the signal-to-noise ratio and making it suitable for further processing.

The most common method of pre-processing biological signals [3,47,48] is to apply one or more filters aiming to reduce the effect of the noise sources on the spectral content of the EEG signal. Digital filtering is easy to implement and apply on the signals and improves the quality of the recordings. However, this approach has disadvantages including the inability to remove noise content within the EEG spectrum and applying filters removes a portion of the total energy of the signal. EOG artefacts are the most troublesome due to their high frequency as to their large amplitude compared to the EEG [48]. So far three main methodologies are proposed for dealing with EOG artefacts. According to the first category, a blind source separation algorithm can be applied so as to decompose EEG signals into statistically independent components (ICs) and then the artefactual components can be removed manually or automatically [49]. The remaining components can be recomposed to form clean EEG signals. Although, this methodology seems very simple and clear, it has one major drawback as the ICs contain also a portion of EEG derived activity, which is negatively correlated with the number of electrodes, and it is removed as well during the artefact rejection procedure. The second methodology is based on regression analysis and adaptive filters, where the regression-based algorithms are used to a) compute the backward propagation coefficients in order to define the relationship between the EEG and EOG channels, and finally, b) subtract the proper amount of EOG activity from the EEG signals. In order to be able to apply a regression-based technique, reference signals (like EOG) should be recorded simultaneously with EEGs. The main problem of this methodology is that the recorded EOG signals contain also information oriented from the cerebral activity, mainly of the pre-frontal cortex, so this activity is going to be also deleted from the EEG signals.

Moreover, independent component analysis (ICA) can be applied to a multichannel signal which is decomposed into ideally ICs equal to the number of EEG channels. Taking into account that artefactual ICs may contain also brain activity, which is positively correlated with the number of artefactual IC detected, a subset of these components which represent only noise sources can be rejected, i.e. not used in reconstructing the EEG signal, thereby rendering the reconstructed EEG suitable for further analysis. As above, this methodology does not come without drawbacks, which include that the initial EEG recordings must be consisted of a significant number of channels, e.g. 19 electrodes, the resulting components may not clearly represent noise sources and removing them may severely affect the reconstructed signal.

Another issue that arises with regards to EEG pre-processing is that depending on the recording device and environment some noise sources may be more prominent than others while others may be almost absent. Hence, it is unlikely that a concise pre-processing pipeline can be developed which will cover all recording scenarios.

2.2.2 Inverse problem solution

After completing the pre-processing pipeline, the artefact free data are ready to be analyzed. There are two possible levels where the analysis can be performed. The sensor level (analysis of the electrodes' signal) and the source level (cortical regions' signal). Computing the activations of the cortical regions based on the electrodes' signal is a complex mathematical problem, namely the inverse problem. The name comes from the fact that we start with the results (EEG electrodes' signal) and we try to estimate the parameters (sources) which caused them. The inverse problem does not have a unique solution, as more than one combination of sources can result to the same signal. Furthermore, the solution is also unstable, in terms of sensitivity to small changes. In order to constrain the solution, a number of assumptions need to be made. In particular, the neural sources are modelled as dipoles, existing only on the cortical regions of the brain.

Mathematically, the inverse problem can be summarized in the following equation:

$$M = GD + n \tag{2.1}$$

M denotes the recorded data matrix, G describes the current flow for an electrode through every dipole, D is the dipole magnitudes matrix (the one that needs to be defined in the inverse solution) and n represents the noise matrix.

The cortical dipoles, thought to be the primary sources of the EEG signal, can be specified by six parameters: three spatial coordinates x, y, z and three momentum factors, the strength d and two orientation angles θ and φ. In order to reduce these components, a number of constraints can be assumed, each leading to different mathematical models, depending on which parameter(s) remain fixed (position, magnitude, orientation). The main mathematical models across the literature are [50]: a single dipole with non-fixed, unknown defying parameters, a fixed number of dipoles with fixed unknown position and orientation, a number of dipoles with fixed known positions and a varying number of dipoles with constrained parameters. There are many factors affecting the accuracy of the source localization, including source and head modelling errors and EEG noise [51].

Across the international literature, a variety of methods for source signal estimation have been proposed. They can be divided in two main categories: non-parametric and parametric methods. The non-parametric methods are known as distributed source models or inverse solutions. The main assumption of these models is the existence of a number of dipoles with fixed locations and in some cases fixed orientations distributed in the cortex, normally aligned. The estimation of amplitude and orientation of the dipoles, combined with the lack of spatial estimation, makes the whole problem linear. When it comes to the parametric methods, the assumption involves a few dipoles with unknown position and orientation. The estimation of the location of the dipoles puts this problem in the non-linear category. They are also known as equivalent current dipole methods or concentrated source or spatiotemporal dipole fit models.

The non-parametric methods are based on [9]:

- The Bayesian Method [52]. It is based on computing an estimator that maximizes the posterior distribution. The estimator is described in (2.2):

$$x = \max_{x} [p(x \mid y)] \tag{2.2}$$

where p is the probability density. In this broad category fall the following mathematic solutions:

- *The Gaussian or normal density function.* It can be used if the posterior density follows a Gaussian or normal distribution. In this approach

$$p(x \mid y) \propto \exp\left(-\mathrm{Tr}\left(X^T \cdot \sigma^1 \cdot X\right)\right) \tag{2.3}$$

where Tr is the trace of a matrix, $X = Kx - y$ (K a compact linear operator), σ^1 the data covariance matrix.
- *The general Normal density function.* It poses a more general approach, proportionate to the Gaussian density function. Here,

$$p(x \mid y) \propto \exp\left(-\mathrm{Tr}\left((X-\mu)^T \cdot \sigma^1 \cdot (X-\mu)\right)\right) \tag{2.4}$$

- *Non-Gaussian priors.* This specific method includes in the computation of p entropy metrics and L_p norms with $p < 2$.
- *Regularization methods.* These methods aim at computing the best approximation to $Kx = y$, x^δ with the assumption that only y^δ an approximation to y (real noisy data) is known.
- *Minimum norm estimates (MNE).* This method is suitable in cases where the activity of the dipoles is extended to many cortical areas. Here,

$$\widehat{D}_{\mathrm{MNE}} = RG^T\left(GRG^T + C\right)^{-1}M \tag{2.5}$$

where $\widehat{D}_{MNE}$ an estimate of D, C the covariance matrix of n and R the covariance matrix of D.
- *Weighted minimum norm estimates (WMNE).* The above method favours weak and surface sources. The WMNE introduces a weight matrix to compensate for this weakness.
- *MNE with FOCUSS (Focal underdetermined system solution).* This recursive procedure was developed in order to provide focal resolution to linear estimators on non-parametric models. It provides more localization accuracy on cortical dipoles, in comparison with MNE.
- *Low-resolution electrical tomography (LORETA).* It combines the Laplacian operator with the lead-field normalization. It is not ideal for focal source estimation. The inverse solution matrix is

$$\widehat{D}_{LOR} = \left(B\Delta^T\Delta B\right)^{-1}G^T\left(G\left(B\Delta^T\Delta B\right)^{-1}G^T + aI_N\right)^{-1}M \tag{2.6}$$

where B is the diagonal matrix for the column normalization of G, a is a regularization parameter and Δ the Laplacian operator.
- *LORETA with FOCUSS.* This method is proportional to MNE with FOCUSS, only it uses LORETA computations. It increases the focal resolution of the LORETA approach.
- *Standardized low-resolution brain electromagnetic tomography (sLORETA).* In this method, the localization is based on the standardized current density. It does not involve the Laplacian operator:

$$\widehat{D}^{T}_{MNE,l}\left\{\mid S_{\widehat{D}}\mid_{ll}\right\}^{-1}\widehat{D}_{MNE,l} \tag{2.7}$$

where $S_{\widehat{D}}$ is the variance of the estimated current density.

- ○ *Variable-resolution electrical tomography activity (VARETA).* A weighted minimum norm solution. Requires very fine grid spacing. In the solution grid, a spatial variability of the regularization parameters is observed:

$$\widehat{D}_{VAR} = arg\min_{D,\Lambda}\left(\|M - GD\|^2 + \|\Lambda L_3 \cdot W \cdot D\|^2 + \tau^2 + \|L \cdot \ln(\Lambda) - a\|^2\right) \tag{2.8}$$

where L a non-singular univariate discrete Laplacian, $L_3 = L\hat{}I_{3\times 3}$, Λ a diagonal matrix used for normalization and τ a parameter which controls the amount of smoothness in the grid.

- ○ *Quadratic regularization and spatial regularization (S-MAP) using dipole intensity gradients.* Quadratic regularization uses dipole intensity gradients to introduce rise to smooth variations of the solution. S-MAP is a modification of it:

$$\widehat{D}_{QR} = \left(G^T G + a\nabla^T\nabla\right)^{-1} G^T M \tag{2.9}$$

- ○ *Spatio-temporal regularization (ST-MAP).* Takes into account the parameter of time. Dipole magnitudes are considered slowly evolving in respect to the sampling frequency.
- ○ *Spatio-temporal modelling.* A highly complex approach. The inverse problem is solved with the use of Kalman filtering. The problem is subjected to high dimensionality.

- The Backus–Gilbert method [53]. In this method,

$$\widehat{D}_{BG} = TM \tag{2.10}$$

where T is an approximate inverse operator of G. In this method, $\widehat{D}_{BG}$ is closest to D inside the brain in at least square sense.

- The weighted resolution optimization [54]. An extension to the Backus–Gilbert method.
  - ○ LAURA. In this method, the assumptions made in order to proceed to the mathematical formalism, incorporate biophysical laws, apart from mathematical ones:

$$\widehat{D}_{\mathrm{LAURA}} = W_j G^T\left(G W_j^{-1} G^T + a I_N\right)^{-1} M \tag{2.11}$$

where W_j a weighted matrix constructed specifically for the calculation.

- The shrinking methods and the multi-resolution methods. These set of methods are based on the application of iterations to the obtained solutions.
  - *S-MAP with iterative focusing.* A modified version of spatial regularization. Can be used when the cortical surface is subjected to sparse spatial sampling and it can recover the focal sources.
  - *Shrinking LORETA-FOCUSS.* This solution combines LORETA and FOCUS, while it increases source resolution and decreases computational time.
  - *Standardized shrinking LORETA-FOCUSS (SSLOFO).* This combination of FOCUS and sLORETA, derives primary regions based on their activity and localizes multiple sources inside them.
  - *Adaptive standardized LORETA/FOCUSS (ALF).* Similar to the other methods described above. The main difference is that this solution requires the computation of 6%–11% of the matrix G.

In the case of parametric approaches, the main mathematical methods on which the most approaches are based, are:

- The non-linear least-squares problem. In order to find the location and the momentum of the dipoles, the global minimum of the residual energy is calculated in this method.
- The beamforming approaches. A set of approaches where the number of the dipoles modelling the cortical area of the brain must not be assumed before the calculations. The neural activity can be estimated at any chosen location of the brain, no matter the geometry of the source.
- The brain electric source analysis (BESA). The dipoles are assumed to have fixed orientations and positions for a given time-point. A continuous distribution of such time-points is hypothesized.
- The subspace techniques. In these techniques, the EEG data are processed before the localization of the dipoles.
  - *Multiple-signal classification algorithm (MUSIC).* A version of the spatio-temporal approach. It has many modifications and extensions.
  - *FINES subspace algorithm.* This approach does not employ the whole subspace for the estimation of the location of the sources, but rather it uses a projection of the subspace spanned by a few vectors.
- Simulated annealing and finite elements. Instead of using measured potential differences, this method uses finite element formulations in a two-dimensional space.
- Computational intelligence algorithms. The localization problem is considered an optimization problem.
  - *Neural networks.* The optimal location and momentum of the dipoles is calculated with the use of neural networks.
  - *Genetic algorithms.* An alternative to using neural networks for the estimation of the cortical activations.

All these solutions pose complex mathematical problems. When it comes to implementing on of them, a variety of custom batch-scripts can be found online. Also, a number of these methods can be computed through the MATLAB® environment (LORETA, sLORETA, WMN and SLF). A number of toolboxes either for MATLAB (e.g. Brainstorm) or for Python (e.g. MNE) are designed specifically for this type of analysis. Finally, standalone desktop applications also exist for implementing LORETA and sLORETA, as these solutions are the most popular ones given their accuracy and simplicity compared to the rest (Figure 2.1).

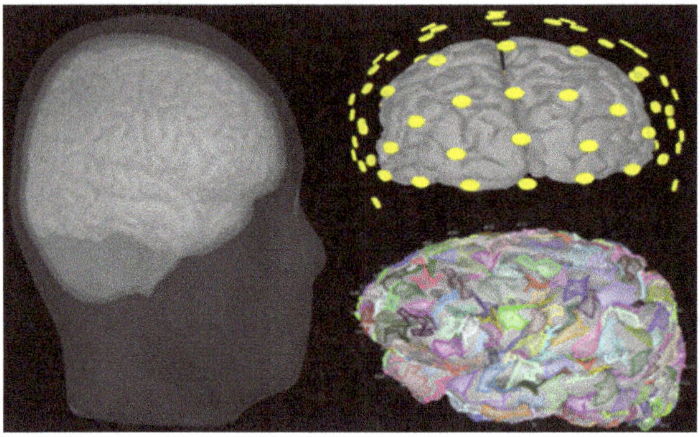

Figure 2.1 LORETA standalone

2.2.3 Principles of FC and NBS analysis

Regions of the human brain often share functional properties. More specifically, spatially remote regions demonstrate temporal correlations. This phenomenon is called FC [55]. The study of FC is of upmost importance when trying to understand the mechanisms of the brain and predict the outcomes of environmental variations, interventions and pathological alterations.

The estimation of FC is a complex mathematical process, since the electrical signals produced and distributed across the brain are four-dimensional, consisting of a 3D spatial and a temporal component [56]. This mathematical solution can be applied on EEG or functional magnetic resonance imaging (fMRI) data. The present study focuses on the analysis of the first category of data (EEG).

After estimating the cortical activations of the selected regions of interest (ROIs), choosing the appropriate connectivity metric is the most critical step of the procedure. In the international literature, a wide variety of connectivity metrics is presented. The metrics can be divided in two broad categories, directed and non-directed. The directed metrics quantify the direction of the interaction, while the non-directed do not represent the direction of the influence. Furthermore, the metrics can be furtherly categorized based on whether they quantify linear or non-linear interactions [57].

In the neuroscientific literature, the most commonly used measures are Pearson's correlation coefficient (COR), cross-correlation function (XCOR), magnitude squared coherence (COH) and phase slope index (PSI). COR [58] detects only linear, non-directed interactions in the time domain between two signals originating from two sources (brain regions) at zero lag. XCOR [59] estimates linear, non-directed correlations between two signals, as a function of time. COH [60] measures linear, non-directed interactions between two time-series as a function of frequency. PSI [61], quantifies the linear, directed correlation between two signals as time-delay between them.

Another wide category of correlation metrics is the phase synchronization (PS) indices, which take into account the PS of the oscillations involved. The most commonly encountered PS measures are the phase-locking value (PLV), the phase-lag index (PLI), the weighted phase-lag index (WPLI), the ρ index (RHO) and the directionality phase indexes (DPI). PLV [62] measures the non-linear, non-directed distribution of the relative phase of two signals over a unit circle. PLI [63] is an alternative measure to PLV. It is sensitive against the presence of common sources as it discards phase differences close to zero or π. WPLI [64] is related to PLI, but WPLI reduces sensitivity to the present noise sources and increases the statistical power when detecting PS. RHO is based on Shannon entropy [65] and it measures the deviation of the phase distribution from the uniform distribution. DPI [66] analyzes the temporal evolution of the phase derivative. It derives directionality by analyzing the relative phase [66].

In case the (sub) systems under investigation are connected (one being a function of the other), the most appropriate metrics for deriving synchronization are the generalized synchronization (GS) Indexes. The most popular index amongst them is the SL. The concept of this metric relies on the fact that if the signals arising from two different sources are correlated, simultaneously occurring patterns will be presented [67]. Other alternative GS metrics are the S index, the H index, the N index, M index and L index.

Granger causality indexes are a set of metrics which detect correlation between time series based on the hypothesis that if two time series have a common source, then each one of them can forecast the other [68]. In this category, fall the classical linear Granger causality (GC), the partial directed coherence (PDC) and the direct transfer function (DTF). GC is based on the classic mathematical approach of causality. PDC [69] is based on the same principles, but the measurements take place in the frequency domain. Finally, DTF [70] is described similarly to PDC, with small mathematic changes.

The final category of connectivity metrics is based on Shannon entropy [71], the so-called information theory metrics. By far, the most popular measures are mutual information (MI), partial mutual information (PMI), transfer entropy (TE) and partial transfer entropy (PTE). MI [72] can be defined as a measure of the entropic correlation between two time series. PMI [73] correlates two signals with the same mathematical formula as MI, while it discards the possibility that the shared information between them arises from a third signal. TE [74] measures the amount of directed information transfer between two time series, while PTE [75] uses the same calculation but again discards the possibility of a third source causing the correlation between them.

The aforementioned metrics are just a small sample of those available in the literature. New metrics continue to arise, deriving from novel mathematical analysis pipelines. Choosing the appropriate metric for the estimation of FC of a specific dataset is not always an easy process. Usually, there is not a correct and a false choice. The procedure goes through trial and error and one must always keep in mind that many analyses have been published dealing with this conundrum [57,76,77].

After choosing the desired FC metric, the next step is its implementation. The two available options are through a batch-script or through an already existing toolbox. The first option requires the user to have an advanced understanding on mathematics and based on the background of the metric to construct the code. The easiest and safest choice is the second one; the toolbox. A variety of toolboxes are available online for calculating various metrics and extracting FC. Most of them are designed to operate via the MATLAB software. The most popular are HERMES toolbox [78] (Figure 2.2), eConnectome [79], FC Toolbox [80] and BrainNetVis [81].

The outcome of the synchronization metric calculation is an **n** × **n** matrix, where **n** denotes either the number of electrodes used in the EEG for a sensor-level analysis, or the number of the selected ROIs. The matrix can be symmetric or asymmetric, depending on the chosen metric. The main diagonal usually is either 1 or 0, portraying the perfect synchronization. The rest of the cells can take any number, depending again on the metric (Figure 2.3).

The final step to the FC analysis is the statistical analysis. The edges with a statistically significant connectivity value need to be identified, in order to

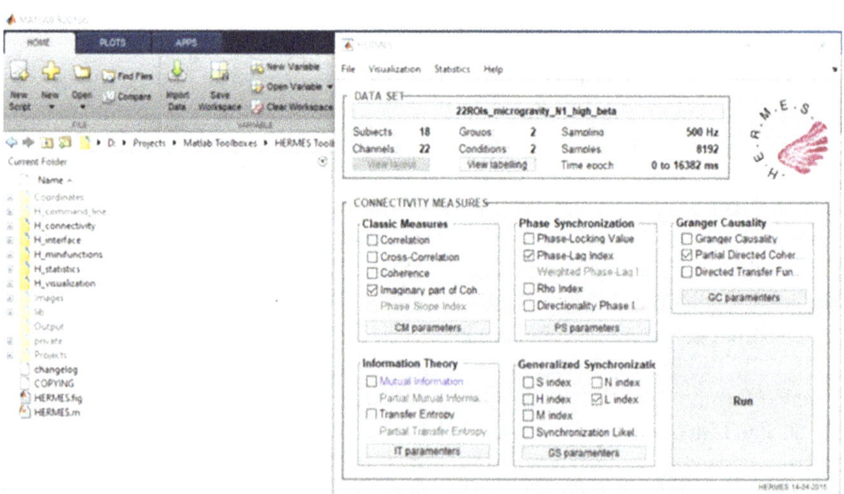

Figure 2.2 The HERMES toolbox operating in MATLAB environment

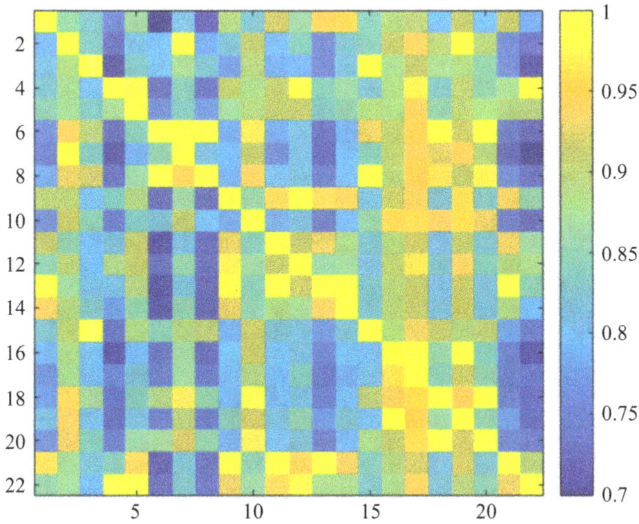

Figure 2.3 Connectivity matrix for 22 ROIs. The metric depictured is mutual information

construct the network. In order to perform this analysis, a large number of EEG data (usually around 20) is required for producing reliable results. There are many programs available for performing statistical analysis on matrix algebra. In the present chapter, we will present NBS [27], a MATLAB toolbox. It offers the possibility of three statistical tests: A one-sample test, a t-test and an F-test. The one-sample test can be used when comparing a sample with a population with known characteristics. The t-test is ideal when comparing means of features under relaxed conditions. Finally, the F-test (ANOVA) is used when the data are grouped into more than two categories. The t-test and the F-test are incorporated by the general linear model, a statistical model ideal when more than one dependent value is presented. In this model, $Y = XB + U$, where Y is the measurements matrix (FC matrix), X the design matrix, B is the matrix containing the parameters which will be estimated (the statistically significant edges) and U is the noise matrix.

In all the above statistical tests (one-sample test, t-test and F-test), the lower 'p-value' needs to be determined in order to calculate the results. The 'p-value' is quantifying the statistical significance of the null (default) hypothesis [82]. It needs to be as low as possible for the results to have statistical value. The highest acceptable amount of it is considered 0.05.

The construction of the design matrix is the most challenging step when performing statistical analyses. It is used to define the general linear model used. It is an $m \times n$ matrix. m is the number of observations and n the number of predictors. It contains explanatory variables regarding the dataset. Depending on the statistical test used, the design matrices vary. An example of a designed matrix can be found in Figure 2.4.

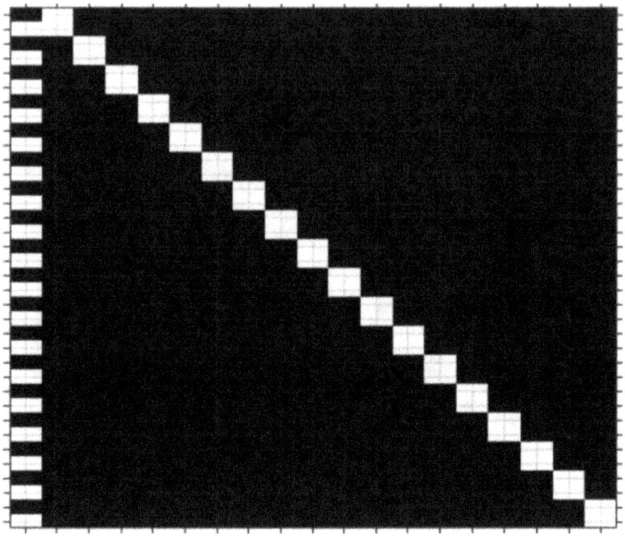

Figure 2.4 Designed matrix used for an F-test in a dataset consisting of 36 EEGs (number of rows); (18 participants, 2 observations each). They were divided in two groups (second column time x group) and the EEGs where obtained in two time-instances (first column)

The NBS toolbox identifies the statistically significant edges for the given ROIs and produces a network based on them (Figure 2.5). The networks identified by the toolbox can be subjected to further analysis via the MATLAB software to extract results regarding the type of interaction (why those edges are statistically significant) and the time-instance which caused it.

2.2.4 Graph theory analysis of functional brain networks

In Mathematics, graphs are structures used for modelling pairs of relationships between objects [83]. Each graph consists of nodes and edges connecting them. When analyzing EEG signals, the nodes represent cortical areas of the brain, while the edges represent pathways for information exchange between them, as they are quantified by the FC of these areas. Each graph can be represented by an $n \times n$ matrix, where n denotes the number of nodes. Every cell has a value which represents the intensity of the FC and can be quantified by the width of the edge connecting the corresponding edges.

Graphs can be divided in directed and undirected. Directed are the ones where the edges have direction from one node to the other, or in terms of linear algebra. On the other hand, graphs where the connections are bi-directional are call undirected. There and the matrix appears symmetric. Furthermore, graphs can be divided to binary and weighted. The matrix cells of binary graphs can get the value ‘0’ if

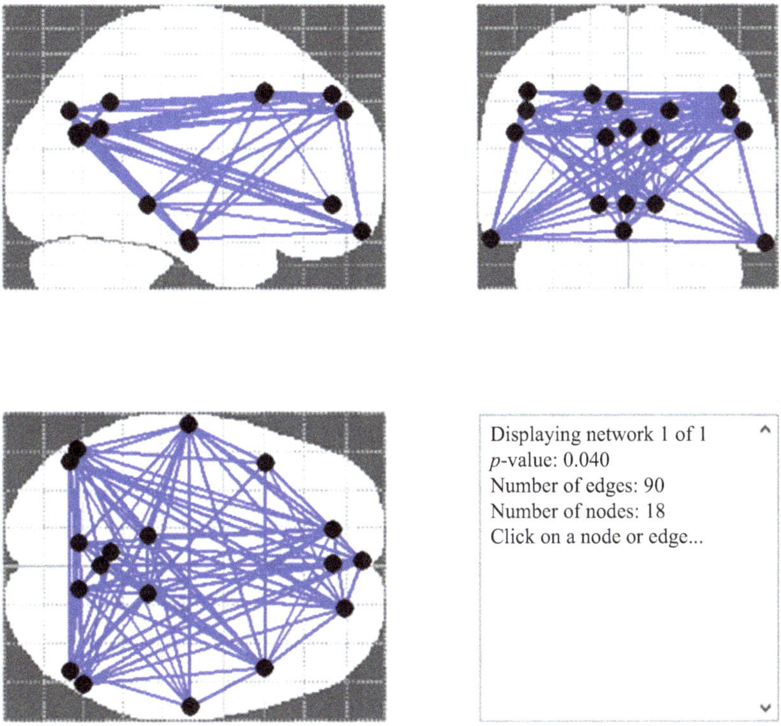

Figure 2.5 Network consisting of statistically significant edges, as they are identified by NBS

there is no connection between the corresponding nodes, or '1' if there is one. Weighted graphs can get any value in their cells.

Every graph has some characteristic metrics which determine their behaviour and in the case of EEG analysis the effectiveness of information transfer within them. These metrics can be divided in global, which concern the total graph, and local which concern each node separately.

- **Global metrics:**
  - *Density:* A dense graph, is a graph where the number of edges is close to the maximum number of edges they can hold. On the other hand, a sparse graph is one with a small number of edges compared to their maximum capacity. In an undirected graph, density (*D*) can be calculated by

$$D = \frac{2|E|}{|V|(|V| - 1)} \tag{2.12}$$

while in a directed one

$$D = \frac{|E|}{|V|(|V| - 1)} \tag{2.13}$$

where E denoted the number of edges and V the number of nodes in the graph [84].

- *Efficiency:* A metric which quantifies the ability of the network to transfer information efficiently within their nodes [85]. Global efficiency refers to the whole network and can be calculated by

$$E(G) = \frac{1}{n(n-1)} \sum_{i \neq j \in G} \frac{1}{d(i,j)} \tag{2.14}$$

where n the number of nodes and $d(i, j)$ the shortest path between nodes i and j.

- *Characteristic path length:* The median value of the mean values of the shortest paths connecting each node to the rest of them. For a graph G, characteristic path length can be calculated by

$$L = \frac{\sum_{x,y \in V(G)} d_G(x,y)}{n(n-1)} = \frac{\sum_{y \in V(G)} \sum_{y \in V(G)\{y\}} d_G(x,y)}{\frac{n}{2}} \tag{2.15}$$

where $d_G(x, y)$ the distance between nodes x and y.

- *Clustering coefficient:* This metric quantifies the degree in which the nodes of a given graph tend to create clusters. In most real networks, the nodes tend to create groups characterized by a large number of edges [86]. The global clustering coefficient is based on the triplets created by the nodes within the graph. A triplet is a set of three nodes connected by two (open triplets) or three (closed triplets) undirected edges. Clustering coefficient can be calculated by

$$C = \frac{\text{number of closed triplets}}{\text{number of all triplets}}. \tag{2.16}$$

- *Modularity:* This metric was designed in order to measure the ability of the network to be divided in modules. Networks with high modularity have dense connections between the nodes belonging to the same module but have sparse connections between nodes belonging to different modules. Modularity can be calculated by

$$Q = \frac{1}{l_n} \sum_{x,y} \left[A_{xy} - \frac{k_x k_y}{l_n} \right] \frac{s_x s_y + 1}{2} \tag{2.17}$$

[87], where l_n the number of modules, x and y two different nodes with degrees k_x, k_y, respectively, A_{xy} the number of edges and s_x,s_y property variables of the nodes.

- *Small worldness:* Any network with the property of small worldness can be represented by a graph where the nodes are not neighbours, but the neighbours of any node are most likely each other's neighbours. In such a graph, transition from one node to another can be performed through a small number of edges. The network is characterized by the ability of information exchange and processing. For a network to be considered small world, σ must be over 1:

$$\sigma = \frac{\frac{C}{C_r}}{\frac{L}{L_r}} \tag{2.18}$$

where C the clustering coefficient, C_r the clustering coefficient of a random network, L the characteristic path length and L_r the characteristic path length of a random network.

- **Local characteristics:**
  - *Clustering coefficient:* The local clustering coefficient quantifies, for a given node, how close are their neighbours in creating a cluster, i.e. a complete graph. For directed graphs, the local clustering coefficient is given by

$$C_i = \frac{\left|\left\{e_{jk}:v_i, v_k \in N_i, e_{jk} \in E\right\}\right|}{k_i(k_i - 1)} \tag{2.19}$$

while for an undirected graph

$$C_i = \frac{2\left|\left\{e_{jk}:v_i, v_k \in N_i, e_{jk} \in E\right\}\right|}{k_i(k_i - 1)} \tag{2.20}$$

where e_{jk} the edge connecting the nodes v_j and v_k, N_i the neighbourhood and k_i the neighbours of v_j.

  - *Participation coefficient:* This metric is used for the quantification of the connection between the number of edges connecting a node with others outside the cluster and the total number of edges belonging to the given node. Participation coefficient for a node I belonging to the cluster S_i is given by [88]:

$$P_i = 1 - \sum\nolimits_S \left(\frac{K_{S_i}}{K_i}\right)^2 \tag{2.21}$$

where K_{S_i} the total number of edges in the cluster and K_i the total edges of the graph.

  - *Within module z-score:* It represents the extent to which a node is connected with the other nodes of the same cluster. It can be calculated by

$$Z_i = \frac{K_i - K_{S_i}}{\sigma_{S_i}} \tag{2.22}$$

where K_{S_i} the grade of each node in the community S_i, K_{S_i} the mean grade of all the nodes in this cluster and σ_{S_i} the standard deviation.

Representing any given cortical network as a graph is an easy way of estimating the FC within them. The aforementioned metrics form a complete picture regarding the properties of the network and the efficiency for information exchange between them. Graph theory is a powerful tool and can be used for studying networks, evaluating interventions and understanding the pathways of communication between cortical areas. It can be implemented through the brain connectivity toolbox (BCT) in MATLAB environment and can be furtherly subjected to statistical analysis or machine-learning techniques.

2.2.5 Biomedical signal-processing application on sleep analysis through PSG data

New trends in biomedical signal processing have been recently proposed in sleep analysis of EEG signals using advanced pre-processing and subsequent sleep staging pipelines, applied on the entire PSG data. It is well-known that sleep analysis of EEG signals can be optimally originated using a pre-processing and subsequent sleep staging pipeline, applied on the entire PSG data derived from each participant. Scientific papers [3] and [43] propose the aforementioned process to be based on two novel methods of FC estimation, i.e. SL and RWE, which are comparatively investigated for automatic sleep staging through manually pre-processed EEG recordings. For instance, in [2], 32 channel PSG devices are used to perform entire night sleep recordings at five different time of the experimental instances. The recordings include 19 EEG, 2 ECG, 4 EOG and 2 chin EMG electrodes.

Subsequently, the raw data of the biological signals (EEG, ECG, EMG and EOG) are pre-processed in a distinct way, in order to remove content irrelevant to brain activity (powerline interference, body movements, noise due to electrode malfunction or incorrect placement), interference from other biological signals, such as EOG, EMG, ECG and noise from other unknown sources. EEG signals where first filtered using digital Butterworth filters with the following order:

- High-pass (cut-off at 0.5 Hz)
- Low-pass (cut-off at 50 Hz)
- Band-stop (cut-off at the range of 47–53 Hz)
- Band-stop (cut-off at the range of 97–103 Hz)

The above filters in combination with a common average re-reference for the EEG signals only result in the rejection of the greater amount of noise. Subsequently, ICA is used in a MATLAB environment to visually identify artefactual sources such as blinks, muscle artefacts, bad electrode placement, etc. The noisy components are visually identified by two experienced neurophysiologists. The full pre-processing pipeline is detailed in Chriskos *et al.* [3].

Then, reconstruction of cortical analysis is conducted. Cleaned sensor raw data are converted into cortical data by using sLORETA inverse problem solution [7].

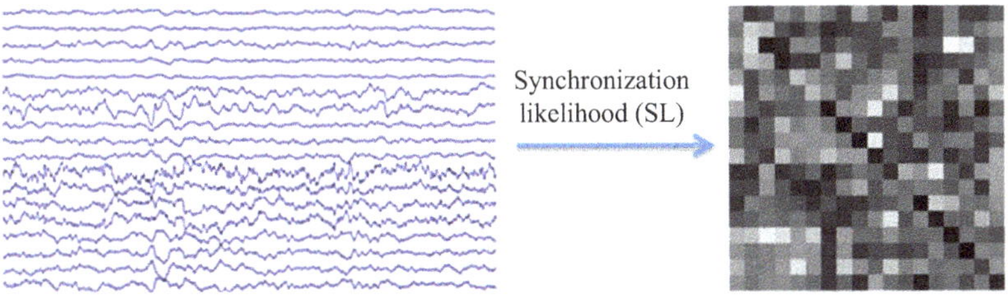

Figure 2.6 Transforming the EEG sensor recordings (left image) to a SL matrix (right image). High-intensity pixels (white) represent strong electrode pair synchronization, while low-intensity pixels (black) denote minimal electrode pair synchronization

This step initially includes generic head modelling by applying Open MEEG Boundary Elements Method (BEM head model). The cortex is modelled through 15,000 dipoles of fixed orientation. Then, their activity is estimated and 20 regions belonging to the visual, auditory, sensorimotor, default-mode and executive resting state network are defined based on *a priori* knowledge [89]. Each node activity is computed as all the dipoles average activity containing in that specific area. Brain storm software is utilized for the cortical reconstruction analysis [90]. SL values are extracted from the cortical reconstruction for each NERM stage 2 (N2) epoch to provide insight into the brain's FC [91] (Figure 2.6). These values quantify the synchronization degree observed between the time series recorded among EEG channels. For a given EEG channel k and at time i the SL value compared to all other channels is given by:

$$S_{k,i} = \frac{1}{2(w_2 - w_1)} \sum_{\substack{j=1 \\ w_1<|i-j|<w_2}}^{N} S_{k,i,j} \tag{2.23}$$

where w_1 denotes the Theiler correction window, w_2 is the window for time resolution sharpening and $S_{k,i,j}$ is the SL value for channel k at time pair i, j which is calculated through:

$$S_{k,i,j} = \begin{cases} \dfrac{V_{i,j-1}}{N-1} & \text{if} \left|x_{k,i} - x_{k,j}\right| < e_{k,i} \\ 0 & \text{if} \left|x_{k,i} - x_{k,j}\right| \geq e_{k,i} \end{cases} \tag{2.24}$$

where $V_{i,j}$ is the number of channels closer to each other than distance $e_{k,i}$, N is the number of EEG channels and $x_{k,i}$ denotes the time-delay embedded vector of x_k. These values are stored in an $N \times N$ symmetric matrix, each cell of which contains SL value between each electrode pair [4].

The calculation of SL results in a symmetrical matrix of 22×22 squares. For each specified brain region pair, matrix cells comprise all SL values containing 0 and 1 to express no synchronization and perfect synchronization of the brain region, respectively. 1 is primarily seen in the SL matrix's main diagonal projecting the activation of each brain region being synchronous with itself. These values are omitted as they do not provide any valuable information. For the extraction of the graph metrics which provide an overview of the function of the brain during individual sleep stage, the SL values must be treated as the adjacency matrix of a weighted un-directional graph with no self-loops. Finally, graph theory analysis of functional brain networks is applied as described in Section 2.2.4.

2.2.6 Biomedical signal-processing application on psychiatric EEG data

EEG can be extremely useful in order to understand various psychiatric disorders. It is commonly used by physicians when the symptoms of the disease do not match the characteristics of any known syndrome. In many cases, abnormalities of the EEG are common for a given disease. Quantitative EEG can identify pathological deviations while performing differential diagnosis among the various psychiatric conditions. It is often used in order to out rule organic factors when it comes to brain behavioural and functional abnormalities. Diffuse abnormalities on the EEG usually suggest encephalopathies or neurodegenerative diseases. Most psychiatric disorders are characterized by functional and structural modifications in the human brain. They can be easily identified with the use of EEG, via FC and spectral analysis. Psychiatric disorders like schizophrenia and depression can alter the chemistry of the brain, thus affecting directly its functionality.

In schizophrenia, the complexity values, metrics used for the estimation of brain dynamics, appear to be lower when compared to healthy participants [92]. They can be calculated through resting-state EEG. Graph theory analysis also reveals significant differences between schizophrenic patients and healthy population. Small-worldness, clustering coefficient and characteristic-path length modulations can be directly associated with cognitive performance in these patients [93]. Information interaction between brain-regions is significantly lower in schizophrenic patients, as proved by graph theory analysis [94]. EEG analysis can also be used for the classification of schizophrenic patients through deep learning techniques [95] and for the prediction of performance in various cognitive tasks [96]. Abnormalities in spectral EEG analysis are also typical in this specific spectrum of disorders [97–99]. Gamma power appears to be elevated, while delta, theta and alpha 2 activity are decreased.

A successful application of biomedical signal processing on the psychiatric EEG data is described in [99]. Herein, potential alterations in resting state and gamma activity linked to specific brain regions are investigated in two different experimental groups (76 healthy and 60 participants with schizophrenia). Resting state with closed eyes state for 2 min is applied using a high-density EEG-system (256 channels and 512 Hz sampling rate), referenced to the vertex. A band-pass-filtered on-line ranging from 0.1 to 100 Hz represents the signal. Eye movement records are collected through EOG for artefact recognition and removal. Using EEGLAB toolbox in the MATLAB environment [100], EEG is re-referenced to the common average potential, and filtered off-line of a 1 to100 Hz band-pass using zero-phase shift-forward and reverse-IIR Butterworth-filter. For removing electrical interference from the 50 Hz line, a 48–52Hz Parks-McClellan stop-band Notch-filter in ERPLAB is applied [101]. Data from EEG recordings are checked manually and artefacts associated to non-brain including muscle contractions and artefacts linked movements are cleared away. In this way, the cleaning process ends up to an average of 94.35 (standard deviation (SD) = 16.36) and 94.53 (SD = 14.67) EEG information, accordingly, for healthy participants and schizophrenic patients. Subsequently, ADJUST toolbox uses an ICA technique to cut out the remaining artefacts related to muscle and eye movement [102]. Actually, ICA is a commonly used technique to rectify EEG signal as it can efficiently reject ocular artefacts that may have been ignored in manual and threshold artefact deletion processes [103]. Next, the assessment is carried out on 60 s of cleaned EEG data based on its reliability in EEG power spectra in the resting state [104]. Furthermore, FFT is performed on the rectified EEG data in order to extract absolute power based on Welch's method, by applying a 2-s, 75% overlap window [105] and 0.5 Hz frequency bin resolution. Gamma band power is estimated in 31–48 Hz, by summing the absolute power in the given frequency range on the frequency bins. For the summed frequency band, log10 transform is calculated, which yields a better approximation of normal distribution compared to raw data. In patients with schizophrenia, the absolute authority is considerably increased in statistical analysis relative to healthy participants (false discovery rate corrected $p < 0.05$). The clustered changes in specific brain areas such as front-central and posterior are favourably correlated with the psychopathology measured severity that positive and negative symptoms scale detects. Changes in gamma activity may result in disrupted large-scale brain network coordination.

Therefore, the enhanced gamma activity in specific brain areas can lead to disturbances in schizophrenia in temporary coordination of task-free resting-state networks. Enhanced gamma power is positively associated with psychopathology indicating that modified gamma activity contributes to the presentation of symptoms.

In depression, EEG analysis can be used for the detection of the syndrome with high accuracy [106,107]. Spectral analysis can also serve as a predictor to treatment outcome. Specifically, theta density has been associated with the response to anti-depressant medication [108]. Resting state EEG is differentiated in patients suffering from major depressive disorder [109]. They demonstrate increased theta and alpha activity in various cortical regions. The brain's FC is also altered due to depression. Studies have reported increases in FC when compared with healthy participants [110].

Alterations in brain's functionality are common in other psychiatric disorders as well, like bipolar, ADHD, schizoaffective, personality and anxiety disorder. Higher absolute theta power is present in bipolar and ADHD disorder [111]. Patients suffering from personality disorder demonstrate increased intermittent rhythmic delta and theta activity (IRDA/IRTA) [112]. An increase in FC in the cerebral regions of the brain is found in patients suffering from generalized anxiety disorder [113].

To summarize, EEG analysis can prove useful when studying psychiatric disorders. It can be used for identifying some disorders and in order to predict the outcome of various treatments. It can also be used for studying and understanding these disorders and their underlying physiological mechanisms. EEG has been used as an assessing tool for mental disorders since the 1930s [114,115]. In spite of all the data pointing towards specific disorders based on the EEG morphology, a normal EEG cannot guarantee the absence of organic disorders. Also, abnormalities in the EEG can also be present in normal cases. Alterations in the EEG can also be caused due to psychotropic medication used for the treatment of mental disorders.

2.3 Open frontiers and conclusions

Overall, future directions on the field of neuroscience should be focused on efficacy accurate data signal analysis and processing research derived pipelines which could easily and automatically bridge research, clinical applications to commercial exploitation. The goal of precision medicine is to create optimally aimed and timed interventions customized to an individual's neurological disease. This has become plausible owing to the rapidly increasing understanding of machine learning that is based on the finest application of signal analysis and processing approaches [42] to tailored neurological disease treatments while providing precise medicine. New techniques for enhancing the collection, storage, cleaning, handling and interpretation of the signal information are still being created to improve current approaches in precision medicine.

However, many difficulties still need to be addressed in neuroscience for the accurate results in precision medicine. For example, data interoperability that addresses the ability of systems and services that generate and exchange data to have clear, shared expectations for the content, context, and meaning of that data without any limitations on neurophysiological data execution or access. For instance, future research in the automatic classification of the sleep stage should focus on experimenting with a greater sample size and extracting and choosing finer characteristics. The use of big data should be a goal to build trust concerning the results ensued from deep learning methodologies, thus precision medicine. Big data in neuroscience can be best described as rapid data increase in the volume, variety and speed of available data information, not only in neuroscience but also in other research fields. Applying larger datasets through signal analysis and processing approaches play a pivotal role in neuroscience. They secure better data validation which can lead to accurate individualized suggestions in predictive medicine in order to prevent or significantly decrease the impact of certain pathological conditions. Beside the aforementioned challenging issues that larger datasets can address, they can also advance current infrastructures which facilitate data sharing and protection, among (neuro) scientists while enhancing research collaboration. While creating user-friendly infrastructures will allow even users with minimal technology skills to explore, navigate, and use scientific data and services provided by those environments.

To that end, future work should focus on the (a) design and implementation of secure and protected systems, (b) advancement of cloud based web applications, (c) facilitation of easy data deployment, (d) reusability and sharing of tools with adaptability to changing requirements and (e) empowering of researchers to share new or advanced functionalities through publications in order to improve precision medicine in the field of neuroscience.

References

[1] Luck S.J. *An Introduction to the Event-Related Potential Technique*. Boston, MA: MIT Press, 2014

[2] Frantzidis C.A., Dimitriadou C.K., Chriskos P., *et al.* 'Cortical connectivity analysis for assessing the impact of microgravity and the efficacy of reactive sledge jumps countermeasure to NREM 2 sleep'. *Acta Astronautica*. 2018;**169**:579–89

[3] Chriskos P., Frantzidis C.A., Gkivogkli P.T., Bamidis P.D., and Kourtidou-Papadeli C. 'Achieving accurate automatic sleep staging on manually pre-processed EEG data through synchronization feature extraction and graph metrics'. *Frontiers in Human Neuroscience*. 2018;**12**:110

[4] Stam C.J., Breakspear M., van Walsum A.M.V.C., and van Dijk B.W. 'Nonlinear synchronization in EEG and whole-head MEG recordings of healthy subjects'. *Human Brain Mapping*. 2003;**19**(2):63–78

[5] Frantzidis C.A., Ladas A.K.I., Vivas A.B., Tsolaki M., and Bamidis P.D. 'Cognitive and physical training for the elderly: evaluating outcome efficacy by means of neurophysiological synchronization'. *International Journal of Psychophysiology*. 2014;**93**(1):1–11

[6] Rosso O.A., Blanco S., Yordanova J., *et al.* 'Wavelet entropy: a new tool for analysis of short duration brain electrical signals'. *Journal of Neuroscience Methods*. 2001;**105**(1):65–75

[7] Frantzidis C.A., Vivas A.B., Tsolaki A., Klados M.A., Tsolaki M., and Bamidis P.D. 'Functional disorganization of small-world brain networks in mild Alzheimer's disease and amnestic mild cognitive impairment: an EEG study using relative wavelet entropy (RWE)'. *Frontiers in Aging Neuroscience*. 2014;**6**:224

[8] Awan F.G., Saleem O., and Kiran A. 'Recent trends and advances in solving the inverse problem for EEG source localization'. *Inverse Problems in Science and Engineering*. 2019;**27**(11):1521–36

[9] Grech R., Cassar T., Muscat J., *et al.* 'Review on solving the inverse problem in EEG source analysis'. *Journal of Neuroengineering and Rehabilitation*. 2008;**5**(1):25

[10] Pascual-Marqui R.D. 'Review of methods for solving the EEG inverse problem'. *International Journal of Bioelectromagnetism*. 1999;**1**(1):75–86

[11] Wendel K., Väisänen O., Malmivuo J., *et al.* 'EEG/MEG source imaging: methods, challenges, and open issues'. *Computational Intelligence and Neuroscience*. 2009;**2009**:13

[12] Clermont G., and Zenker S. 'The inverse problem in mathematical biology'. *Mathematical Biosciences*. 2015;**260**:11–15

[13] Hauk O., Wakeman D.G., and Henson R. 'Comparison of noise-normalized minimum norm estimates for MEG analysis using multiple resolution metrics'. *NeuroImage*. 2011;**54**(3):1966–74

[14] Pascual-Marqui R.D., Esslen M., Kochi K., and Lehmann D. 'Functional imaging with low-resolution brain electromagnetic tomography (LORETA): a review'. *Methods and Findings in Experimental and Clinical Pharmacology*. 2002;**24**(Suppl C):91–95

[15] Pascual-Marqui R.D., Lehmann D., Koenig T., *et al.* 'Low resolution brain electromagnetic tomography (LORETA) functional imaging in acute, neuroleptic-naive, first-episode, productive schizophrenia'. *Psychiatry Research: Neuroimaging*. 1999;**90**(3):169–79

[16] Pettie M., Oakly A., Harper D.N., and Ellenbroek B.A. 'A genetic deletion of the serotonin transporter differentially influences the behavioural effects of MDMA'. *Journal of Psychopharmacology*. 2019;**33**(3):355–63

[17] Friston K.J., Frith C.D., Liddle P.F., and Frackowiak R.S.J. 'Functional connectivity: the principal-component analysis of large (PET) data sets'. *Journal of Cerebral Blood Flow and Metabolism*. 1993;**13**(1):5–14

[18] Horwitz B. 'The elusive concept of brain connectivity'. *NeuroImage*. 2003;**19**(2):466–70

[19] Klados M.A., Pandria N., Micheloyannis S., Margulies D., and Bamidis P.D. 'Math anxiety: brain cortical network changes in anticipation of doing mathematics'. *International Journal of Psychophysiology*.2017;**122**:24–31

[20] Pezoulas V.C., Zervakis M., Michelogiannis S., and Klados M.A. 'Resting-state functional connectivity and network analysis of cerebellum with respect to IQ and gender'. *Frontiers in Human Neuroscience*. 2017;**11**:189

[21] Zilidou V.I., Frantzidis C.A., Romanopoulou E.D., Paraskevopoulos E., Douka S., and Bamidis P.D. 'Functional re-organization of cortical networks of senior citizens after a 24-week traditional dance program'. *Frontiers in Aging Neuroscience*. 2018;**10**:422

[22] Athanasiou A., Klados M.A., Styliadis C., Foroglou N., Polyzoidis K., and Bamidis P.D. 'Investigating the role of alpha and beta rhythms in functional motor networks'. *Neuroscience*. 2018;**378**:54–70

[23] Klados M.A., Paraskevopoulos E., Pandria N., and Bamidis P.D. 'The impact of math anxiety on working memory: a cortical activations and cortical functional connectivity EEG study'. *IEEE Access*. 2019;**7**:15027–39

[24] Messé A., Rudrauf D., Giron A., and Marrelec G. 'Predicting functional connectivity from structural connectivity via computational models using MRI: an extensive comparison study'. *NeuroImage*. 2015;**111**:65–75

[25] Ellenbroek B., and Youn J. 'Schizophrenia'. In *Gene-Environment Interactions in Psychiatry*. 2016, pp. 289–322

[26] Stephan K.E., and Friston K.J. 'Functional Connectivity'. In *Encyclopedia of Neuroscience*. 2010, pp. 391–97

[27] Zalesky A., Fornito A., and Bullmore E.T. 'Network-based statistic: identifying differences in brain networks'. *NeuroImage* 2010;**53**(4):1197–207

[28] Bullmore E.T., Suckling J., Overmeyer S., Rabe-Hesketh S., Taylor E., and Brammer M.J. 'Global, voxel, and cluster tests, by theory and permutation, for a difference between two groups of structural MR images of the brain'. *IEEE Transactions on Medical Imaging*. 1999;**18**(1):32–42

[29] Nichols T.E., and Holmes A.P. 'Nonparametric permutation tests for functional neuroimaging: a primer with examples'. *Human Brain Mapping*. 2002;**15**(1):1–25

[30] Fornito A., Zalesky A., Pantelis C., and Bullmore E.T. 'Schizophrenia, neuroimaging and connectomics'. *NeuroImage*. 2012;**62**(4):2296–314

[31] Long Z., Duan X., Wang Y., *et al.* 'Disrupted structural connectivity network in treatment-naive depression'. *Progress in Neuro-Psychopharmacology and Biological Psychiatry*. 2015;**56**:18–26

[32] Li W., Mai X., and Liu C. 'The default mode network and social understanding of others: what do brain connectivity studies tell us'. *Frontiers in Human Neuroscience*. 2014;**8**:74

[33] Cao Q., Shu N., An L., *et al.* 'Probabilistic diffusion tractography and graph theory analysis reveal abnormal white matter structural connectivity networks in drug-naive boys with attention deficit/hyperactivity disorder'. *Journal of Neuroscience*. 2013;**33**(26):10676–87

[34] Hong S.B., Zalesky A., Fornito A., *et al.* 'Connectomic disturbances in attention-deficit/hyperactivity disorder: a whole-brain tractography analysis'. *Biological Psychiatry*. 2014;**76**(8):656–63

[35] Mulders P.C., van Eijndhoven P.F., and Beckmann C.F. Identifying Large-Scale Neural Networks Using fMRI. Systems Neuroscience in Depression 2016 Jan 1 (pp. 209-237). Academic Press

[36] Bullmore E., and Sporns O. 'Complex brain networks: graph theoretical analysis of structural and functional systems'. *Nature Reviews Neuroscience*. 2009;**10**(3):186

[37] Dwyer D.B., Harrison B.J., Yücel M., *et al.* Adolescent Cognitive Control: Brain Network Dynamics. InStress: Concepts, Cognition, Emotion, and Behavior 2016 Jan 1 (pp. 177–185). Academic Press

[38] Delorme A., and Makeig S. 'EEGLAB: an open source toolbox for analysis of single-trial EEG dynamics (pdf, 0.7 MB)'. *Journal of Neuroscience Methods*. 2004;**134**(1):9–21

[39] Pezoulas V.C., Athanasiou A., Nolte G., *et al.* FCLAB: An EEGLAB module for performing functional connectivity analysis on single-subject EEG data. IEEE EMBS International Conference on Biomedical and Health Informatics (BHI) 2018 Mar 4 (pp. 96–99)

[40] Wallander M.A., Johansson S., Ruigómez A., Rodríguez L.A.G., and Jones R. 'Morbidity associated with sleep disorders in primary care: a longitudinal

cohort study'. *Primary Care Companion to the Journal of Clinical Psychiatry.* 2007;**9**(5):338

[41] Skeldon A.C., Dijk D.J., and Derks G. 'Mathematical models for sleep-wake dynamics: comparison of the two-process model and a mutual inhibition neuronal model'. *PloS One.* 2014;**9**(8):e103877

[42] Chriskos P., Frantzidis C.A., Gkivogkli P.T., Bamidis P.D., and Kourtidou-Papadeli C. 'Automatic Sleep Staging Employing Convolutional Neural Networks and Cortical Connectivity Images'. *IEEE transactions on neural networks and learning systems.* 2019

[43] Frantzidis C.A., Kontana E., Karkala A., *et al.* 'Current trends and future perspectives of space neuroscience towards preparation for interplanetary missions'. *Neurology India.* 2019;**67**(8):182

[44] Kerr M.P., Mensah S., Besag F., *et al.* 'International consensus clinical practice statements for the treatment of neuropsychiatric conditions associated with epilepsy'. *Epilepsia.* 2011;**52**(11):2133–8

[45] Kanner A.M. 'Depression and epilepsy: a bidirectional relation?'. *Epilepsia.* 2011;**52**:21–7

[46] Chang Y.T., Chen P.C., Tsai I.J., *et al.* 'Bidirectional relation between schizophrenia and epilepsy: a population-based retrospective cohort study'. *Epilepsia.* 2011;**52**(11):2036–42

[47] Klados M.A., Papadelis C.L., and Bamidis P.D. REG-ICA: a new hybrid method for eog artifact rejection. 9th International Conference on Information Technology and Applications in Biomedicine 2009 Nov 4 (pp. 1–4)

[48] Klados M.A., Papadelis C., Braun C., and Bamidis P.D. 'REG-ICA: a hybrid methodology combining blind source separation and regression techniques for the rejection of ocular artifacts'. *Biomedical Signal Processing and Control.* 2011;**6**(3): 291–300

[49] Klados M.A., Bratsas C., Frantzidis C., Papadelis C.L., and Bamidis P.D. A Kurtosis-based automatic system using naïve bayesian classifier to identify ICA components contaminated by EOG or ECG artifacts. InXII Mediterranean Conference on Medical and Biological Engineering and Computing 2010 (pp. 49–52). Springer, Berlin, Heidelberg

[50] De Munck J.C. 'The estimation of time varying dipoles on the basis of evoked potentials'. *Electroencephalography and Clinical Neurophysiology/Evoked Potentials Section.* 1990;**77**(2):156–60

[51] Whittingstall K., Stroink G., Gates L., Connolly J.F., and Finley A. 'Effects of dipole position, orientation and noise on the accuracy of EEG source localization'. *Biomedical Engineering Online.* 2003;**2**(1):14

[52] Baillet S., and Garnero L. 'A Bayesian approach to introducing anatomo-functional priors in the EEG/MEG inverse problem'. *IEEE Transactions on Biomedical Engineering.* 1997;**44**(5):374–85

[53] Riera J.J., Valdes P.A., Fuentes M.E., and Oharriz Y. Explicit Backus and Gilbert EEG inverse solution for spherical symmetry. Department of Neurophysics, Cuban Neuroscience Center, Havana, Cuba. 2002

[54] Grave de Peralta Menendez R., Hauk O., Gonzalez Andino S., Vogt H., and Michel C. 'Linear inverse solutions with optimal resolution kernels applied to electromagnetic tomography'. *Human Brain Mapping*. 1997;**5**(6):454–67

[55] Biswal B.B., Kylen J.V., and Hyde J.S. 'Simultaneous assessment of flow and BOLD signals in resting-state functional connectivity maps'. *NMR in Biomedicine*. 1997;**10**(4–5):165–70

[56] Beauchene C., Roy S., Moran R., Leonessa A., and Abaid N. 'Comparing brain connectivity metrics: a didactic tutorial with a toy model and experimental data'. *Journal of Neural Engineering*. 2018;**15**(5):056031

[57] Bastos A.M., and Schoffelen J.M. 'A tutorial review of functional connectivity analysis methods and their interpretational pitfalls'. *Frontiers in Systems Neuroscience*. 2016;**9**:175

[58] Pearson K. Notes on regression and inheritance in the case of two parents. *Proceedings of the Royal Society of London*. 1895;**58**(347–352):240–42

[59] Bracewell R. *Pentagram Notation for Cross Correlation. The Fourier Transform and its Applications*. New York: McGraw-Hill, 1965;**46**:243

[60] French C.C., and Beaumont J.G. 'A critical review of EEG coherence studies of hemisphere function'. *International Journal of Psychophysiology*. 1984;**1**(3):241–54

[61] Nolte G., Bai O., Wheaton L., Mari Z., Vorbach S., and Hallet M. 'Identifying true brain interaction from EEG data using the imaginary part of coherency'. *Clinical Neurophysiology*. 2004;**115**(10):2292–307

[62] Lachaux J.P., Rodriguez E., Martinerie J., and Varela F.J. 'Measuring phase synchrony in brain signals'. *Human Brain Mapping*. 1999;**8**(4):194–208

[63] Stam C.J., Nolte G., and Daffertshofer A. 'Phase lag index: assessment of functional connectivity from multi channel EEG and MEG with diminished bias from common sources'. *Human Brain Mapping*, 2007;**28**(11):1178–93

[64] Vinck M., Oostenveld R., Van Wingerden M., Battaglia F., and Pennartz C. M. 'An improved index of phase-synchronization for electrophysiological data in the presence of volume-conduction, noise and sample-size bias'. *NeuroImage*. 2011;**55**(4):1548–65

[65] Tass P., Rosenblum M.G., Weule J., *et al*. 'Detection of n:m phase locking from noisy data: application to magnetoencephalography'. *Physical Review Letters*. 1998;**81**(15):291–4

[66] Rosenblum M., and Pikovsky A. 'Detecting direction of coupling in interacting oscillators'. *Physical Review E*. 2001;**64**(4):045202

[67] Stam C.J., and Van Dijk B.W. 'Synchronization likelihood: an unbiased measure of generalized synchronization in multivariate data sets'. *Physica D: Nonlinear Phenomena*. 2002;**163**(3–4):236–51

[68] Granger C.W. 'Investigating causal relations by econometric models and cross-spectral methods'. *Econometrica: Journal of the Econometric Society*. 1969;424–38

[69] Baccalá L.A., and Sameshima K. 'Partial directed coherence: a new concept in neural structure determination'. *Biological Cybernetics*. 2001;**84**(6): 463–74

[70] Kamiński M.J., and Blinowska K.J. 'A new method of the description of the information flow in the brain structures'. *Biological Cybernetics.* 1991;**65**(3): 203–10

[71] Shannon C.E. 'A mathematical theory of communication'. *Bell System Technical Journal.* 1948;**27**(3):379–423

[72] Cover T.M., and Thomas J.A. *Elements of Information Theory*. New York, NY: John Wiley & Sons, 2012

[73] Frenzel S., and Pompe B. 'Partial mutual information for coupling analysis of multivariate time series'. *Physical Review Letters.* 2007;**99**(20):204101

[74] Schreiber T. 'Measuring information transfer'. *Physical Review Letters.* 2000;**85**(2):461–4

[75] Papana A., Kugiumtzis D., and Larsson P.G. 'Detection of direct causal effects and application to epileptic electroencephalogram analysis'. *International Journal of Bifurcation and Chaos.* 2012;**22**(09):1250222

[76] Fraschini M., Lai M., and Didaci L. 'Comparison of functional connectivity metrics using an unsupervised approach: a source resting-state EEG study'. *Journal of Integrative Neuroscience.* 2018;**17**(4):393–6

[77] Zhang Z., Telesford Q.K., Giusti C., Lim K.O., and Bassett D.S. 'Choosing wavelet methods, filters, and lengths for functional brain network construction'. *PloS One.* 2016;**11**(6):e0157243

[78] Niso G., Bruña R., Pereda E., *et al.* 'HERMES: towards an integrated toolbox to characterize functional and effective brain connectivity'. *Neuroinformatics.* 2013;**11**(4):405–34

[79] He B., Dai Y., Astolfi L., Babiloni F., Yuan H., and Yang L. 'eConnectome: a MATLAB toolbox for mapping and imaging of brain functional connectivity'. *Journal of Neuroscience Methods.* 2011;**195**(2):261–9

[80] Zhou D., Thompson W.K., and Siegle G. 'MATLAB toolbox for functional connectivity'. *NeuroImage.* 2009;**47**(4):1590–607

[81] Christodoulou E.G., Sakkalis V., Tsiaras V., and Tollis I.G. 'BrainNetVis: an open-access tool to effectively quantify and visualize brain networks'. *Computational intelligence and neuroscience.* 2011

[82] Wasserstein R.L., and Lazar N.A. 'The ASA's statement on p-values: context, process, and purpose'. *The American Statistician.* 2016;**70**(2): 129–33

[83] Tutte W.T. *Graph Theory*, Cambridge: Cambridge University Press, 2001, p. 30

[84] Humphries M.D., and Gurney K. 'Network "small-world-ness": a quantitative method for determining canonical network equivalence'. *PloS One.* **3** (4):e0002051

[85] Latora V., and Marchiori M. 'Efficient behavior of small-world networks'. *Physical Review Letters.* 2001;**87**(19):198701

[86] Watts D.J., and Strogatz S.H. 'Collective dynamics of "small-world" networks'. *Nature.* 1998;**393**(6684):440

[87] Newman M.E. 'Modularity and community structure in networks'. *Proceedings of the National Academy of Sciences.* 2006;**103**(23):8577–82

[88] Kwon Y. *4.2.4: Partition and Distribution Coefficients. Handbook of Essential Pharmacokinetics, Pharmacodynamics and Drug Metabolism for Industrial Scientists*. New York: Kluwer Academic/Plenum Publishers; 2001. p. 44

[89] Tagliazucchi E., von Wegner F., Morzelewski A., Borisov S., Jahnke K., and Laufs H. 'Automatic sleep staging using fMRI functional connectivity data'. *NeuroImage*. 2012;**63**(1):63–72

[90] Tadel F., Baillet S., Mosher J.C., Pantazis D., and Leahy R.M. 'Brainstorm: a user-friendly application for MEG/EEG analysis'. *Computational Intelligence and Neuroscience*. 2011;**2011**:8

[91] Shao Y., Wang L., Ye E., *et al*. 'Decreased thalamocortical functional connectivity after 36 hours of total sleep deprivation: evidence from resting state FMRI'. *PLoS One*. 2013;**8**(10):e78830

[92] Akar S.A., Kara S., Latifoğlu F., and Bilgiç V. 'Analysis of the complexity measures in the EEG of schizophrenia patients'. *International Journal of Neural Systems*. 2016;**26**(02):1650008

[93] Gomez-Pilar J., Lubeiro A., Poza J., *et al*. 'Functional EEG network analysis in schizophrenia: evidence of larger segregation and deficit of modulation'. *Progress in Neuro-Psychopharmacology and Biological Psychiatry*. 2017;**76**:116–23

[94] Yin Z., Li J., Zhang Y., Ren A., Von Meneen K.M., and Huang L. 'Functional brain network analysis of schizophrenic patients with positive and negative syndrome based on mutual information of EEG time series'. *Biomedical Signal Processing and Control*. 2017;**31**:331–8

[95] Chu L., Qiu R., Liu H., Ling Z., Zhang T., and Wang J. 'Individual recognition in schizophrenia using deep learning methods with random forest and voting classifiers: Insights from resting state EEG streams'. 2017;arXiv preprint arXiv:1707.03467

[96] Johannesen J.K., Bi J., Jiang R., Kenney J.G., and Chen C.M.A. 'Machine learning identification of EEG features predicting working memory performance in schizophrenia and healthy adults'. *Neuropsychiatric Electrophysiology*. 2016;**2**(1):3

[97] Ludwig S., Spitzer B., Jacobs A.M., Sekutowicz M., Sterzer P., and Blankenburg F. 'Spectral EEG abnormalities during vibrotactile encoding and quantitative working memory processing in schizophrenia'. *NeuroImage: Clinical*. 2016;**11**:578–87

[98] Kim J.W., Lee Y.S., Han D.H., Min K.J., Lee J., and Lee K. 'Diagnostic utility of quantitative EEG in un-medicated schizophrenia'. *Neuroscience Letters*. 2015;**589**:126–31

[99] Baradits M., Kakuszi B., Bálint S., *et al*. 'Alterations in resting-state gamma activity in patients with schizophrenia: a high-density EEG study'. *European Archives of Psychiatry and Clinical Neuroscience*. 2019;**269**(4):429–37

[100] Delorme A., Mullen T., Kothe C., *et al*. 'EEGLAB, SIFT, NFT, BCILAB, and ERICA: new tools for advanced EEG processing'. *Computational Intelligence and Neuroscience*. 2011;**2011**:10

[101] Lopez-Calderon J., and Luck S.J. 'ERPLAB: an open-source toolbox for the analysis of event-related potentials'. *Frontiers in Human Neuroscience.* 2014;**8**:213

[102] Mognon A., Jovicich J., Bruzzone L., and Buiatti M. 'ADJUST: an automatic EEG artifact detector based on the joint use of spatial and temporal features'. *Psychophysiology.* 2011;**48**(2):229–40

[103] Delorme A., Palmer J., Onton J., Oostenveld R., and Makeig S. 'Independent EEG sources are dipolar'. *PloS One.* 2012;**7**(2):e30135

[104] Lund T.R., Sponheim S.R., Iacono W.G., and Clementz B.A. 'Internal consistency reliability of resting EEG power spectra in schizophrenic and normal subjects'. *Psychophysiology.* 1995;**32**(1):66–71

[105] Thatcher R.W., Palmero-Soler E., North D.M., and Biver C.J. 'Intelligence and EEG measures of information flow: efficiency and homeostatic neuroplasticity'. *Scientific Reports.* 2016;**6**:38890

[106] Bachmann M., Lass J., and Hinrikus H. 'Single channel EEG analysis for detection of depression'. *Biomedical Signal Processing and Control.* 2017;**31**:391–7

[107] Acharya U.R., Sudarshan V.K., Adeli H., Santhosh J., Koh J.E., and Adeli A. 'Computer-aided diagnosis of depression using EEG signals'. *European Neurology.* 2015;**73**(5–6):329–36

[108] Arns M., Etkin A., Hegerl U., *et al.* 'Frontal and rostral anterior cingulate (rACC) theta EEG in depression: implications for treatment outcome?'. *European Neuropsychopharmacology.* 2015;**25**(8):1190–200

[109] Koo P.C., Thome J., Berger C., Foley P., and Hoeppner J. 'Current source density analysis of resting state EEG in depression: a review'. *Journal of Neural Transmission.* 2017;**124**(1):109–18

[110] Li Y., Cao D., Wei L., Tang Y., and Wang J. 'Abnormal functional connectivity of EEG gamma band in patients with depression during emotional face processing'. *Clinical Neurophysiology.* 2015;**126**(11):2078–89

[111] Rommel A.S., Kitsune G.L., Michelini G., *et al.* 'Commonalities in EEG spectral power abnormalities between women with ADHD and women with bipolar disorder during rest and cognitive performance'. *Brain Topography.* 2016;**29**(6):856–66

[112] Van Elst L.T., Fleck M., Bartels S., *et al.* 'Increased prevalence of intermittent rhythmic delta or theta activity (IRDA/IRTA) in the electroencephalograms (EEGs) of patients with borderline personality disorder'. *Frontiers in Behavioral Neuroscience.* 2016;**10**:12

[113] Wang Y., Chai F., Zhang H., *et al.* 'Cortical functional activity in patients with generalized anxiety disorder'. *BMC Psychiatry.* 2016;**16**(1):217

[114] Berger H. 'Das Elektrenkephalogram des Menschen. 12 Mittlg. (12th report)'. *Arch Psychiatr Nervenkr.* 1937;**106**:165–87

[115] Da Silv F.L. *Electroencephalography: Basic Principles, Clinical Applications, and Related Fields.* Philadelphia, PA: Lippincott Williams & Wilkins, 2005

Chapter 3

Neuroimage acquisition and analysis

Thomas Schultz[1]

Neuroimaging has facilitated major advances in understanding the human brain, and provides invaluable guidance for diagnosis and therapy of disease. This chapter provides a broad overview of human *in-vivo* neuroimaging techniques, and introduces algorithms for neuroimage processing and analysis. Covered modalities include structural, functional, and diffusion MRI (dMRI), positron emission tomography (PET), and magnetoencephalography (MEG). We discuss state-of-the-art algorithms for image registration and segmentation, for statistical analysis, and for predictive modeling.

3.1 Introduction

In recent years, the range, availability, and quality of techniques for imaging the structure and function of the human brain has rapidly increased. However, obtaining scientific insight, diagnostic, or therapeutic guidance from neuroimaging data requires a complex pipeline in which image acquisition is only the initial step, and in which computational techniques play a key role.

This chapter provides an overview of that pipeline. In particular, Section 3.2 covers image acquisition with the most widely used modalities, such as magnetic resonance imaging (MRI), its variants functional and dMRI, PET, as well as electroencephalography (EEG) and MEG. The rest of the chapter covers image processing and data analysis. This starts in Section 3.3 with image registration, which is required to bring the images taken at different times, with different modalities, or from different subjects, into anatomical correspondence. Section 3.4 covers different approaches to image segmentation, from classical deformable models and Markov random fields (MRFs) to modern convolutional neural networks (CNNs). Section 3.5 introduces the most important statistical methods that are used for inference, and Section 3.6 highlights some recent approaches to integrating neuroimaging data into predictive models. Finally, Section 3.7 provides an outlook on potential future developments.

[1]B-IT and Department of Computer Science, University of Bonn, Bonn, Germany

3.2 Neuroimaging modalities

The popularity of MRI for neuroimaging is due to the facts that it does not involve ionizing radiation, it achieves a wide range of contrasts without the need to inject contrast agents, and it is well suited for imaging soft tissues such as the brain. It is also fast enough for functional imaging, and flexible in the sense that it can be sensitized to different physical tissue properties.

We will begin with a brief introduction to MRI. Textbooks such as [1] are available for readers with a deeper interest in its physical foundations. Two separate subsections will be devoted to specific variants of MRI that are particularly relevant to neuroimaging: in functional MRI (fMRI), changes associated with blood flow are used to map neural activity [2]. In dMRI, the effects of structures at the cellular scale on water self-diffusion are observed in order to infer information about tissue properties that are far below the image resolution [3].

We will also cover two alternative imaging modalities that are used to map brain function: PET uses radioactive tracers that take part in metabolic activity, and therefore allows for more specific investigations. MEG measures the electromagnetic fields generated by neural activity, and provides higher temporal resolution than fMRI.

3.2.1 Magnetic resonance imaging

In vivo brain MRI is based on the magnetic moment of hydrogen nuclei. MR scanners generate a strong magnetic field, which is referred to as the B_0 field. It causes the nuclei to align either in a parallel or an anti-parallel configuration, corresponding to slightly lower or higher energy levels, respectively. At equilibrium, a tiny excess of nuclei is in the lower energy state. The individual nuclei rotate around an axis that itself rotates (precesses) around B_0. The precession frequency is proportional to the strength of the external magnetic field, and is called the Larmor frequency. Initially, the phases of this precession are random. Therefore, averaging over a volume element (voxel) leads to a net magnetization vector that is aligned with the direction of the B_0 field (longitudinal axis). The contributions from different nuclei in the transverse plane cancel out.

The MR signal is generated by lifting some of the nuclei into their higher energy state. This requires electromagnetic radiation whose frequency needs to match the Larmor frequency, to achieve *magnetic resonance.* This electromagnetic wave is referred to as a radiofrequency (RF) pulse, since the Larmor frequency at typical B_0 field strengths is in the same range as radio transmissions. Magnetic resonance brings the nuclei in phase with the RF pulse. This generates a transverse component in the net magnetization, which rotates at the Larmor frequency. An RF pulse with suitable amplitude and duration can cause the longitudinal part of the net magnetization to vanish. This is referred to as a 90° pulse, since the net magnetization has been flipped into the transverse plane.

After the RF pulse has been turned off, the net magnetization gradually returns to its equilibrium state: the longitudinal part returns, and the transverse part vanishes. Both happen exponentially, with time constants T_1 and T_2^*, respectively. The MR signal is generated by the rotating transverse magnetization. After its initial decay, the signal is made to return either using magnetic field gradients (gradient recalled echo) or a 180° RF pulse (spin echo). In the latter case, the magnitude of the echo is determined by another time constant, T_2. Different biological tissues, such as gray and white matter in the brain, have different relaxation times, which gives rise to the MR image contrast between them.

Generating an image requires knowing the spatial origin of the signal, but the mechanisms described so far do not yet allow us to distinguish between parts of the signal that arise at different locations. This is achieved by modulating the static B_0 field with magnetic field gradients at different times during the measurement sequence. Since the Larmor frequency is proportional to the magnetic field strength, applying a field gradient causes the Larmor frequency to vary along its direction. Therefore, applying an RF pulse with limited bandwidth while a spatial gradient is active will only cause magnetic resonance in a corresponding two-dimensional slice, perpendicular to the gradient. This is known as slice selection, and is used to acquire three-dimensional volumes one slice after the other.

Two additional mechanisms, frequency encoding and phase encoding, are required to encode spatial position within two-dimensional slices. To achieve frequency encoding, a magnetic field gradient is applied during signal readout (i.e., during the echo). The resulting signal is still a mixture from the entire selected slice. However, position along the direction of the frequency encoding gradient can now be recovered by analyzing the spectrum, since differences in position have been encoded as differences in frequency. Phase encoding recovers the remaining dimension, by applying a corresponding gradient between excitation and readout. This encodes spatial position in the phase of the signal.

After a variable substitution, the effects of frequency and phase encoding can be interpreted as a two-dimensional Fourier transformation of the desired slice image. In this context, the Fourier domain is referred to as k-space. MR images are reconstructed by first filling the corresponding k-space matrix with MR measurements, and then computing its inverse Fourier transform, as illustrated in Figure 3.1. High-quality anatomical MRI acquires a single line of k-space per excitation, making it relatively time consuming. For functional and dMRI, echo planar imaging acquires a full 2D slice with a single excitation, at the cost of reduced resolution and increased image noise.

3.2.2 Functional MRI

Comparing repeated MR scans taken while the brain performed different functions reveals that neural activity leads to small differences in MR image intensity in the corresponding brain region. This effect is now widely used to investigate with MRI not only the anatomical structure, but also the functional parcellation of the brain.

Variations in MR signal intensity have been attributed to differences in the magnetic properties of oxygenated and deoxygenated blood: oxygenated blood is diamagnetic, i.e., it weakly counteracts the local magnetic field. Deoxygenated

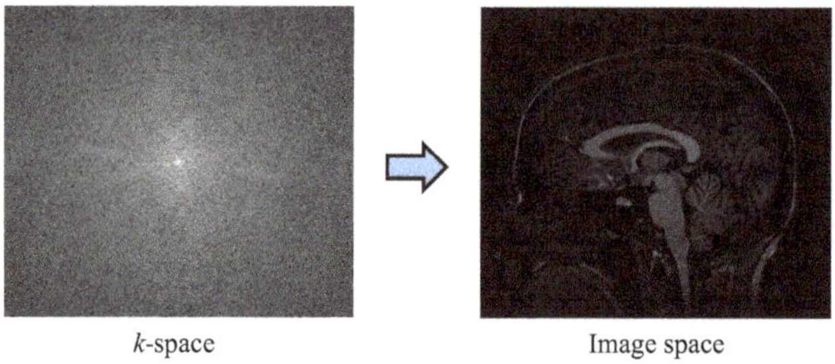

Figure 3.1 MRI encodes spatial position as frequency and phase. Consequently, images are acquired in k-space, and have to be reconstructed using the inverse Fourier transformation

blood, however, slightly enhances the field, decreasing T_2^*, and thereby attenuating the MR signal. This is referred to as the blood oxygenation level dependent (BOLD) effect.

Local blood oxygenation depends on neural activity through rather complex mechanisms of neurovascular coupling. This makes fMRI an indirect way of observing neural activity. The relationship between activity and signal change is modeled with a hemodynamic response function (HRF), which is measured empirically by averaging over many trials in which a short stimulus is presented, and the time course of the resulting signal change is observed. The exact shape of the HRF differs between subjects and brain regions. In general, a brief and weak "initial dip" is followed by a much stronger and longer increase in signal intensity, before the signal returns to the baseline after going through a final "undershoot" phase. The initial dip is explained by the onset of increased oxygen consumption; the subsequent increase is explained by an increased supply of oxygenated blood to the activated area, which overcompensates the additional need [2].

In order to achieve a temporal resolution on the order of a few seconds, fMRI is typically performed using echo planar imaging, an MR imaging sequence that acquires full two-dimensional slice images with a single spin excitation, leading to much faster imaging at the cost of reduced spatial resolution.

There are two main variants of fMRI: task-based fMRI studies the response to presenting certain stimuli, or asking subjects to perform specific tasks while in the scanner. In this case, the expected time course of the BOLD response is known a priori, and brain regions whose signal correlates significantly with it can be highlighted, as in Figure 3.2(a). In resting-state fMRI, measurements are taken over an extended period of time during which the subject is awake, but relaxed and instructed not to think about anything in particular. In the resulting data, significant correlations are observed between certain brain regions. They are taken as evidence of the corresponding regions being connected functionally. Statistical tools for the analysis of task-based and resting-state fMRI will be the topics of Sections 3.5.3 and 3.5.4, respectively.

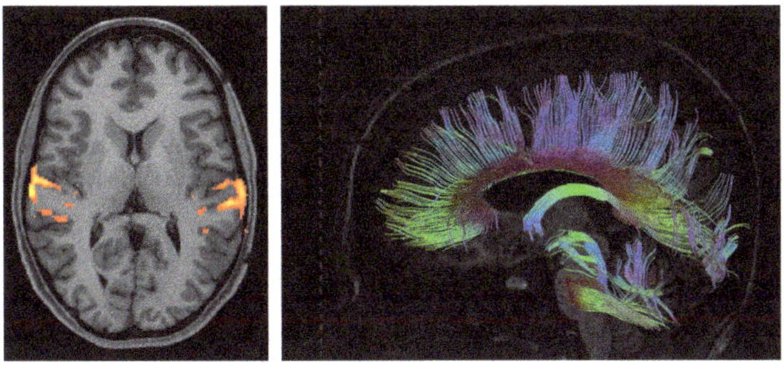

(a) fMRI activations (b) dMRI tractography

Figure 3.2 Functional MRI highlights regions of increased blood flow in the brain, overlaid in red/yellow on a structural MRI scan in (a). dMRI permits reconstruction of major white matter tracts. Some of them are shown in color in (b), again in the context of a structural scan

3.2.3 Diffusion MRI

The spatial resolution of *in-vivo* human brain MRI is on the order of 1 mm. Therefore, MR images do not directly reveal cellular structures, which are on the much smaller scale of micrometers. dMRI can be seen as a superresolution technique which uses MRI to acquire information on the microscale. This is achieved by observing the natural self-diffusion (Brownian heat motion) of water molecules. During diffusion times on the order of milliseconds, molecular motion is restricted and hindered by cellular structures. Therefore, observing it provides a means to probe microstructure that is not visible in the image directly [3].

In order to take diffusion MR measurements, additional magnetic field gradients are added to a standard MR sequence, resulting in a diffusion weighting. Consequently, molecular motion along a pre-specified gradient direction will lead to an attenuation of the MR signal. Comparing the resulting intensity to a measurement without diffusion weighting then provides an estimate of the apparent diffusivity in the given direction. Taking repeated measurements with varying directions reveals the anisotropy (directional dependence) of diffusivities. Diffusion tensor imaging (DTI) is one particular variant of dMRI. It models anisotropic diffusion with a 3×3 symmetric positive definite diffusion tensor. It is particularly informative in the white matter tracts that connect different brain regions: they contain bundles of coherently oriented axons, which lead to high apparent diffusivities along the tract direction, and decreased diffusivities perpendicular to it.

Information encoded in the diffusion tensor includes the mean diffusivity, which has been found to be informative for the diagnosis of stroke. The tensor also reveals the extent to which diffusivity is anisotropic, which correlates, among other things, with the density and myelination of neural fibers. Finally, it indicates the principal direction, in which diffusivity is greatest. Assuming fiber orientation is sufficiently homogeneous within the voxel, this principal direction can serve as its estimate. The integration of curves that are everywhere tangential to the principal direction has been found to result in plausible three-dimensional reconstructions of major tracts, and has been referred to as tractography [4]. It is illustrated in Figure 3.2(b), and is now used in surgery planning, as well as in scientific studies.

Due to its limited requirements, diffusion tensor imaging remains the most popular variant of dMRI in a clinical context. However, much more advanced acquisition schemes and mathematical models are now available, and are used in large-scale initiatives such as the Human Connectome Project [5], or the Rhineland Study [6]. One goal of advanced dMRI is to increase angular resolution, since a substantial fraction of white matter voxels has been found to contain multiple bundles in different directions [7], which cannot be modeled with standard DTI. Other extensions that reveal additional information include the use of diffusion encodings at varying strengths [8], diffusion times [9], and more complicated diffusion gradient waveforms that better distinguish between components in heterogeneous voxels [10].

3.2.4 Positron emission tomography

Molecular imaging visualizes activity at the molecular and cellular level. A popular example is PET, which is based on injecting into the subject biologically active molecules that have been modified to become radioactive. Such radioactive tracers aggregate in a characteristic fashion. For example, the widely used fludeoxyglucose F 18 indicates regions of high metabolic activity. Tracers used in PET emit positrons during their radioactive decay. After traveling a small distance, these positrons encounter their antiparticles, i.e., electrons. This leads to their annihilation and emission of a pair of photons at 511 keV, in opposite directions.

A ring of detector blocks around the patient registers coinciding photon pairs, and localizes the annihilation event to the line segment connecting the corresponding detectors, the line of response. From a list of such detections, the three-dimensional activity distribution can be reconstructed computationally. It represents the image acquired by PET. Quantitative reconstructions (in units of kBq/ml) are made possible by calibration with respect to a phantom with known activity. In medical practice, such values are further converted to standardized uptake values, which are normalized with respect to patient weight and injected dose [11].

Proper quantification of PET has to account for the attenuation which occurs on the path from the photon source (within the patient) to the detector through interactions with bones and tissues. Suitable attenuation maps can be derived from computed tomography (CT). Moreover, CT provides anatomical context, which is helpful for correctly interpreting the activities visualized by PET. Acquiring PET and CT on separate scanners results in the difficult problem of correctly aligning the two scans. This is greatly reduced by hybrid PET/CT, which integrates both modalities. Concerns about the combined radiation dose of PET and CT, and the limited soft tissue contrast provided by CT, have led to the development of PET/MR. However, unlike CT, MRI does not quantify the attenuation of electromagnetic radiation. Therefore, deriving attenuation maps from its results requires a more indirect strategy, e.g., using machine learning [12].

Compared to MRI, PET offers increased flexibility due to the availability of different and often highly specific tracers. For example, certain tracers indicate amyloid deposition, which goes along with Alzheimer's disease [13]. With suitable tracers, PET can also be used for activation studies. However, fMRI provides greater spatial and temporal resolution, and it does not require radioactive tracers. Consequently, it can be conducted without access to a cyclotron, and it does not expose the subject to ionizing radiation. The main reason why one might still consider PET for activation studies is its reduced sensitivity to subject motion [14].

3.2.5 EEG and MEG

Neural information processing involves small electric currents, which can be monitored with two related techniques. The first of them is EEG, which records the brain's electrical activity with multiple electrodes that are attached to the scalp. Compared to fMRI and PET, EEG provides a more direct measure of neural activity, as well as a much higher temporal resolution. However, brain tissue and skull act as a low-pass filter which significantly reduces the spatial resolution that can be obtained with scalp EEG [15], to an extent that it is sometimes not even classified as an "imaging" technique [16].

The electrical activity in the brain also produces magnetic fields. They pass through the skull relatively undisturbed, but are so weak that measuring them requires a magnetically shielded room, as well as highly sensitive detectors. Arrays of superconducting quantum interference devices (SQUIDs) have made it possible to measure the magnetic fields associated with neural activity from outside the skull, and form the technical basis of MEG.

EEG and MEG can be used to record spontaneous neural activity, or to capture event-related activity, such as the response to a sensory, motor, or cognitive event. In the latter case, it is common to repeat the experiment several times, and to average the results in order to improve the signal-to-noise ratio. Source localization is the task of computationally inferring the three-dimensional origin of a signal that was measured on the two-dimensional surface of the skull. It is easier when given MEG data compared to EEG, because the magnetic fields suffer from less distortion. However, even in this case, it is impossible to recover a unique solution for this inverse problem, unless additional assumptions are made concerning the number of sources (e.g., a single dipole) or their location (e.g., on the cortex) [17].

Magnetic source imaging (MSI) integrates data from MEG and MRI in order to constrain the source localization, and to visualize the resulting estimate within its anatomical context. Most applications of MSI have been for fundamental research, but some clinical applications have also been established [17].

3.3 Image registration

Image registration is an important step in many neuroimage analysis pipelines. It brings brain images into spatial alignment, so that anatomical structures are in correspondence when the images are overlaid (cf. Figure 3.3). Registration is

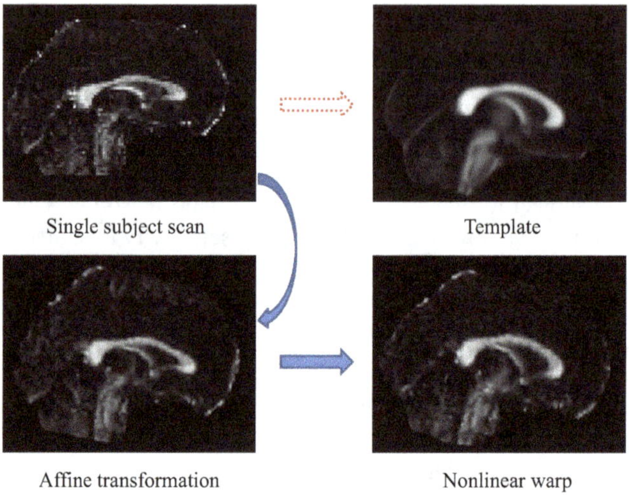

Figure 3.3 Aligning a single subject to a template is a typical use case of image registration. It is often done in two steps: an affine transformation results in an overall alignment of position, size, and rotation. A nonlinear warp is required to compensate for local shape differences

required to compare images between subjects, from different modalities, or taken at different points in time.

Registration is typically formalized as a nonlinear optimization problem, in which a cost function quantifies the extent to which a successful alignment has been achieved. Often, an optimization algorithm transforms one image until the cost with respect to a second ("reference") image is minimized. In other cases, all involved images are transformed simultaneously, so as not to bias the result toward any of the original images. It depends on the specific application which cost function is optimal, and which set of transformations should be considered. In this section, we will discuss the most widely used choices, as well as standard spaces that have been established via co-registration of a large number of brains.

3.3.1 *Cost functions for registration*

Since the goal of image registration is to bring anatomical structures into alignment, it might seem natural to localize specific structures in all involved images, and to deform the images so that the distances between corresponding structures are minimized. However, manual annotation of anatomical landmarks is labor intensive, and might not be fully reproducible; marking them automatically is difficult and error-prone. Therefore, we will focus on a widely used alternative, which is to quantify the achieved alignment directly based on the image intensities.

Defining an intensity-based objective function for image registration is easiest if we can assume that correctly aligned voxels will have similar intensity values, possibly up to some rescaling. Often, this is a reasonable assumption when the involved images have been acquired by the same modality. In this case, simple cost functions can be effective, such as the least-squares cost, which averages the squared differences between pixel values, or the normalized cross-correlation, which remains unaffected by a global rescaling of intensities. Slowly varying changes in overall image intensity, as they for example arise due to magnetic field inhomogeneities in MRI, can be compensated for by subtracting the local mean from images before comparing them; this is the main idea behind local cross-correlation.

A more sophisticated approach is required to successfully deal with cases in which the same structures might have a different visual appearance, as it often happens when imaging them with different modalities. In those cases, it is common to assess the degree of alignment with information theoretic measures, based on the individual and joint distributions of intensities. The intensity distribution of a single image X can be captured in a histogram: it partitions the overall range of values into equally sized intervals ("bins"), and records the frequencies with which voxel intensities fall into each bin. From the histogram, the entropy $H(X)$ can be computed. It can be interpreted as a measure of the image's information content. It attains its minimum when all voxels share the same intensity, and its maximum when all bins are equally populated.

Similarly, when comparing two images X and Y on their overlapping domain, a 2D histogram records the frequency with which certain intensities in X co-occur with specific intensities in Y. From this, their joint entropy $H(X, Y)$ can be computed. Even if intensities in the two images are not the same after correct alignment, registration should maximize their mutual dependence: it should make it easier to predict values in Y from values in X. Mutual information is an information theoretic measure that quantifies just that: the amount of information that we gain about the intensity in a given voxel of Y from observing the corresponding voxel of X. Formally, the mutual information of X and Y is defined from their individual and joint entropies as $I(X, Y) = H(X) + H(Y) - H(X, Y)$ [18]. It has been found that, in cases where the region of overlap between the images might vary during registration, a modified definition that is referred to as normalized mutual information $NMI(X, Y) = (H(X) + H(Y))/H(X, Y)$, provides a more robust alternative to $I(X, Y)$ [19].

3.3.2 Linear registration

It depends on the specific use case of image registration which spatial transformations should be considered in order to align the involved images. In the simplest case, such as when compensating for head motion within an imaging session, only rigid transformations make sense. They encompass global rotation and translation, and amount to six degrees of freedom. To compensate for differences in resolution, one might additionally allow for uniform scaling, leading to the set of shape-preserving similarity transformations. Adding nonuniform scaling and shear, which can help to compensate for certain image distortions, results in the 12-dimensional space of affine transformations.

Rotation, scaling, and shear are linear transformations. In three dimensions, they can thus be represented as 3×3 transformation matrices, and any combination of them can be multiplied out into a single matrix. Since translations are a very fundamental operation for image registration, it is computationally convenient to express them as matrices as well. This becomes possible when using homogeneous coordinates, which represent three-dimensional points using four coordinates, so that Cartesian coordinates $\mathbf{p}_c = (x, y, z)^T$ correspond to homogeneous coordinates $\mathbf{p}_h = (wx, wy, wz, w)^T$ with $w \in \mathbb{R}\backslash\{0\}$. Any non-zero scaling of $\mathbf{p}_h$ still represents the same point, and scaling by $1/w$ results in a canonical representation $\mathbf{p}_h = (x, y, z, 1)^T$. When using homogeneous coordinates, all affine transformations $\mathbf{p}'_c = \mathbf{L}\mathbf{p}_c + \mathbf{t}$ can be expressed as a single matrix-vector product $\mathbf{p}'_h = \mathbf{M}\mathbf{p}_h$, where

$$\mathbf{M} = \begin{pmatrix} L_{xx} & L_{xy} & L_{xz} & t_x \\ L_{yx} & L_{yy} & L_{yz} & t_y \\ L_{zx} & L_{zy} & L_{zz} & t_z \\ 0 & 0 & 0 & 1 \end{pmatrix} \tag{3.1}$$

with L_{ij} and t_k being the coefficients of linear transformation matrix $\mathbf{L}$ and translation vector $\mathbf{t}$, respectively. Registration algorithms that only apply affine transformations, or a subset of them, are often said to perform "linear registration". When using this terminology, we should keep in mind that they typically include translations which, as we just discussed, are not linear in the original Cartesian space. Despite this, they can be expressed as linear transformations when using homogeneous coordinates.

Now that we have a reference image Y, moving image X, cost function C, and a space S_T of allowable transformations in place, we can formulate registration as the following optimization problem, which aims to find the optimal transformation T^* in order to align X with Y:

$$T^* = \arg\min_{T \in S_T} C(Y, T(X)) \tag{3.2}$$

Some of the above-mentioned objective functions, such as normalized cross-correlation or mutual information, require maximization rather than minimization. However, we can easily turn them into a cost by scaling them by -1. A large variety of optimization algorithms have been applied to image registration, and a full survey is beyond our scope. Since C is typically non-convex, we lack efficient algorithms that guarantee finding its global minimum. Instead, local optimization is commonly employed. This implies the risk of terminating in a suboptimum, and is an important reason why registration methods can fail to provide a suitable alignment.

Systematically initializing the local optimization in different ways, and retaining the result with the lowest cost, greatly reduces the risk of misregistration. An important factor in the practical success of widely used software packages for image registration is that they implement sophisticated multistage search strategies that empirically provide a favorable tradeoff between computational speed and quality of the final result. In particular, in order to reduce the computational effort, and the prevalence of local optima, it is often beneficial to first solve a simplified version of the problem, and increase the complexity gradually [20].

First, a rough alignment is often achieved on a coarse scale, i.e., at reduced image resolution. Methods might also start with a reduced search space, in which only a subset of transformations is permitted. For example, when the goal is to apply a full affine transformation, the search might be limited to similarity transformations initially. Moreover, the resolution at which the search space is sampled may be increased once a substantial part of it has been pruned.

Once an initial run of the optimization algorithm has finished, another strategy that helps escape local optima is to randomly perturb the parameters of the transformation it has found, to re-initialize the optimization with the resulting values, and to check whether that results in a lower cost.

Many local optima in image registration are caused by the image contents, and cannot easily be avoided. However, others arise from discontinuities in the computation of the cost function, e.g., when a voxel leaves the region of overlap between the involved images, or when, due to interpolation, an intensity value passes from one histogram bin to another. Registration can be improved by avoiding such local optima via apodization, e.g., smoothly reducing the impact of voxels as they approach the boundary of the domain [20].

3.3.3 Nonlinear registration

The variability between brains of different subjects, and even when observing individual brains during maturation or aging, is not adequately captured by the space of affine transformations. It requires handling local deformations, which can be modeled as a displacement field. In d dimensions, it is a vector field $\mathbf{u} : \Omega \subset \mathbb{R}^d \rightarrow \mathbb{R}^d$ that is defined on the same domain as the moving image X, and matches $X(\mathbf{x} + \mathbf{u}(\mathbf{x}))$ to the fixed image $Y(\mathbf{x})$ [21]. This allows us to model arbitrary, even nonlinear transformations, and thus leads to nonlinear image registration. It is typically performed in a separate step, after a coarse alignment has been achieved with linear registration.

Deformation fields can also model extreme and implausible deformations. Therefore, they are usually subject to constraints or regularization. One approach is to represent the deformation field with splines that are defined on a sufficiently coarse grid to enforce a minimum amount of smoothness [21,22].

A second strategy derives a regularizer $R(\mathbf{u})$ that penalizes less plausible transformations. A modified optimization problem

$$\mathbf{u}^* = \arg\min_{\mathrm{u}} \; C(Y(\mathbf{x}), X(\mathbf{x} + \mathbf{u}(\mathbf{x}))) + \lambda R(\mathbf{u}) \tag{3.3}$$

is then formulated. It simultaneously minimizes the intensity-based cost C, in order to achieve a good alignment, and the regularizer R, to maintain a plausible transformation. Typically, more complex transformations allow for a better alignment, but are less regular, so these two objectives compete. A suitable trade-off between them has to be found by tuning the regularization weight λ. One popular choice is to derive $R(\mathbf{u})$ from the energy of linear elasticity, leading to the family of elastic registration methods [23].

The optical flow denotes the velocities with which local intensity patterns move in a temporal image sequence. The task of estimating optical flow is similar to image registration in that both aim to establish correspondences; however, optical flow methods do so more locally, and usually based on more specific assumptions [24]. The Demons algorithm [25] is a classic nonlinear registration method. It is based on iteratively computing, regularizing, and accumulating optical flow; its regularizer has been shown to approximate linear elasticity.

Elastic regularizers have been found to prevent large displacements, which are, however, needed to model phenomena such as brain atrophy. Therefore, diffeomorphic registration has been proposed as an alternative. It constrains $\mathbf{u}(\mathbf{x})$ to the group of diffeomorphisms [26]: this means that it should be differentiable, invertible, and its inverse should also be differentiable. This constraint is now widely used, and has produced plausible results in many cases. However, it has been noted that it might oversimplify the true variability in the brains of different individuals [27].

When working with displacement vector fields, it is difficult to ensure that they remain diffeomorphic. In particular, the Euclidean mean of displacement vector fields that represent diffeomorphisms is not necessarily invertible. For this reason, it is more convenient to represent transformations as smooth time-dependent [28] or stationary [29] *velocity* vector fields, whose flow describes the corresponding displacement. The flow can be computed by numerically integrating an ordinary differential equation. Deformations obtained in this way are guaranteed to be diffeomorphic: their inverse is easily found by scaling all velocities by -1, and re-computing the flow.

Many diffeomorphic image registration methods are symmetric. This means that, rather than transforming a moving image toward a fixed reference, they treat all involved images identically while simultaneously transforming them toward their common mean [27]. This guarantees that the registration result, and the downstream analysis, will not depend on the order of the subjects, which is often arbitrary.

Most nonlinear registration methods solve a difficult optimization problem, and are consequently computationally expensive. A recently explored alternative is to replace optimization with prediction. For example, one might design and train a neural network that takes a pair of patches from the involved images as its input, and outputs the parameters of the transformation that is required to register them. One such approach has reported a $35\times$ speedup over a comparable optimization-based method [30].

3.3.4 Standard spaces and templates

Image registration is frequently used to normalize images from different brains, i.e., to transform them into a common reference space. This establishes a stereotactic coordinate system (or "atlas") in which the position of a specific anatomical structure can unambiguously be specified with three numbers. One widely used atlas was established by, and is named after, the French neurosurgeon Jean Talairach, in collaboration with Gabor Szikla and Pierre Tournoux [31].

The Talairach atlas is based on the dissection of one single brain. Subsequent efforts have defined atlases that are more representative of the general population, based on MRI of a larger sample. Widely used examples are the MNI atlas, which was derived from 305 subjects at the Montreal Neurological Institute [32], as well as an atlas that was developed by the International Consortium for Brain Mapping (ICBM) [33].

Templates are 3D images that show an "average" brain for the given population. In case of the MNI atlas, these are young, normal, right-handed subjects. The brains in specific age groups or diseases can systematically differ from such a reference. In such cases, it is often beneficial to create a custom template that is specific to the group under investigation. It has been shown to be advantageous to use "symmetric" template construction, which avoids having to select one of the subjects as an initial reference. Instead, all images are treated identically while alternating between registering them to a preliminary template, and improving the template [34].

3.4 Image segmentation

Image segmentation is the task of partitioning a given image into connected regions that belong to the same object, or consist of the same material or tissue type. In the context of neuroimaging, a frequent use case of segmentation is to extract the brain from a volumetric image, i.e., to remove skull and surrounding tissues. Other uses include defining regions of interest for further analysis, or measuring the volume of specific structures or tissue types such as gray or white matter.

The most basic segmentation methods simply group together voxels with similar intensity levels in the image. Advanced methods handle more difficult segmentation problems by combining image information with prior assumptions about properties of the expected segmentation, such as smoothness, or even sophisticated prior knowledge about the shape and appearance of the segmented object.

Among the basic segmentation methods are ones that determine suitable thresholds for separating foreground and background. For example, *Otsu's method* picks a threshold that maximizes the ratio of between-class and within-class intensity variation [35]. Image noise can lead to irregularities, such as isolated pixels, in the results of thresholding. To some extent, applying *image smoothing* as a pre-process or *morphological operations,* such as erosion and dilation with a given structure element as a post-process, can remove such artifacts [36]. However, achieving reliable results usually requires integrating the regularization, and possibly more complex prior knowledge, into the segmentation method itself. We will

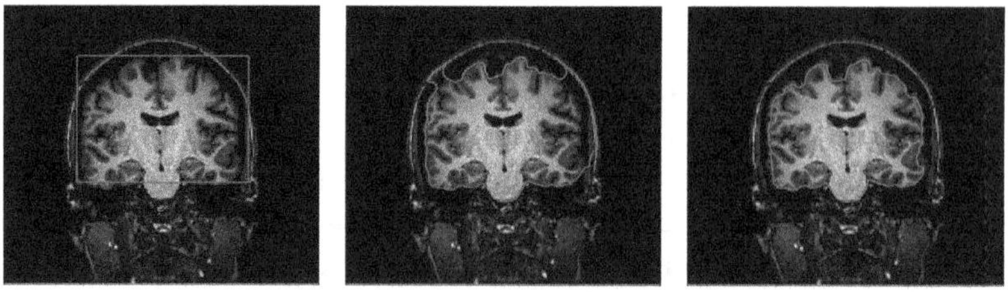

Figure 3.4 Deformable models for image segmentation start with a rough initial outline (left) and continuously deform it (center) so that it aligns with image structures, while remaining smooth. Ideally, the desired segmentation is achieved when the contour stops moving (right)

cover three families of segmentation methods that achieve this in different ways: deformable models, MRFs, and CNN.

3.4.1 Deformable models

Deformable models are one popular class of segmentation methods that balance information from the image against smoothness of the resulting segmentation boundary. These two criteria are formalized as forces that act on an initial contour that is placed on the image: external forces pull the contour toward image features such as edges or specific intensity levels. Internal forces deform it to become more regular, e.g., by reducing its length or curvature. They can also represent more sophisticated prior knowledge, such as a statistical shape model. The way the contour moves under these forces has led to the alternative names *snakes* or *active contours* for this approach [37]. As illustrated in Figure 3.4, the final segmentation is given by a steady state in which the two types of forces balance out.

There are two main ways to design deformable model-based segmentation algorithms: a pragmatic one is to specify the forces directly. A more mathematically elegant one is to derive them from an energy functional that assigns a cost to any possible segmentation result. It usually involves an external (data) term, which penalizes misalignment with the desired image structure, and an internal (regularization) term which penalizes implausible contours. Corresponding external and internal forces can be derived using the calculus of variations such that, at each point in time, the contour deforms in such a way that the energy decreases most rapidly.

This provides transparency in the sense that the energy functional states the properties of the desired segmentation in exact mathematical language, and often quite concisely. A simple example would be maximizing the magnitude of the image intensity gradient along the curve, while minimizing its length. The fact that deformable contours implement a gradient descent type of optimization implies that they might converge in a local suboptimum. This means that the final segmentation depends not just on the energy functional, but also on the initialization of the contour.

How to represent the evolving contour is another important choice when implementing deformable models. For simplicity, we will consider the two-dimensional case, in which the contour amounts to a curve. An explicit representation will represent this curve in parametric form $\mathbf{v}(s) : [s_{\min}, s_{\max}] \subset \mathbb{R} \to \mathbb{R}^2$ with curve parameter s. It will be discretized with line segments connecting vertices at discrete positions along the curve. This allows for efficient computation, since external and internal forces only need to be computed for, and applied to, the vertices. However, an explicit refinement is required to avoid artifacts in cases where neighboring vertices move too far away from each other. Correctly dealing with topological changes, such as merging, disconnecting, or vanishing contours, requires substantial implementation effort.

For these reasons, *level set methods* [38,39] represent the segmentation boundary as the zero level set of a height field $\varphi(x,y) : \Omega \subset \mathbb{R}^2 \to \mathbb{R}$. In this case, the curve is given implicitly, as the set of points $\{(x,y)|\varphi(x,y) = 0\}$, and is discretized by sampling φ at all pixel positions. For all practical purposes, this can represent arbitrarily complex shapes, and it requires no specific treatment of topological changes as the curve evolves. The signed distance transform is used to convert the given initial contour to a height field representation, which is then again iteratively updated in a way that minimizes an energy functional describing desirable properties of the final segmentation. Finally, a standard isocontouring algorithm can be used to recover the final segmentation boundary. A drawback of level set methods is their increased computational cost: in a straightforward implementation, the value at each pixel has to be updated in each iteration. Substantial speedups can be achieved by exploiting the fact that only values close to the boundary are relevant, which has been implemented in fast marching and narrow band methods [38].

3.4.2 Markov random fields

MRFs are the basis of another popular group of image segmentation methods [40]. They achieve a probabilistic segmentation by inferring the posterior probability $P(L|I)$ of region labels L given image I within a Bayesian framework:

$$P(L|I) = \frac{P(I|L) \cdot P(L)}{P(I)}, \tag{3.4}$$

where the likelihood $P(I|L)$ specifies how well the observed image intensities I agree with the assumed labels L. As illustrated in Figure 3.5, MRFs are widely used to segment different tissue types, such as gray matter, white matter, and cerebrospinal fluid in the brain [41]. A simple choice is to model $P(I|L)$ with Gaussians that reflect the mean and variance of intensity levels that correspond to the different tissues.

The prior probability $P(L)$ specifies how probable a labeling L is without considering any information from the image. The assumption that spatially smooth labelings are more plausible than ones with many isolated voxels introduces spatial dependencies in the prior: if many neighboring voxels have a certain label, this

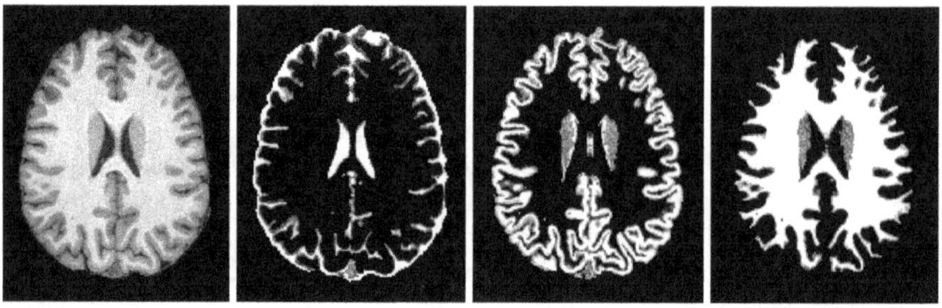

Figure 3.5 A classical application of MRFs in neuroimaging is to segment input MR images into tissue maps that indicate the presence of CSF, gray matter, or white matter, respectively

increases our expectation that the current voxel should share the same label. MRFs maintain computational feasibility by restricting such dependencies to pairwise interactions between immediate neighbors. This implies that the joint probability $P(L)$ over the labels of all voxels factorizes into probabilities over immediate neighbors [40].

Rather than working with the involved probabilities directly, it is more common to work with their negative logarithms. This turns products of probabilities into sums of potentials. Evaluating the likelihood $P(I|L)$ for a given voxel involves only one label, from that same voxel. Therefore, the corresponding potential is referred to as a unary potential. Potentials from the factorized representation of $P(L)$ involve labels from two neighboring voxels, and are thus referred to as binary potentials. When looking for a labeling that maximizes $P(L|I)$, the denominator in (3.4) can be neglected, since it does not depend on L. Therefore, maximizing $P(L|I)$ is equivalent to minimizing a corresponding energy, which is the overall sum of unary and binary potentials. This also establishes a link to the energy functionals used in deformable models, with the sum of unary potentials taking the role of the external energy, binary potentials the role of the internal energy.

Segmentation with MRFs is typically implemented with the expectation-maximization (EM) algorithm: given an initial estimate of the intensity distributions corresponding to the different classes, the expectation step first computes a maximum a posteriori (MAP) estimate of the labels. This means that it finds a hard labeling that maximizes (3.4). Afterwards, for each voxel, the prior can be evaluated by conditioning on the MAP labels, and a probabilistic (soft) labeling can be computed from (3.4). In the maximization step, parameters of the individual class distributions, such as means and variances of tissue-specific intensities, are updated. These E and M steps are alternated until convergence.

Since the labels of different voxels depend on each other through the binary potentials, MAP estimation has to determine them simultaneously. Deriving a graph from the image so that the optimal label assignment corresponds to a maximum flow (minimum cut) is a popular approach to solving this problem, and provides an exact solution under certain conditions [40].

Two extensions to the basic MRF segmentation algorithm are widely considered when using it for tissue segmentation. First, MR images often suffer from low-frequency variations in the overall image intensity, which arise from inhomogeneities in the magnetic field, referred to as bias fields. Their effect is multiplicative, and can be strong enough to compromise correct tissue segmentation when left uncorrected. Estimating bias fields is simplified by working on log-transformed intensities, which turns the multiplicative factor into a linear offset. In the expectation step of the EM algorithm, the bias field can then be estimated as a spatially smoothed version of the residuals, i.e., the differences between the observed intensities and the predictions based on the current label assignments. In all subsequent steps, the effect of the estimated bias field can then be removed [42]. In order to obtain useful bias field estimates also in non-tissue classes, such as cerebrospinal fluid or air, their intensities are better described by uniform than by Gaussian distributions [43].

A second extension of MRF segmentation, which can be helpful especially in cases of severe bias fields, is to account for additional prior knowledge in the form of an atlas which indicates where the different tissue types are expected in a normal brain. Such an atlas can be formed by normalizing probabilistic segmentation results from a group of reference subjects to a standard space, and averaging them. Using the atlas for segmentation requires registration of the new subject to the same space, using the algorithms discussed in Section 3.3.

3.4.3 Convolutional neural networks

Deep neural networks have received much attention in recent years, since they made it possible to solve difficult problems in computer vision and medical image analysis, sometimes at accuracies that are on par with, or even above, what is achieved by human experts. Their strength derives from their ability to learn from a large number of training images. In the context of image segmentation, training them on a sufficient number of representative examples in which a correct segmentation is given, and which include cases that pose challenges to traditional methods, such as weak contrast, or the presence of pathologies, neural networks will learn to handle them more robustly. Even though training neural networks is computationally expensive, applying the final model is often faster than previous state-of-the-art methods, sometimes by orders of magnitudes [44].

CNNs are a specific type of neural network that are particularly popular for medical image segmentation. They include convolutional layers, which apply convolution filters to their input, and whose weights are determined by the above-mentioned training process [45]. CNNs combine such convolutions with activation layers, which apply a nonlinear function to each voxel of the image, and stack multiple layers into deep architectures that generate a feature hierarchy: layers close to the input image extract low-level features, such as edges, while neurons in higher layers are activated more specifically by certain textures, or semantic image components.

Individual neurons in a CNN only receive input from a limited region in the input volume, which is referred to as their receptive field. In a single standard convolution, the receptive field equals the size of the convolution filter. Recognizing more complex objects in an image requires a sufficiently large receptive field, but enlarging it via the size or number of subsequent convolutions increases computational effort, as well as the number of parameters in the network, which determines not only the memory requirements, but also affects the number of training samples required for training.

Therefore, different strategies are employed to enlarge the receptive field without increasing the number of model parameters. In strided convolutions, neighboring voxels in the output are obtained by shifting the convolution kernel by more than one voxel. This makes the receptive field grow more rapidly in subsequent convolutions. In dilated (or atrous) convolution, the filter weights are not applied to immediately adjacent voxels. Rather, gaps are left between them, which also leads to a greater receptive field. Finally, pooling downsamples the image, e.g., by taking the maximum within a certain block. This also makes the network more robust with respect to small shifts.

All these strategies effectively reduce the image resolution. In the context of image segmentation, this is problematic, since the final labeling is typically desired at the original resolution. Ideally, the steps described so far lead to an accurate description of the image contents. However, it will have a coarse resolution, and restoring a sharp labeling from it is not a simple task. Among several solutions that have been proposed for this, the U-Net has become especially popular in medical image segmentation [46]. Its name derives from its U-shaped architecture, in which an encoder branch which consists of multiple convolutional, activation, and pooling layers, is connected to a mirror symmetric decoder branch, in which up-convolutions with learnable weights take the places of the pooling layers in the encoder. In addition to the upsampled feature maps, convolutions in the decoder branch are given access to the original feature maps from the same level in the encoder path. These retain fine structures from the original image, which helps to achieve a meaningful segmentation at the full resolution.

3.5 Statistical testing

Statistical hypothesis testing plays a central role in neuroimage analysis. Depending on the specific context, it might be used to address questions such as "Which cortical areas are involved in different aspects of speech?" [47], or "In which white matter tracts is integrity affected by a certain neurological disease?" [48]. For a general introduction to hypothesis testing, interested readers are invited to consult relevant textbooks [49].

In this section, we will first discuss the problem of familywise error, which is shared by almost all statistical testing in neuroimaging. We will then turn to methods for specific use cases: based on structural MRI, voxel-based morphometry (VBM) performs a comparison of regional gray matter volumes, either between different groups, or longitudinally (over time). fMRI and dMRI were introduced in Sections 3.2.2 and 3.2.3, respectively, and require their own specific modeling.

3.5.1 Statistical parametric maps and familywise errors

Neuroimaging data is commonly analyzed in a mass univariate fashion, i.e., a separate statistical test is performed at each voxel. This results in a statistical parametric map (SPM), in which the value at each voxel follows, under the null hypothesis, a specific null distribution, such as Student's t distribution [50].

In statistical testing, a type I error occurs when a true null hypothesis is incorrectly rejected. This amounts to reporting a false finding. In neuroimaging, the level α, indicating the conditional probability of such an error under the null hypothesis, is generally set to $\alpha = 5\%$. Obviously, performing a large number of tests, one for each voxel, substantially increases the risk that at least one of them will result in a type I error. The conditional probability of committing a type I error at least once under the null hypothesis is referred to as the familywise error rate α_{FW}. In general, we would like to limit it to $\alpha_{\mathrm{FW}} = 5\%$.

Standard approaches for controlling familywise error, such as Šidák or Bonferroni correction, reduce the α level of individual tests in order to guarantee the desired α_{FW}. However, when applied in neuroimaging, they are overly conservative, since they assume that individual tests are independent from each other while in reality, neighboring voxels are often highly correlated. This leads to a dramatic increase of type II errors, i.e., an inability to detect true deviations from the null hypothesis.

Several approaches have been developed to correct for multiple comparisons while accounting for spatial correlations. A classical one is based on random field theory (RFT) [51]. It assumes that the SPM follows a specific distribution, such as t, χ^2, or F, and it involves estimating the smoothness of the image to determine the severity of the multiple testing problem [52]. RFT can be applied in two modes: in voxelwise mode, each voxel's significance is determined separately. In clusterwise mode, the SPM is first thresholded to obtain clusters, i.e., connected components of super-threshold voxels. Significance is then determined for clusters as a whole, based on the number of voxels contained in them. An empirical study found voxelwise RFT inference to be valid, but clusterwise RFT to be invalid, since it resulted in familywise error rates far above the nominal level in practice [53].

Clusterwise testing is often more sensitive than voxelwise tests. A valid, but more time-consuming way to perform it is via permutation testing [54]. Unlike RFT, it does not attempt to analytically derive the null distribution of cluster volumes under the null hypothesis. Rather, it computes an empirical null distribution by repeatedly permuting the labels (indicating group membership) of the images, and recomputing the test statistic each time. To account for familywise errors, the largest value over the full 3D volume is recorded after each iteration.

Finally, a p value is obtained by comparing test statistics corresponding to the correct labels with the recorded values from the permutations.

Limitations of clusterwise testing include the absence of a principled way to select the cluster-forming threshold and the amount of pre-smoothing which is commonly performed before thresholding, as well as the fact that large clusters can make it difficult to precisely localize effects. These concerns are addressed by threshold-free cluster enhancement (TFCE) [55]. It transforms the SPM such that high values are boosted if they are surrounded by a supporting region of high values, while preserving the locations of local maxima. Inference is again done with permutation testing, but on the transformed SPM and at a voxel level. Similar to clusterwise testing, this increases sensitivity by boosting the belief in extended areas of signal. At the same time, it eliminates the need for smoothing or selection of an arbitrary cluster-forming threshold, and leads to better localization than clusterwise testing.

3.5.2 Voxel-based morphometry

Brain morphometry studies how volumes of specific brain structures differ between certain populations, e.g., in relation to brain maturation and aging, to specific abilities, or to disease. Even though it is possible to explicitly segment brain structures from 3D MR images in order to compute their volumes, it remains difficult to reliably achieve such a segmentation automatically. Moreover, such a study would be biased toward findings in the structures that were pre-selected for segmentation.

For these reasons, VBM has been developed to compare local gray matter volumes throughout the whole brain [56]. It first brings images from different subjects into anatomical alignment via nonlinear registration, as explained in Section 3.3.3. It then performs a probabilistic tissue type segmentation, which was discussed in Section 3.4.2. This determines the local amount of gray matter, which is much simpler to achieve than a segmentation of specific cortical areas. Gaussian smoothing is applied to the segmentation results. This turns local, per-voxel gray matter concentration to a regional measure, accounting for the fact that the cortical regions of interest span multiple voxels. It also compensates for imperfections in the nonlinear registration, which remain with all current methods. The resulting values are then logit transformed to make them more normally distributed, and finally entered into a statistical analysis as outlined in the previous section.

Often, a modulation step is added to account for volumetric distortions, i.e., the fact that nonlinear registration might locally compress or expand volumes to achieve a better match. Studies that interpret gray matter probabilities from a tissue type segmentation as a measure of the absolute amount of local gray matter should modulate them accordingly. To give an example, if three voxels that contain the same amount of gray matter in an individual subject get compressed into a single voxel in standard space, the value in that voxel should be scaled by a factor of three in order to preserve the overall amount of gray matter. When the warp is described as a vector field, the required factor is given by the determinant of its Jacobian matrix. When this modulation step is omitted, results should be interpreted as gray matter *density* (i.e., the regional ratio of gray matter compared to white matter and CSF), as opposed to absolute concentration [57].

Limitations of VBM include a potential for group-specific difficulties in the normalization or segmentation steps, difficulties to deal with cases of severe pathology, and the fact that the explicit smoothing step, as well as smoothness of the underlying registration, limits localization accuracy [58].

3.5.3 Modeling task-based fMRI

In task-based fMRI, experimental tasks or stimuli evoke a BOLD response, as explained in Section 3.2.2. Analysis of the resulting data involves predicting the expected BOLD response, so that we can identify brain regions in which the observed time course correlates significantly with it.

The HRF captures the BOLD response to a short stimulus. The response to repeated or extended stimuli is commonly based on assuming linearity of the hemodynamic response, such that the overall BOLD response is obtained by convolving an indicator function, describing the timings and durations of stimuli, with the HRF. Linearity is known to be violated for very short or tightly spaced stimuli, but these are usually avoided in experimental designs [2]. Even though HRFs are known to vary between subjects and brain regions, a gamma distribution with fixed parameters is often used as a "canonical" model. Sometimes, a second gamma distribution is subtracted to model the post-stimulus undershoot.

Statistical testing in fMRI is commonly done per-voxel with a general linear model. It amounts to a multiple linear regression, in which the predicted BOLD response is entered as the regressor of interest, and it permits converting the corresponding coefficient to a t score. Often, additional coregressors of no interest are added to the model in order to absorb undesired effects. Examples are small temporal shifts between the canonical and the actual HRF, which can be modeled by including the first derivative of the predicted BOLD response, or parameters from motion correction and their derivatives, which can absorb motion artifacts.

The fMRI data undergoes preprocessing before it is entered into the analysis. This commonly includes image registration to correct for head motion during the experiment; spatial smoothing to reduce noise and compensate residual misalignment; and temporal high-pass filtering to remove scanner drift. Sometimes, images are also normalized to a standard space, bias fields are compensated, and slice timings might be corrected, to account for the fact that different 2D slices within the same 3D volume have been acquired at different times.

Studying how the BOLD response differs between groups of subjects requires a second-level analysis. This is commonly done by entering regression coefficients from the general linear model described above into another statistical test, such as a two-sample t test. Two-level analyses involve two types of variances, within and between subjects. Some approaches explicitly estimate within-subject variances and account for them in the second level; others make the simplifying assumption that all subjects share the same within-subject variance [2].

3.5.4 Modeling resting-state fMRI

Many different approaches have been developed for analyzing resting-state fMRI [59]. A widely used one quantifies the amplitude of low frequency fluctuations (ALFF), i.e., the total power of fluctuations in the BOLD signal in a frequency band that is assumed to reflect neural activity, between 0.01 and 0.08 Hz. Statistical testing can be used to detect group mean differences in the resulting values, or variations within individual brains [60].

Independent component analysis (ICA) decomposes the observed spatio-temporal BOLD signal into spatial components that are mixed by corresponding time courses. The underlying mathematical model is quite similar to the general linear model that is used in task-based fMRI. However, in task fMRI, the (temporal) BOLD response is assumed to be known, and only the corresponding spatial activation maps are estimated. In an ICA of resting-state fMRI, a given number of time courses and spatial maps are estimated simultaneously, based on the assumption that spatial maps are statistically independent and non-Gaussian [61]. Certain areas of the brain emerge as independent components consistently across subjects, and correspond to specific functional regions. They are referred to as resting-state networks, and could also be reproduced using other methods, such as clustering [62].

When viewing the brain as a whole as a complex functional network, graph theory provides a tool to analyze its properties. Graph-based analyses of resting-state fMRI rely on parcellating the cortex into discrete areas, each of which is represented by a node in the graph. A measure of functional connectivity between these areas is computed from the correlation between the resting-state fMRI signals in them; if it exceeds a certain threshold, the two nodes will be connected by an edge. The resulting graphs typically represent small-world networks. This means that they contain short paths between most pairs of nodes, which guarantees efficient communication. At the same time, they are strongly clustered: even though two independently selected nodes are unlikely to be neighbors, two neighbors of the same node are likely to have a direct connection. Brain maturation, senescence, and brain disorders have been shown to affect the properties of functional connectivity graphs [63].

Functional connectivity between two regions could indicate that activity in one causes the activity in the other, or it could merely be due to both responding to a common input. Measures of functional connectivity cannot distinguish between these cases, nor establish the direction of causal links. Effective connectivity, estimated with techniques such as dynamic causal models, goes beyond this by providing a directed measure of the influence that one regions exerts over another [64].

3.5.5 Modeling dMRI

As it was explained in Section 3.2.3, dMRI measures water self-diffusion, which is described by a 3×3 matrix in diffusion tensor imaging, or by even more complex mathematical models in advanced dMRI. Only few statistical hypothesis tests are available that fully use this complex information [65]. In practice, it is more common to summarize it with scalar quantities, which can be entered into standard tests. Fractional anisotropy (FA) is particularly widely used. It ranges within [0,1] and reflects diffusion anisotropy: FA = 0 indicates equal diffusivities in all directions; FA = 1 indicates that diffusion is limited to the principal direction.

FA can be analyzed in analogy to the above-described VBM. However, this involves spatial smoothing to compensate residual misalignments. Smoothing decreases spatial resolution and therefore works against the primary goal of dMRI, i.e., to exceed the resolution limit of *in-vivo* brain MRI.

For this reason, alternative techniques have been developed that analyze dMRI data without smoothing. Tract-based spatial statistics (TBSS) [66] has become a de-facto standard. Its main idea is to limit statistical analysis to the center of white matter structures. To this end, FA maps from all subjects are nonlinearly registered and averaged. A medial surface representation of the white matter, the "white matter skeleton", is then extracted from the locations where group-averaged FA is locally maximal. Residual misalignment is removed by a projection step, which searches for the maximal FA value in each individual subject, orthogonally to the skeleton, and projects it to the skeleton. Despite the popularity of TBSS, it suffers from the fact that differently oriented tracts can be impossible to distinguish in FA maps. It has been demonstrated that this may lead to voxels being missed, or assigned to an incorrect part of the skeleton [67].

Along-the-tract statistical analysis is an alternative to TBSS. Here, white matter tracts are reconstructed using tractography, and quantities such as FA are then averaged over tracts, or binned based on their position along the bundle. The latter requires finding a joint parameterization of the bundles from multiple subjects, i.e., establishing anatomical correspondences between them [68,69].

The extent to which two given brain regions are connected by dMRI tractography is referred to as structural connectivity. Graph-based analyses of structural connectivity have been performed in analogy to the ones of functional connectivity described above, and the relationship between structural and functional connectivity has been studied [63,70].

3.6 Predictive modeling

Neuroimaging reveals information about factors such as neurological disease, age, gender, or current neuronal activity. Predictive modeling of neuroimaging data allows us to make individual predictions related to such factors. These predictions may or may not relate to the future: they might attempt to predict the future onset of some disease or future behavior, but they might also aim to support a diagnosis [71], to estimate brain maturity [72], or to "read a subject's mind" in the sense of detecting what class of object they are currently looking at [73], or whether they are currently telling the truth [74].

In this section, we will first introduce some fundamental supervised classification and regression techniques that are widely used in the context of neuroimaging. We will then discuss the extraction of suitable features from neuroimaging data, as well as aspects of feature selection and weighting, parameter tuning, and validation, which are required to obtain accurate and valid results in practice.

3.6.1 Supervised classification and regression

Assume that we are predicting a specific value y. We will refer to it as a label or response. The information based on which we can make the prediction is represented numerically by a feature vector $\mathbf{x}$. Depending on the nature of the label, we distinguish two types of predictive tasks: in classification, the label is discrete. An example would be to judge whether a brain scan indicates a disease, or which one. In regression, the label is continuous. An example would be to infer a subject's age based on observing their brain activity. In supervised learning, a training dataset $\{(\mathbf{x}_i, y_i)\}$ that contains pairs of feature vectors $\mathbf{x}_i$ and corresponding labels y_i is given. The goal is to derive a general function $y = f(\mathbf{x})$ which assigns a label to arbitrary inputs $\mathbf{x}$, including ones that were not part of the training data.

Neuroimaging frequently leads to high-dimensional feature vectors $\mathbf{x} \in \mathbb{R}^p$; more details about their exact derivation will be given in the next subsection. At the same time, the number n of training samples is often rather limited. This "large p, small n" scenario poses a challenge for many supervised learning techniques. Support vector machines (SVMs) are a family of methods that include variants for supervised classification and regression. They deal with this case relatively well, since they have been designed to make predictions based on a sparse subset of the available training data, referred to as support vectors [75]. For this reason, they are very popular in neuroimaging applications.

Other methods for supervised classification and regression have also been used, including linear discriminant analysis (LDA), naive Bayesian classification, k Nearest Neighbors, relevance vector regression, and random forests. A more detailed explanation of these methods is outside of our scope, and can be found in machine learning textbooks [76]. Initial attempts of using deep neural networks for predictive modeling of brain images have also been made [77]. However, the effective use of this paradigm is still limited by the small sample sizes of most studies. More generally, this has been identified as one of the current limiting factors for predicting brain disorders from neuroimaging [78].

3.6.2 Features for predictive modeling

Accurate modeling requires a suitable representation of the input data as a feature vector $\mathbf{x} \in \mathbb{R}^p$. Feature extraction is the process of transforming the measured data into such a vector. A basic strategy is to process the images in the same way as when preparing them for mass univariate testing (cf. Section 3.5). The resulting data values or test statistics from all voxels can then be stacked into a very high-dimensional feature vector [79]. More concise feature vectors have been formed by computing the average over spatial regions, such as uniform blocks [74], functional units [72], or specific sections of white matter tracts [69]. Another alternative, which is specific to dMRI and resting-state fMRI, is to derive features from a connectivity graph, as it was discussed in Section 3.5.4 [80].

The original set of extracted features can be quite large (e.g., $p > 10^5$ [48]). It often contains many features that are irrelevant for predicting the label y, e.g., when detecting a disease that only affects a certain part of the brain. Feature selection is the process of retaining only the most important features. It can greatly increase the achieved accuracy, in particular when only few training samples are available. Feature selection is often a separate step between feature extraction and supervised learning; some techniques treat it as an integral part of the training [81].

Feature selection is made difficult by the fact that features might interact: features that are non-informative in isolation might become useful when combined with each other, and multiple useful features might be redundant. Accounting for such interactions would require considering all possible combinations of features, which is computationally infeasible in almost all practical scenarios. Therefore, heuristics are widely used, such as defining an importance score for individual features, and retaining only the k highest-ranking ones. Classification and regression tasks require different importance scores. Examples include the F score [82], t score [83], and Pearson correlation coefficient [72].

The number of retained features is sometimes set to a fixed value [72]. In other cases, it is determined via cross-validation [83], or based on statistical significance [84]. In practice, feature relevance is a gradual concept. Thus, an alternative to feature selection is to weight features by their continuous importance score. Many supervised techniques permit this [85], and this strategy has been used successfully also in neuroimaging [48].

3.6.3 Hyperparameter tuning and evaluation

Most machine learning models involve parameters, which are adjusted automatically by the training process, as well as hyperparameters, which control aspects of the training process, and have to be adapted to the characteristics of the data to obtain optimal results. A widely used strategy is to systematically train the model with many different hyperparameter values, to estimate the corresponding model accuracy, and to select the setting that achieved the highest estimate. This is valid as long as the three distinct tasks of fitting the parameters, tuning the hyperparameters, and estimating the final accuracy of the model are each performed on independent data. When a sufficient amount of data is available, this can be achieved by splitting it into three sets, which are used for training, validation (estimating accuracy for the purpose of hyperparameter tuning), and estimation of the final accuracy, respectively.

Even though it is preferable to evaluate models on data that was not available during their design and training, this is not always possible, for example when dealing with small datasets. In such cases, it is customary to use cross-validation. It subdivides the available data into n equal parts, which are called “folds”. Accuracy is evaluated by predicting the labels in each fold, using a model that was trained on the respective $n - 1$ remaining folds. When using cross-validation both for tuning hyperparameters and for estimating the final accuracy, it is important to use it in a nested manner: an outer cross-validation loop estimates the final accuracy; an inner loop makes the tentative estimation of accuracy that is required for tuning hyperparameters. Crucially, the inner loop may not use the data in the test fold of the outer loop at any point. Failing to cleanly separate parameter tuning and testing, or performing feature selection on the full data, outside of the cross-validation loop, can lead to unrealistic and greatly inflated estimates of model accuracy [65].

3.7 Outlook

On our tour through the neuroimage acquisition and analysis pipeline, we highlighted several recent research trends and innovations, such as the use of deep neural networks for image processing, advanced models for complex neuroimaging modalities such as functional and dMRI, or the use of machine learning for the predictive modeling of neuroimaging data.

Several of these trends can be expected to continue in the coming years. In particular, we can expect ever more complex and higher-resolution imaging modalities. Neuroimage analysis is likely to continue moving away from traditional mass-univariate to multivariate or higher-level approaches that account for interdependencies between different brain regions, and between quantities captured with different modalities. Machine learning models that assess individual patients based on their imaging data should greatly benefit from the availability of larger cohorts, and from data pooling, which will require data harmonization. Finally, with an increasingly wide use of neuroimaging, and maturation of analysis techniques, integration of image-based biomarkers with other types of data, such as omics data, can be expected to move more into the focus.

References

[1] Webb S. *The Physics of Medical Imaging*. Boca Raton, FL: CRC Press; 1988.

[2] Poldrack RA, Mumford JA, and Nichols TE. *Handbook of Functional MRI Data Analysis*. Cambridge: Cambridge University Press; 2011.

[3] Jones DK (editor). *Diffusion MRI: Theory, Method, and Applications*. Oxford: Oxford University Press; 2011.

[4] Jeurissen B, Descoteaux M, Mori S, *et al.* Diffusion MRI fiber tractography of the brain. NMR in biomedicine. 2018; *Early View*. DOI:10.1002/nbm.3785.

[5] Sotiropoulos SN, Jbabdi S, Xu J, *et al.* Advances in diffusion MRI acquisition and processing in the human connectome project. *NeuroImage*. 2013;80:125–43.

[6] Tobisch A, Stirnberg R, Harms RL, *et al.* Compressed sensing diffusion spectrum imaging for accelerated diffusion microstructure MRI in long-term population imaging. *Frontiers in Neuroscience*. 2018;12:650.

[7] Schultz T. Learning a reliable estimate of the number of fiber directions in diffusion MRI. In: Ayache N *et al.* (editor). *Proc. Medical Image Computing and Computer-Assisted Intervention (MICCAI)* Part III. vol. 7512 of LNCS; 2012. p. 493–500.

[8] Jensen JH, Helpern JA, Ramani A, *et al.* Diffusional kurtosis imaging: the quantification of non-gaussian water diffusion by means of magnetic resonance imaging. *Magnetic Resonance in Medicine*. 2005;53:(6)1432–40.

[9] Assaf Y, Blumenfeld-Katzir T, Yovel Y, *et al.* Axcaliber: a method for measuring axon diameter distribution from diffusion MRI. *Magnetic Resonance in Medicine*. 2008;59(6):1347–54.

[10] Topgaard D. Multidimensional diffusion MRI. *Journal of Magnetic Resonance*. 2017;275:98–113.

[11] Sanders J. Emission tomography. In: Maier A, Steidl S, Christlein V, *et al.* (editors). *Medical Imaging Systems*. vol. 11111 of LNCS. Cham: Springer; 2018. p. 207–36.

[12] Hofmann M, Pichler B, Schölkopf B, *et al.* Towards quantitative PET/MRI: a review of MR-based attenuation correction techniques. *European Journal of Nuclear Medicine and Molecular Imaging*. 2009;36(1):93–104.

[13] Marcus C, Mena E, and Subramaniam RM. Brain PET in the diagnosis of Alzheimer's disease. *Clinical Nuclear Medicine*. 2014;39(10):e413–e26.

[14] Kameyama M, Murakami K, and Jinzaki M. Comparison of [^{15}O] H_2O positron emission tomography and functional magnetic resonance imaging in activation studies. *World Journal of Nuclear Medicine*. 2016;15(1):3–6.

[15] Burle B, Spieser L, Roger C, *et al.* Spatial and temporal resolutions of EEG: is it really black and white? A scalp current density view. *International Journal of Psychophysiology*. 2015;97(3):210–20.

[16] Lewine JD, and Orrison Jr WW. Clinical electroencephalography and event-related potentials. In: *Functional Brain Imaging*. Mosby; 1995. p. 327–86.

[17] Lu ZL, and Kaufman L, (editors). *Magnetic Source Imaging of the Human Brain*. Abingdon: Taylor & Francis; 2004.

[18] Viola P, and Wells III WM. Alignment by maximization of mutual information. *International Journal of Computer Vision*. 1997;24(2):137–54.

[19] Studholme C, Hill DLG, and Hawkes DJ. An overlap invariant entropy measure of 3D medical image alignment. *Pattern Recognition*. 1999;32(1):71–86.

[20] Jenkinson M, Bannister P, Brady M, *et al.* Improved optimization for the robust and accurate linear registration and motion correction of brain images. *NeuroImage*. 2002;17:(2)825–41.

[21] Klein S, Staring M, and Pluim JPW. Evaluation of optimization methods for nonrigid medical image registration using mutual information and B-Splines. *IEEE Trans on Image Processing*. 2007;16(12):2879–90.

[22] Szeliski R, and Coughlan J. Spline-based image registration. *International Journal of Computer Vision*. 1997;22(3):199–218.

[23] Christensen GE, and Johnson HJ. Consistent image registration. *IEEE Trans on Medical Imaging*. 2001;20(7):568–82.

[24] Horn BKP, and Schunck BG. Determining optical flow. *Artificial Intelligence*. 1981;17(1–3):185–203.

[25] Thirion JP. Image matching as a diffusion process: an analogy with Maxwell's demons. *Medical Image Analysis*. 1998;2(3):243–60.

[26] Joshi SC, and Miller MI. Landmark matching via large deformation diffeomorphisms. *IEEE Trans on Image Processing*. 2000;9(8):1357–70.

[27] Avants BB, Epstein CL, Grossman M, *et al.* Symmetric diffeomorphic image registration with cross-correlation: evaluating automated labeling of elderly and neurodegenerative brain. *Medical Image Analysis*. 2008;12(1): 26–41.

[28] Beg MF, Miller MI, Trouvé A, *et al.* Computing large deformation metric mappings via geodesic flows of diffeomorphisms. *International Journal of Computer Vision*. 2005;61(2):139–57.

[29] Arsigny V, Commowick O, Pennec X, *et al.* A log-euclidean framework for statistics on diffeomorphisms. In: *Proc. Medical Image Computing and Computer-Assisted Intervention (MICCAI)*. vol. 4190 of LNCS. Cham: Springer; 2006. p. 924–31.

[30] Yang X, Kwitt R, Styner M, *et al.* Quicksilver: fast predictive image registration – A deep learning approach. *NeuroImage*. 2017;158:378–96.

[31] Talairach J, and Tournoux P. *Co-Planar Stereotaxic Atlas of the Human Brain*. Stuttgart: Thieme; 1988.

[32] Evans AC, Collins DL, Mills SR, *et al.* 3D statistical neuroanatomical models from 305 MRI volumes. In: *Proc. IEEE Nuclear Science Symposium and Medical Imaging Conference*; 1993. p. 1813–17.

[33] Mazziotta JC, Toga AW, Evans A, *et al.* A probabilistic atlas of the human brain: theory and rationale for its development. *NeuroImage*. 1995;2(2): 89–101.

[34] Avants BB, Yushkevich P, Pluta J, *et al.* The optimal template effect in hippocampus studies of diseased populations. *NeuroImage*. 2010;49(3): 2457–66.

[35] Otsu N. A threshold selection method from gray-level histograms. *IEEE Trans on Systems, Man and Cybernetics*. 1979;9(1):62–66.

[36] Sonka M, Hlavac V, and Boyle R. *Image processing, analysis, and machine vision*. 4th ed. Stamford, CT: Cengage; 2013.

[37] Kass M, Witkin A, and Terzopoulos D. Snakes: active contour models. *International Journal of Computer Vision*. 1988;1(4):321–31.

[38] Sethian JA. *Level Set Methods and Fast Marching Methods*. 2nd ed. Cambridge: Cambridge University Press; 2001.

[39] Osher S, and Fedkiw R. Level set methods and dynamic implicit surfaces. vol. 153 of *Applied Mathematical Sciences*. Cham: Springer; 2003.

[40] Prince SJD. *Computer Vision: Models, Learning, and Inference.* Cambridge: Cambridge University Press; 2012.

[41] Zhang Y, Brady M, and Smith S. Segmentation of brain MR images through a hidden Markov random field model and the expectation-maximization algorithm. *IEEE Transactions on Medical Imaging*. 2001;20(1):45–57.

[42] Wells WM, Grimson WEL, Kikinis R, *et al.* Adaptive segmentation of MRI data. *IEEE Transactions on Medical Imaging*. 1996;15(4):429–42.

[43] Guillemaud R, and Brady M. Estimating the bias field of MR images. *IEEE Transactions on Medical Imaging*. 1997;16(3):238–51.

[44] Roy AG, Conjeti S, Navab N, *et al.* QuickNAT: a fully convolutional network for quick and accurate segmentation of neuroanatomy. *NeuroImage*. 2019;186:713–27.

[45] Goodfellow I, Bengio Y, and Courville A. *Deep Learning*. Cambridge, MA: MIT Press; 2016.

[46] Ronneberger O, Fischer P, and Brox T. U-Net: convolutional networks for biomedical image segmentation. In: *Medical Image Computing and Computer-Assisted Intervention (MICCAI)*. vol. 9351 of LNCS. Cham: Springer; 2015. p. 234–41.

[47] Price CJ. A review and synthesis of the first 20 years of PET and fMRI studies of heard speech, spoken language and reading. *NeuroImage*. 2012; 62(2):816–47.

[48] Schmidt-Wilcke T, Cagnoli P, Wang P, *et al.* Diminished white matter integrity in patients with systemic lupus erythematosus. *NeuroImage: Clinical*. 2014;5:291–97.

[49] Reid HM. *Introduction to Statistics*. Thousand Oaks, CA: SAGE Publications; 2013.

[50] Friston KJ, Ashburner JT, Kiebel SJ, *et al.* (editors). Statistical parametric mapping. *The Analysis of Functional Brain Images*. Cambridge, MA: Academic Press; 2007.

[51] Worsley KJ, Marrett S, Neelin P, *et al.* A unified statistical approach for determining significant signals in images of cerebral activation. *Human Brain Mapping*. 1996;4(1):58–73.

[52] Nichols T, and Hayasaka S. Controlling the familywise error rate in functional neuroimaging: a comparative review. *Statistical Methods in Medical Research*. 2003;12(5):419–46.

[53] Eklund A, Nichols TE, and Knutsson H. Cluster failure: why fMRI inferences for spatial extent have inflated false-positive rates. *Proceedings of the National Academy of Sciences*. 2016;113(28):7900–05.

[54] Nichols TE, and Holmes AP. Nonparametric permutation tests for functional neuroimaging: a primer with examples. *Human Brain Mapping*. 2001; 15(1):1–25.

[55] Smith SM, and Nichols TE. Threshold-free cluster enhancement: addressing problems of smoothing, threshold dependence and localisation in cluster inference. *NeuroImage*. 2009;44(1):83–98.

[56] Ashburner J, and Friston KJ. Voxel-based morphometry—the methods. *NeuroImage*. 2000;11(6):805–21.

[57] Good CD, Johnsrude IS, Ashburner J, *et al.* A voxel-based morphometric study of ageing in 465 normal adult human brains. *NeuroImage*. 2001; 14(1):21–36.

[58] Mechelli A, Price C, Friston K, *et al.* Voxel-based morphometry of the human brain: methods and applications. *Current Medical Imaging Reviews.* 2005;1(2):105–13.

[59] Lv H, Wang Z, Tong E, *et al.* Resting-state functional MRI: everything that nonexperts have always wanted to know. *American Journal of Neuroradiology.* 2018.

[60] Yu-Feng Z, Yong H, Chao-Zhe Z, *et al.* Altered baseline brain activity in children with ADHD revealed by resting-state functional MRI. *Brain & Development.* 2007;29(2):83–91.

[61] Beckmann CF, DeLuca M, Devlin JT, *et al.* Investigations into resting-state connectivity using independent component analysis. *Philosophical Transactions of the Royal Society B: Biological Sciences.* 2005;360 (1457):1001–13.

[62] van den Heuvel MP, and Pol HEH. Exploring the brain network: a review on resting-state fMRI functional connectivity. *European Neuropsycho pharmacology.* 2010;20(8):519–34.

[63] Bullmore E, and Sporns O. Complex brain networks: graph theoretical analysis of structural and functional systems. *Nature Reviews Neuroscience.* 2009;10:186–98.

[64] Friston K. Functional and effective connectivity: a review. *Brain Connectivity.* 2011;1(1):13–36.

[65] O'Donnell L, and Schultz T. Statistical and machine learning methods for neuroimaging: examples, challenges, and extensions to diffusion imaging data. In: Hotz I, and Schultz T, (editors). *Visualization and Processing of Higher Order Descriptors for Multi-Valued Data.* Cham: Springer; 2015. p. 293–313.

[66] Smith SM, Jenkinson M, Johansen-Berg H, *et al.* Tract-based spatial statistics: voxelwise analysis of multi-subject diffusion data. *NeuroImage.* 2006;31(4):1487–505.

[67] Bach M, Laun FB, Leemans A, *et al.* Methodological considerations on tract-based spatial statistics (TBSS). *NeuroImage.* 2014;100:358–69.

[68] O'Donnell LJ, Westin CF, and Golby AJ. Tract-based morphometry for white matter group analysis. *NeuroImage.* 2009;45(3):832–44.

[69] Khatami M, Schmidt-Wilcke T, Sundgren PC, *et al.* BundleMAP: anatomically localized classification, regression, and statistical analysis in diffusion MRI. *Pattern Recognition.* 2017;63:593–600.

[70] Honey CJ, Sporns O, Cammoun L, *et al.* Predicting human resting-state functional connectivity from structural connectivity. *Proceedings of the National Academy of Sciences of the United States of America (PNAS).* 2009;106(6):2035–40.

[71] Kambeitz J, Kambeitz-Ilankovic L, Leucht S, *et al.* Detecting neuroimaging biomarkers for schizophrenia: a meta-analysis of multivariate pattern recognition studies. *Neuropsychopharmacology.* 2015;40(7):1742–51.

[72] Dosenbach NUF, Nardos B, Cohen AL, *et al.* Prediction of individual brain maturity using fMRI. *Science.* 2010;329(5997):1358–61.

[73] Cox DD, and Savoy RL. Functional magnetic resonance imaging (fMRI) "brain reading": detecting and classifying distributed patterns of fMRI activity in human visual cortex. *NeuroImage*. 2003;19(2 Pt 1):261–70.

[74] Davatzikos C, Ruparel K, Fan Y, *et al.* Classifying spatial patterns of brain activity with machine learning methods: application to lie detection. *NeuroImage*. 2005;28(3):663–68.

[75] Schölkopf B, and Smola AJ. Learning with Kernels. Cambridge, MA: MIT Press; 2002.

[76] Hastie T, Tibshirani R, and Friedman J. *The Elements of Statistical Learning: Data Mining, Inference, and Prediction*. 2nd ed. *Springer Series in Statistics*. Cham: Springer; 2011.

[77] Cole JH, Poudel RPK, Tsagkrasoulis D, *et al.* Predicting brain age with deep learning from raw imaging data results in a reliable and heritable biomarker. *NeuroImage*. 2017;163:115–24.

[78] Arbabshirani MR, Plis S, Sui J, *et al.* Single subject prediction of brain disorders in neuroimaging: promises and pitfalls. *NeuroImage*. 2017; 145:137–65.

[79] LaConte S, Strother S, Cherkassky V, *et al.* Support vector machines for temporal classification of block design fMRI data. *NeuroImage*. 2005; 26(2):317–29.

[80] Wee CY, Yap PT, Zhang D, *et al.* Identification of MCI individuals using structural and functional connectivity networks. *NeuroImage*. 2012; 59(3):2045–56.

[81] Bradley PS, and Mangasarian OL. Feature selection via concave minimization and support vector machines. In: *Proceedings of the International Conference on Machine Learning (ICML)*; 1998. p. 82–90.

[82] Cui Y, Wen W, Lipnicki DM, *et al.* Automated detection of amnestic mild cognitive impairment in community-dwelling elderly adults: a combined spatial atrophy and white matter alteration approach. *NeuroImage*. 2012; 59(2):1209–17.

[83] Wee CY, Yap PT, Li W, *et al.* Enriched white matter connectivity networks for accurate identification of MCI patients. *NeuroImage*. 2011;54(3): 1812–22.

[84] Dai Z, Yan C, Wang Z, *et al.* Discriminative analysis of early Alzheimer's disease using multi-modal imaging and multi-level characterization with multi-classifier (M3). *NeuroImage*. 2012;59(3):2187–95.

[85] Kpotufe S, Boularias A, Schultz T, *et al.* Gradient weights improve regression and classification. *Journal of Machine Learning Research*. 2016; 17(22):1–34.

Chapter 4

Virtual and augmented reality in neuroscience

Panagiotis E. Antoniou[1], *Alkinoos Athanasiou*[1], *and Panagiotis D. Bamidis*[1]

Neuroscience encompasses the interpretive layers of a person with the environment, implemented as the central and peripheral nervous systems. In neuroscientific research, normal and abnormal neurophysiological function is studied for diagnosis, or altered through treatment by affecting paired or evoked responses from environmental sensory input. With the advent of virtual reality (VR) and augmented reality (AR), as human–computer interaction (HCI) modalities of highly immersiveness and malleability, a new avenue for previously unfeasible environmental input has been realized. This chapter explores the potential of VR/AR in the axes that these modalities are currently being deployed in neuroscience. Virtual reality environments (VREs) are implemented as alternative customizable platforms for brain–computer interface (BCI) applications but are also exploited in neurorehabilitation paradigms. In medical neurosciences, VR/AR have proven added value in the visualization of the nervous system both for operative enhancement and for surgical training. Additionally, in medical education there is a strong case to be made in favor of the "virtual lab" (VL) for neuroanatomy offering added value to teaching of the complexity of the neural structures and networks. Furthermore, the controllability of VREs and VR/AR render them excellent candidates for studying cognition through event related potentials (ERPs). Finally, we explore the combined role of VR/AR and affect in educational and internal learning processes, aiming toward an integrated sensor immersion ecosystem.

4.1 Introduction

Human neuroscience studies, modifies and treats the nervous systems of a person, those that constitute the interpretive interface between human and environment, natural of artificial. In the context or research expeditions in the field and various

[1]Medical Physics Laboratory, School of Medicine, Faculty of Health Sciences, Aristotle University of Thessaloniki, Thessaloniki, Greece

subfields of human neuroscience, it is the structure, function and interactions between them that are studied in normal conditions and abnormal pathologies. Eventually, the field aims to offer novel interpretations of neurophysiological processes, enhancement of physiological function or treatment of pathology; all the while modulating paired and evoked responses from environmental multimodal sensory input constitutes the backbone of non-pharmacological approaches to affecting the nervous systems. In neurotechnology, VR and AR have recently emerged as modalities of highly immersiveness and malleability in the context of HCI and have offered a new and unexplored avenue for immersive environmental input that was otherwise previously unfeasible.

Immersive visualization modalities such as VR/AR have been validated as safe, motivating, affordable and effective approaches for BCI systems [1,2]. Furthermore, the above-mentioned experiential modalities are used as sensory inputs for evoked responses study and research which are inherently linked to contemporary BCIs. Studies, addressed for the potential for neurological rehabilitation, have shown that the motor system and the current state of embodiment (or body schema) are plastically linked [3,4]. Sensorimotor triggers and multisensory inputs are crucial for enabling body ownership and agency while embodiment modifications affect motor responses of limbs and aim at promoting adaptive neuroplasticity and neurological recovery [5–7].

On another approach to VR and AR through the prism of information and communication technologies (ICT) for healthcare and associated education, digital innovations have shown the potential to reduce costs and increase capacity to support learning, growth and address social inequalities as well as to improve diagnostic efficacy and treatment effectiveness. Contemporary medical education, in particular, has progressed extensively toward widely diverse learning resources and healthcare-specific educational activities in the ICT domain and “virtual” patients (VPs) have been already established as an effective and important tool for modern medical education [8]. Further down this pathway, more accurate approaches to the “virtual” term have been also implemented, such as the VL, digital complex simulations readily supported by VR, AR and mixed reality (MR) technologies [9–11]. While the distinctions are somewhat blurred at times, VR can be defined as the substitution of external sensory inputs (mainly visual and audio) with computer-generated ones through the use of a headset device. AR is the superposition of digital content over the real world utilizing either two- or three-dimensional markers in the real-world environment. MR, finally, is similar to AR with the key difference that instead of the real-world marker being a pre-programmed static item or image, the superposition of content is rendered after a 3D mapping of the current environment has been completed.

The remainder of this chapter will attempt to categorize the use of VR and AR in areas of research, albeit sometimes overlapping, explore the paradigms of use in those areas and offer examples and use cases, while discussing advantages and added value against unmet challenges. The areas that will be explored are BCI, neurorehabilitation, operative neurosurgery, neuroanatomy teaching, exploration of cognition and finally affective computing.

4.2 BCI trends

BCI refers to an interactive technology that detects active or passive brain responses pertaining to volition and offers users with an alternative way to translate volition into control of external devices [12]. BCI technology now spans its fifth decade while boasting tremendous improvements and impressive results in portability, reliability, performance and robustness, as well as an ever-increasing array of clinical, experimental and demonstrational applications, ranging from avatar control and word spelling to neurorehabilitation [13]. Nevertheless, even to this day, it can be argued that despite advances, BCI technology has not been integrated into common daily activities due to drawbacks considering unobtrusiveness, ease of use, cost, availability and smooth performance. The combination of VREs—and of AR in particular—with BCIs has been proposed as early as 2004, to be able to "revitalize" the potential of brain controlled technology for everyday use, taking also advantage of the progress in new wireless technologies (such as ultra-low energy Bluetooth) [2]. While such a prospect seems to not have been yet realized, BCI community has indeed caught upon the "trend" and an increasing number of very interesting investigations are now focusing on VR and BCI synergies [14–19].

In the now classic study of Pfurtscheller *et al.* [20], navigation inside a VRE was performed using detection of only two classes of motor imagery in the EEG signals of three subjects. Two VR setups were explored in this study, one using a head-mounted display (V8-HMD, Virtual Research Systems, Inc. Aptos, USA) and the second using full immersion in a VR Cave (ReaCTor, SEOS Ltd., West Sussex, UK). A Cave is a category of VRE that uses wall screen projections that surround the user, in order to create a continuous, navigable virtual setting [21]. Both setups allowed for successful navigation using three central electroencephalography (EEG) electrodes after some training in a classic BCI set on computer display. This study demonstrated the feasibility of navigating a VR urban setting without muscular activity and investigated the motivation effect of VR on BCI performance [20]. Concentration, fatigue, loss of attention and visual interference with moving objects inside the simulation were identified as parameters that affect BCI performance within such a setting. Subjective evaluation from the participants in this study offered some very interesting insight, as they reported positive feelings for the VR setting over the traditional BCI experiment, for the immersion Cave over the HMD as well as increased motivation; this insight needs to be taken into account not only for the design of BCI applications for commercial uses, but also for the use of brain-controlled VR modalities in neurorehabilitation.

In accordance with this insight, the sense of "agency" and the registration of virtual representation of self (avatar) into the body schema is another aspect of BCIs currently under investigation and presents implications for the integration of VR into BCI systems [15]. On the other hand, VREs may not be universally appropriate for every BCI task or user category and the positive or negative effects on BCI performance may be influenced by visual comfort characteristics of the design and the relative comfort experienced by the user [17]. The task-related drawbacks can indeed by addressed by gamification of BCI tasks (Figure 4.1) which can induce synergistic effects with VR regarding immersiveness [16,22]. Moreover tailored design of the tasks, such as progressive training [16] and hybrid BCI systems with appropriate combination of EEG features for different control elements [22], should be considered in order to overcome task-based and medium-based discomfort and discouragement that could hurt performance.

While the previous directions of research consider VR addition to BCI applications for neurophysiological research, BCI optimization and neurorehabilitation applications, an interesting approach would also be to consider adding BCI functionality to VR applications for improvement of a system's performance. Such an approach could be inferred by the study of Brouwer *et al.* [23]. The authors investigated a passive EEG-based BCI, that was used to predict intention of head rotation in order to proactively improve the quality of streaming in the VR headset. The EEG sensors were incorporated on the VR headset and the results from 11 participants showed that a neural network could timely and correctly classify the intention and direction of head turn and allow for retrieving and streaming higher quality imagery, thus making a case for possible improvement of immersiveness of the VR experience [23]. While this approach could help popularization of BCI technology for commercial uses, retroactively, an increase of immersiveness could be expected to lead to better performance for more "traditional" approaches. At the same time, flexibility and modularization of BCI applications should be considered desired properties of BCI design architecture; those properties could be combined with VR environments to allow end-users intuitive control of devices beyond those applied in research and tread into the domain of Internet of Things (IoT), as proposed and demonstrated by [24].

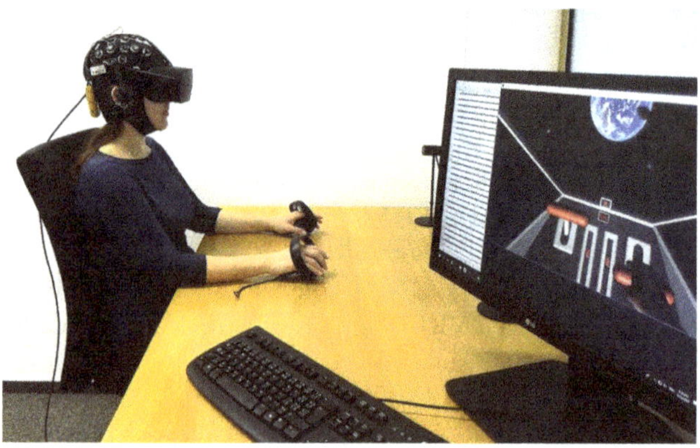

Figure 4.1 Use case example of an EEG-based BCI, using a VR HMD where the participants performs hands motor imagery. Figure from [16]

4.3 Neurorehabilitation and neurotherapy

Implementation of VREs in rehabilitation of neurological disorders, although still at very early stages, has opened up some very promising prospects for clinicians and researchers alike, as this technology has demonstrated real potential for added benefit to traditional rehabilitation practice and for enhancing neurological recovery [25,26]. Traumatic brain injury (TBI) [27,28], especially vestibular disfunction in mild injury (mTBI) [29], motor rehabilitation following stroke [18,19,30,31] and, recently, even spinal cord injury (SCI) [32,33] fall into the scope of VRE-assisted treatments.

TBI is an important cause of neurological disability with variety in severity and phenotypes among survivors ranging from mTBI, with minimal to no residual neurological symptoms, to chronic severe "vegetative" states. VR technology has been tested mostly in neurorehabilitation for mild TBI, that nonetheless features symptoms may hamper patients up to the degree of functional disability [28,29]. Furthermore, VR is also being discussed as a potential life-changing technology, possibly able to modify outcomes in cases of moderate TBI [27] targeting affective states, deficits in cognition and behavior, as well as degraded motor skills. In the case of mTBI, VRE applications have demonstrated clinical success, for example improvement in gait speed, weight shift, visual and physical motion tolerance in the context of post-concussion residual vestibular dysfunction [28,29]. In these studies the intervention employed an immersive VRE featuring Cave-like characteristics using a motion platform as control method for navigating the environment, belt treadmill, motion capture system and 2D virtual graphics projected on the surrounding screen (CAREN, Motek Medical BV, Amsterdam, The Netherlands). The study demonstrated feasibility for such a setup and measurable improvement in clinical outcomes and should be interesting to be validated against outcomes in traditional rehabilitation for vestibular dysfunction [29].

Stroke has perhaps been the most popular indication for incorporating VR into neurorehabilitation protocols so far [18,19,30,31]. The sense embodiment that, as already discussed, is enhanced by immersiveness of VR techniques can aid the recruitment of neuroplasticity and induce partial motor recovery even in chronic phases after the stroke using intensive training protocols [31]. Multimodal feedback plays a key role in the illusion construct and the combination of visual and vibrotactile input shows important synergy in immersive VR-based neurorehabilitation [18,19]. At the patient end of the loop, the appropriateness of control modalities are still not clearly defined as BCIs, electromyography (EMG) biofeedback and repetitive transcranial magnetic stimulation (rTMS) pose as suitable candidates and their relative roles are still under investigation [18,19,30]. As with BCI development, VR should be combined instead of antagonizing other modalities, offering its key comparative advantage that is immersiveness. Combining motor imagery BCI with stereoscopic VR presentation and rTMS (Figure 4.2) showed motor improvement for three patients a year post-stroke that was validated both with fMRI investigation showing significant desired neuroplasticity but also against a

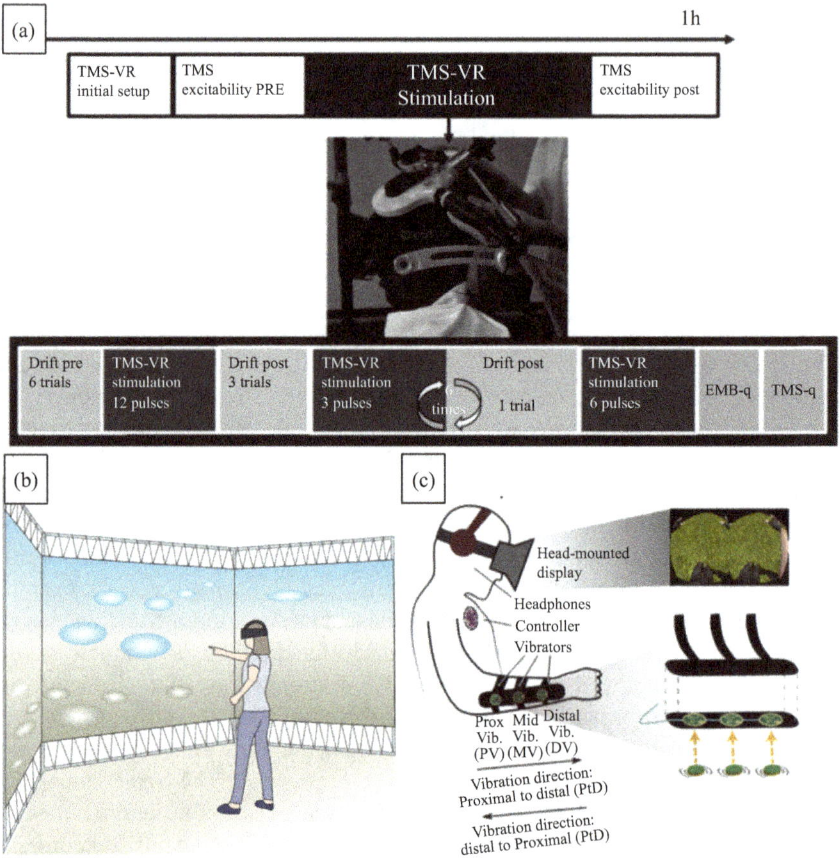

Figure 4.2 (a) VR has been increasingly chosen in multimodal neurorehabilitation protocols, often combined with other novel technologies, such as transcranial magnetic stimulation (TMS). Here, [3] present such a protocol aimed to induce rubber hand embodiment. (b) [25] discuss that HMDs and VR Caves or open VR displays both present relative advantages and disadvantages, depending on the scenario. (c) VR HMD here used for integrating visual with tactile feedback illusion, virtually corresponding to the feet of paraplegic patients and delivered to the forearm, in a ground-breaking study from [32,33]

sham intervention group which had no significant neuroimaging alterations [30]. The importance of this study lies not only with the seamless integration of VR functionality into the experiment but also with the neurophysiological validation, an aspect that technology-based interventions often lack.

Rehabilitation of SCI patients remains a significant medical, engineering and societal challenge, especially in chronic complete injury that causes irreversible paraplegia or tetraplegia depending on the level of injury. Only during the most recent years, ground-breaking advances in science and technology have offered a light of hope for those patients in the form of spinal neurostimulation implants and exoskeletons that often aim to bypass the neurological level of injury (NLI). It is now accepted that a percentage of ascending and descending white matter tracts that cross the injury remain structurally intact, offering the neural basis for neuroengineering and neurorehabilitation to investigate promoting dormant neuroplasticity in the sensorimotor system of those patients [32,33]. It has been demonstrated that long-term training in multimodal man–machine interfaces does indeed have the ability to produce synergistic effect and promote said dormant neuroplasticity and offer a measurable degree of neurological recovery for those patients. In the line of work from [32,33] patients with chronic complete SCI were trained in different combinations of modalities that included robotic gait systems, treadmills, body-weight support gait systems and more importantly BCI-controlled exoskeletons and BCI controlled rich immersive VR environment. VR through a commercial HMD (Occulus Rift) was used in combination with the other technologies to promote the sense of embodiment and agency. The researchers deployed novel strategies, such as integrating visual with tactile feedback into the virtual "illusion" (Figure 4.2) and delivered to the forearm tactile signals that corresponded to the feet of paraplegic patients (no sense of touch is preserved normally in complete injury below the NLI). Not only did the researcher demonstrate the feasibility and usability of their strategies and of successfully incorporating VR paradigms in neurorehabilitation protocols for SCI, but they also reported significant clinical outcomes: patients previously were classified as chronic complete SCI had their disability status transformed to incomplete injury, showing sensory and—most importantly—motor recovery below the NLI. As forefront neurorehabilitation researchers and centers further investigate synergistic effects of neurorobotics, BCIs, exoskeletons and VR technology, important issues may arise regarding good practices of such approaches. These issues include availability, task design and deliverance as well as expertise from the part of the researchers, possible adverse effects and maladaptive plasticity, but also tolerance, acceptance and strain associated with these novel technologies from the part of the patients. Nevertheless, this technology in general seems safe and well-accepted, while its financial feasibility and cost-effectiveness should be determined by clinical and research outcomes, as suggested by [26].

The applications and usefulness of VR in the context of nonpharmacological treatments should be expected to grow beyond the scope of neurorehabilitation as well. Immersiveness affects motivation that has been show to play a role in efficacy of neurofeedback (NF) [34,35] (an EEG biofeedback technique used to regulate brain activity), that is used in a variety of indications including attention deficit-hyperactivity disorder [36] and refractory epilepsy [37] and is being currently investigated for others, including addiction [38]. A recent study explored emotion regulation using real-time functional magnetic resonance imaging (rtfMRI)-based NF, while the visual cues were created using the Unity 3D asset Autumn Nature Pack [39]. The authors describe using a "virtual environment" BCI, that in actuality consisted of 3D-realistic graphics projected on 2D monitor, and demonstrated increased engagement of users while recruiting associated brain regions. The setup, as was realized in Unity, can be considered only a step away to be usable in a more accurate VR Environment and the results regarding motivation, engagement and immersiveness would be interesting to be comparatively investigated. As VR and AR have been also demonstrated as very useful tools in psychotherapy due to their properties of immersiveness, their combined use with BCIs and NF could be explored for the multimodal treatment of psychiatric conditions such as schizophrenia [40].

4.4 Operative virtual guidance and neurosurgical education

Neurosurgical specialty is eager to test and develop novel technologies that have a potential to improve outcomes (albeit not so eager to adopt such technologies in standard practice). VR and AR have gained gradual popularity with neurosurgeons regarding mainly education and pre-surgical planning but also novel intraoperative uses of VR and AR have been also investigated, especially for assistance with neuro-oncologic operations [41].

Basic AR functionality on the other hand, has become a common feature in modern neurosurgical microscopes, allowing surgeons to use AR as a standard tool in number of operations [42]. These devices nowadays can be registered to traditional neuronavigation protocols and they also can import mages from patient examinations, such as computed tomography (CT), magnetic resonance imaging (MRI) and magnetic resonance angiography (MRA). Accordingly, since the examinations are co-registered to the patient anatomical features in a 3D neuronavigation space, the microscope can overlay obscure anatomical and pathological features onto the surgeon's bifocal vision, creating a dynamically augmented surgical field (Figure 4.3) [43,44]. Elements such as obscure vasculature, a non-delineating tumor's relative position below the cerebral surface or a very small aneurysm can appear on the projected field, allowing better planning, targeted guidance, individualized approach and increased safety for the surgeon. Already a wide variety of cerebral and cerebrovascular pathologies have been reportedly treated in AR-assisted operations, covering a large portion of brain operations, making a case for the need for comparative assessment of its efficiency against traditional non-AR-assisted surgery [42].

Furthermore, sophisticated AR techniques have been developed. One such technique allows image guidance directly overlaid on the patient operating field (and not just through the microscope), after a co-registration of physical objects and patient imaging scans in a similar fashion to neuronavigation, as is the case described in [45]. The authors here presented a novel and reliable approach to AR that offers layer to layer (on skin, skull or brain) rendering of the pathology and continuous guidance during all surgical steps that was successfully tested on patient cases. Although still not as popularized as the AR functionality of modern

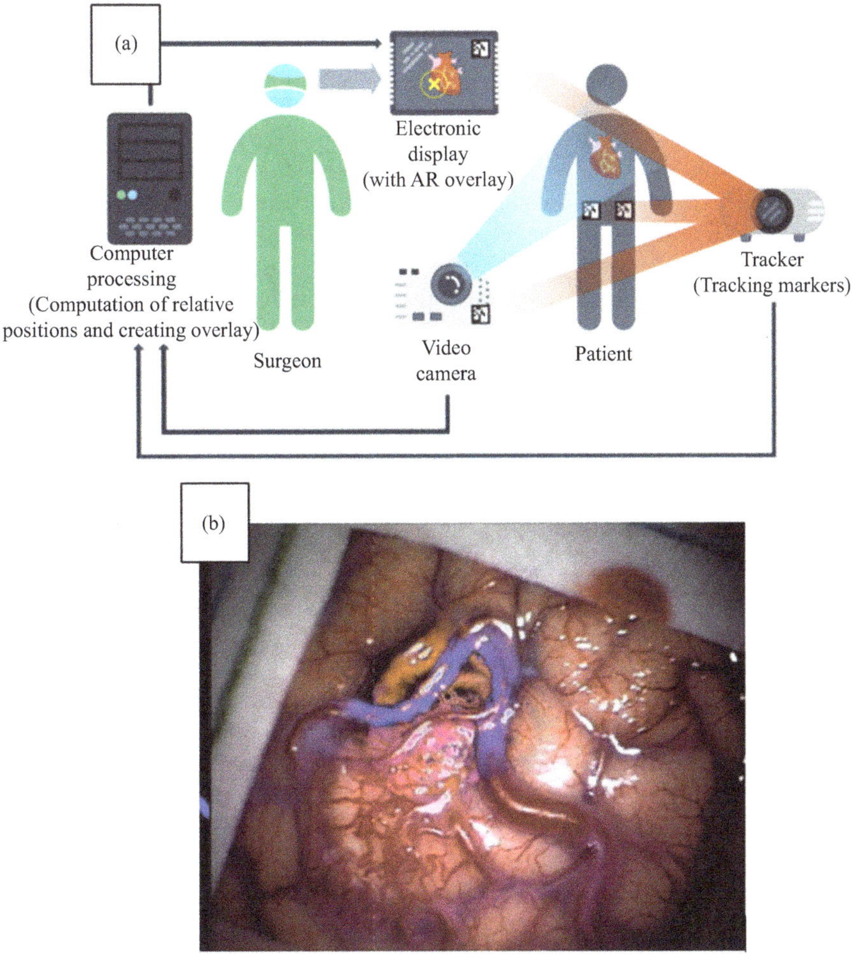

Figure 4.3 *(a) AR overlay has demonstrated added value to some surgical specialties, most notably Neurological Surgery. Hereby is presented a basic schema of AR integration to the operational setup, as shown in [43]. (b) Example of a neurosurgical operative field using AR functionality to enhance anatomical perception of cerebral vasculature, as shown in [44]*

neurosurgical microscopes, this methodology can be considered a more accurate application of AR, being unrestricted to the medium (binocular view) and possibly more dynamic to operation progress. Indeed, critique that microscope AR functionality should not be considered "true" AR has been expressed [46]. Still various AR approaches are indeed gaining popularity in neurosurgical practice, including commercial HMD AR (HoloLens) for placement of pedicle screw in cadavers [46].

Regarding further ready-made VR technologies that concern neurological surgery, evaluation of patient neuroimaging scans can be facilitated by 3D exploration of DICOM data using a commercially available HMD, in lieu of traditional 2D display-based DICOM viewers, as has been also demonstrated [47], and could as well become a practice standard in the following years. The creation of VR ready immersive 3D cerebrovascular models for education and clinical applications has been facilitated by the boost in power and lower costs of graphics processors in the recent years [48]. Commercially available surgical simulators, enabled by VR headsets, are also becoming more widespread in neurosurgical training and their steady advancements have transformed them, from a design novelty, into a validated tool that is able to differentiate levels of ability in specific neurosurgical skills [49]. Hybrid education settings, combining visual supervision and VR-enabled simulation can provide added value in training and evaluation of neurosurgical trainees as demonstrated by [50] and VR simulation has been established at least as a useful addition to state-of-the-art neurosurgery training programs, able to provide exposure to techniques prior to cadaveric studies or live surgeries, volume of training and self-paced learning [51].

4.5 Virtual reality, the virtual laboratory and the case for neuroanatomy

Medical education has as of late centered around strengthening experiential methods like virtual worlds [52,53] or AR and VR [54]. Such endeavors aim to draw in the students in the broad, various and important points of their educational plan. On this experiential front, numerous concepts have been realized. One of them is the VL. VLs use a diverse number of media, for example, 3d models and video, so as to supplant genuine research facility interactions. A VL would comprise of a few computerized simulations and would be augmented by thematic discussions and video showings, or even, as a coordinated effort, complex reenactments [55]. Such intelligent conditions encourage self-guided learning (for example repeating content, accessing it at off hours). That way, students keep up activity and extend commitment in the learning procedure while interactivity sharpens lab abilities that move past just information transfer. Such lab aptitudes, in medicine, reduce shortcomings of contemporary medicinal educational program. Hands on research center methods are typically not accessible to preclinical students because of cost, time or health hazard limitations [56,57]. This leaves them lacking in clinical and lab abilities [58].

A methodology for implementing the VL has been the use of virtual augmented and, recently mixed, reality (VR/AR/MR) technologies with the emergence of devices such as the Microsoft HoloLens [59]. These facilitate completely untethered, holographic encounters so as to enable the learner in novel manners. They mix optics and sensors to convey consistent 3D content links with this present reality. The differentiations sometimes blur but VR is the substitution of outside sensory information sources (predominantly visual and sound) with computer-generated ones using a headset. AR is the superposition of digital content over this present reality using either two-dimensional or three-dimensional markers in the real world. Blended reality is like AR with a key difference. Rather than the markers being prearranged static things or pictures, the digital superposition is done after a 3D mapping of the real-world space has been finished. Along these lines, 3D objects can very well behave like normal objects (e.g., 3D models standing on tables, 2D pictures "hanging" on dividers, etc.). There are evidence that such innovations altogether augment instructive effect of a learning episode and can accordingly influence instructive results [60]. Acknowledged models incorporate experiential world investigation [61], physics and chemistry [62–64], as well as fusion of such modalities in virtual scenarios [54]. It is the immediacy of these modalities that can engage the learner, but also help her absorb instructive material, and in this way avoid conceptual fallacies [65].

Especially in neuroanatomy, MR becomes highly relevant. The functionally varied neural pathways that all consist of the same kind of tissue are not able to be viewed in real cadavers since they are not anatomically differentiated. Through V/MR, series of diverse showcases can be crafted, so that the student learns, visually, in a virtual human body, location and functional connectivity of neuronal groups and pathways. In that context V/MR is a versatile educational modality, especially in this highly sensitive medical domain. It provides the capacity for highly impactful visualizations that can convey knowledge greater than which can be seen in real medical objects (e.g., patients, cadavers, etc.).

4.6 Event related potentials (ERPs) from virtual stimuli

The controllability of virtual environments and virtual reality makes it an excellent candidate for use in studying cognition. It expands the bounds of possible event related potentials (ERP) experiments by providing complex, dynamic environments in order to study decision making in cognition without sacrificing environmental control.

There have been several experimental setups to leverage these capacities. Even in the 1990s there has been created a flexible system for real-time EEG collection and analysis from within virtual environments. The ability to give quick feedback enables it to be used in BCl research, aimed at helping individuals with severe motor deficits to become more independent. BCl work, even back in the 1990s has shown the feasibility of online averaging and biofeedback methods in order to choose characters or move a cursor on a computer screen with up to 95% accuracy while sitting still and concentrating on the screen [66,67]. Works such as [68] dramatically extended BCl by allowing evoked potentials to propel the user through alternate virtual environments. For instance, a client could pick a virtual space from a menu of rooms, explore it naturally in the HMD, and afterwards turn on the sound system. As appeared in [69], the P3 evoked potential might be utilized for a BCI that picks characters on a PC screen. Found by [70,71] and widely examined [72], P3 is a positive waveform happening roughly 300–500 ms after an in-visit task-pertinent upgrade. Basic yet significant analyses expecting subjects to stop or go at virtual traffic lights evokes this ERP. The contingent negative variety, of an ERP that happens going before a normal improvement, happens at slow down lights. Feasibility trial of recognition in the noisy virtual condition of a BCI controlled virtual vehicle's brakes provided, on average, very good precision [68].

Down the road equally impactful experiments attempted to test through ERP study whether seeing one's body affects the spatial attentional link between vision and touch and to determine at which point the effect takes place. Based on the findings of [73], it was suggested that seeing one's hands would make the tactile non-targets more distracting, and thus lead to response inhibition indexed by an enhanced N200 component [74].

This was affirmed in a test that utilized VR headset to make different conditions of visual embodiment in the subject (no visible hands, visible virtual hands, visible real hands). Subjects reacted with increasingly negative N200 to tactile distractors when they either emerged from the same modality or were introduced on the same side as the objective. This impact was more pronounced in the VR hands and control conditions than in the VR no-hands condition. In opposition to expectations, the presence of hands had no impact on visual-evoked N200. In spite of this, visual N200 was in like manner increasingly sensitive to target-like distractors, demonstrating an inhibitory job like what was seen in somatosensory-evoked potentials (SEPs). Along these lines, visual target-like distractors resulted to increased response in visual-evoked potentials, no extra inhibitory exertion was required when (virtual or genuine) hands were available. The observed disparity among visual and material N200 is fascinating as it demonstrates that visual presence of one's body influences the tuning of tactile spatial attention more than it influences the tuning of attention to visual spatial senses. The finding is significant as it affirms and broadens previous ERP evidence which proposes that the visual body as opposed to ambient visual data (e.g., the lab space) is utilized for remapping tactile stimuli [75]. Additionally, it was shown that embodied presence influences sensory-perceptual procedures as well as later executive function. The main advantage of controllability in a virtual environment was obvious in such experiments since it provided the option to tentatively adjust the environmental sensory input [76] in order to explore whether substantial presence would impact late attention and memory-related procedures. As opposed to previous components, both visual and material P3 were selectively responsible of targeted stimuli. Likewise viewing condition influenced P3, albeit just in SEPs. The modulation of attention at the P3 territory was the same between differing view conditions, proposing that both target and non-target tactile stimuli brought about improved P3 if the hands were visible. The finding of P3's specific responsiveness agrees with prior literature demonstrating that cross-modal interaction in spatial attention is absent at the later ranges [77].

Such examinations additionally affirmed the dependability of ERPs acquired in VR conditions with signal-to-noise ratio (SNR) investigations. Looking at the SNRs of viewing conditions, the VR conditions had a stronger P3 signal and less noise than what was seen in a non- HMD control condition. The finding was interesting as it was expected that HMD would induce electrical impedance in the EEG signal. The higher SNR in the VR conditions could be because of the head motion limitations that the HMD enforces, thus removing head movement and ocular artifacts. On the contrary, in the control condition, subjects without an HMD may have triggered them to move more uninhibitedly thus causing more such artifacts. In conclusion it has been demonstrated by the previous, amongst others, experimental studies, that VR HMDs did not unfavorably influence the ERPs, which at the time suggested an important role for VR headsets in ERP research without compromising EEG recording quality.

This led to the current state of the art that has allowed full-fledged ERP studies to take place in order to validate the VR modality as an equally valid experimental setup for full scale ERP studies. High quality EEG data has been proven to be able to be acquired with time-locked VR stimulus delivery with temporal resolution at the millisecond level. A within-subjects analysis revealed no significant differences in the EEG measures between Desktop and VR AMP acquisitions [78].

These studies demonstrated that integrated solutions are viable and can prove useful for numerous research and clinical applications. However, there are still issues to be addressed prior to the widespread acceptance of VR-EEG. Usual VR headsets have a non-trivial weight and exert significant pressure on the face, causing neck strain and overall discomfort. Ergonomics solutions such as engaging only for 30 min for were proposed along with regular intermittent breaks between each such interval. VR headsets are still bulky and feature exposed wiring, thus limiting mobility while wearing the device. These issues are in fact getting addressed with technological progress, while integration solutions are becoming available. Specifically, solutions that integrate EEG electrodes in the chassis of the VR headset are underway [79], along with improvements in the weight and volume of the device.

As such, VR headsets seem to be important tools for such further research because they allow full immersion of the user without core changes in the neurological signatures of capacities like attention, learning and memory. A core potential application of VR, as it offers 3-D realistic environments, is the measurement of optic flow (OF). OF is the overall sense of perception and integrative understanding of the visual field as someone is navigating a physical environment. More specifically, radial optic flow (ROF) is the perception of motion in the periphery of a central visual field especially during changes in scenery from self-directed movement or from objects moving away or toward the viewer. ROF is getting attention because it has been associated with Alzheimer's disease (AD), or other neurodegenerative diseases such as amnestic mild cognitive impairment (aMCI) [80,81]. OF-based psychophysical tests provided the interesting result of differentiating AD vs aMCI as OF has been proven to be impaired in AD but not in aMCI. Neurophysiological studies demonstrated perceptual, cognitive and sensory (early components P100 and N130), ERP component differences for tasks relating to OF in AD as well as variations in cognitive ERP components for aMCI during varied experimental OF stimuli-based paradigms [82,83]. A core find across AD patient studies is longer latency and reduced amplitude of the P200 with one study of aMCI patients correlating the inverse of the latency of the P200 with mini-mental state examination (MMSE) scores, proposing links to cognitive decline. These ERP changes can be linked with navigational abilities deficits in AD. However, not all MCI or AD patients demonstrated ERP changes while there have also been several reports of older adults with no neurodegenerative disease evidence that presented ERP change in OF and impaired navigational capabilities [84]. Given these facts it becomes apparent that VR ERP studies can play a crucial role in exploring these preliminary diverse results.

4.7 Toward an integrated sensor immersion ecosystem

The role of emotions and mood is crucial in the educational process, as well as in the internal learning process itself, since it directly taps into concepts such as attention, motivation and interest [85]. It has been suggested [86] that learning mostly happens as an emotional process, thus the interaction of learning and affect is invaluable for discerning and facilitating the former. Also, links between affective states and learning have been established before. A learner's emotional state will consistently affect how they process new knowledge. It has been reported [87] that veteran learning experts have a positive impact on their learner's education because they recognize and respond to their emotional states. Affective states like curiosity have been identified as an indicator of motivation [88]. These constitute a driver for learning, as they can motivate and affectively engage learners, involving them more in the process and removing stress and anger, or frustration, providing pleasure and engagement while lowering boredom [89,90].

Biosignals such as blood volume pressure (BVP), external body temperature, electro-dermal activity (EDA), heart rate (HR) and EEG are physiological markers that can characterize the human emotional state [91]. Tracking and analyzing these marker recordings can determine the emotional state, a process that can be accomplished by devices that use various physicochemical processes to achieve these recordings, called biosensors [92–94]. Extracting affect and emotional state from the analysis of these recordings has been the research interest of several studies. EEG biosignals classification for emotion detection, utilizing the Valence-Arousal model of affect classification have been explored [95], while EEG and EDA biosignals, more specifically, have successfully detected stress (e.g., during indoor mobility [96]).

In the context of today's untethered MR technologies, experiential training that integrate advanced sensors such as portable EEGs and smart watch devices can capture information about the affective state of the user based on her biosignals. High fidelity area scanning with inertial and depth sensors allow for mapping and understanding of the physical places, spaces and things around the user. This untethered integration of sensors and "sense projectors" can integrate into a real-time affective analytics (RTAAs) digital ecosystem. Such ecosystems are able to implement RTAAs in authentic learning environments ("in the wild") during MR experiential educational episodes, for example in neuroanatomy training that is, as mentioned previously, highly appropriate for MR immersive resources. Experimental scenarios include recordings through a wearable unit. EEG devices, and wearable multisensory smart wristband are just a few of the options for integrating sensors at the time of this chapter's writing (e.g., [97,98]). The recording of such biosignals can provide significant objective indications of a person's affective state relevant to its educational capacity.

Increment of theta power is reported to encourage both episodic and working memory just as the encoding of new information [99,100]. In the frontal-midline brain areas, theta activity additionally identified with creativity, attention and focus [101–105]. In accordance with this, increased theta rhythm has been observed in the frontal-midline regions, during high cognitive demands tasks that do require higher working memory use [106,107] and a high attention threshold [108,109]. Moreover, commitment to judgment calls, or attention demanding tasks has been found to prompt alpha power suppression [110]. Raised EDA and HR are indications of high arousal, independent of valence [111,112].

The technical barriers for this endeavor are yet significant. While the wrist-wearable sensors are unobtrusive and easy to wear, EEG recordings when wearing a highly sophisticated electronic device like the Microsoft HoloLens are not easily obtainable. The literature has significant findings regarding EEG and VR, but usually these involve virtual environments (3D environments projected in a 2D screen) instead of true immersive VR that requires a dedicated headset [113]. In other cases, successful endeavors incorporating EEG to VR scenarios require expensive and cumbersome simulation rooms [114]. However, the fact of the matter remains that given the state of current technologies, such experimental setups are viable as proofs of application.

Similar endeavors, for affective analytics in non-technology heavy medical education episodes have already been initially explored [115]. However, technology exists, as described above, for a first proof of concept configuration that is realistically unconfined and applicable in a simple educational setting for emotion detection in technology heavy, V/MR medical education episodes (Figure 4.3). The full spectrum application of this work is the integration of such a sensors and devices ecosystem in an intelligent tutoring systems (ITSs).

ITSs are computer systems that provide immediate and customized feedback and instruction to learners [116], without human teacher involvement. ITSs typically aim to replicate, albeit in an automated fashion, the benefits personalized, one-on-one educational experiences from a single or multiple teaching entities (e.g., lectures tutoring) or even to motivate self-learning (e.g., online homework). ITS implementations aim to leverage individual learner characteristics and needs, analyzing and/or anticipating their emotional shifts and behaviors during the educational process. That way these systems can, tailor tutoring instructions upon inferences on strengths and weaknesses, handle content adjustment, suggest additional work and in general improve the learning level of subjects [117]. In that context, the integration of emotion detection for content/experience customization is not a new endeavor. Affective tutoring systems (ATSs) use bias-free physiological expressions of emotion (facial expressions, eye-tracking, EEG, bio-sensors

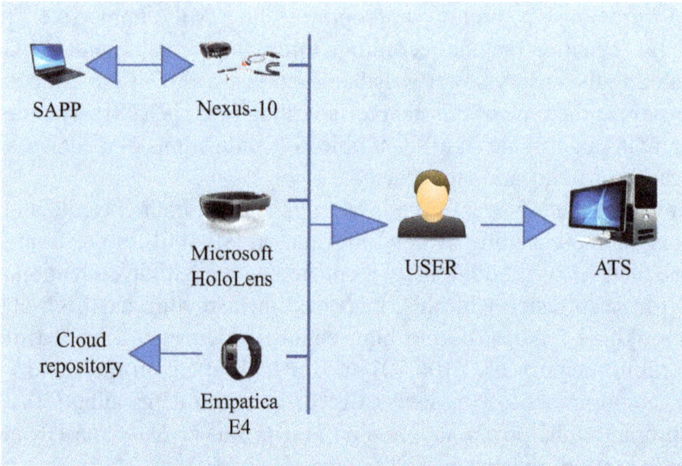

Figure 4.4 A suggested integrated sensor immersion ecosystem with existing platforms and technologies. The user is experiencing a mixed reality educational episode in a platform like the Microsoft HoloLens. A signal acquisition and post processing (SAPP) PC Unit is driving a Nexus 10 portable EEG device. An Empatica E4 wearable biosignal sensor records heart rate and other biosignals storing them in the cloud. An activity and timestamp synchronization (ATS) PC Unit synchronizes and timestamps all-time series as they are recorded

etc.) used "to detect and analyze the emotional state of a learner and respond in a way similar to human tutors" [118], hence being able to adapt to the affective state of students [119]. However, the integration of "untethered" sensors and devices (Emotiv EEG, Empatica wearables, The HoloLens MR platform) can integrate into an "in the wild" affect aware ITS (Figure 4.4).

4.8 Conclusions

The advent of VR and AR, as HCI modalities of highly immersiveness and malleability, has paved ways for previously unfeasible environmental input in the field of Neuroscience. A number of approaches, proposals and success stories were presented in the areas of BCIs, neurorehabilitation, operative neurosurgery, neuroanatomy teaching, exploration of cognition and affective computing. We have discussed the added value of virtual paradigms to these areas as well as the challenges and drawbacks that need to be addressed. Steps that need to be followed for the advancement of the field include expansion of indications, exploration of neural basis, appropriate task design, development of new applications, feasibility and comparative validation. Use of "virtual" and "augmented" reality terms sometimes refer to functionality and do not fully comply with the definitions but nonetheless demonstrate immersiveness and engagement for this technology. The role of VR/AR in neuroscience should be also explored alongside important synergies with other emerging neurotechnologies, possible examples being ATS in affective computing and medical education [115], 3D rapid prototyping in the educational and training paradigms [120] or micro- and nanofabrication technologies for the creation of advanced central interfaces and peripheral neuroprostheses with desired properties [121]. It is our conviction that, as the paradigms of BCIs and neurorehabilitation [32,122] have shown, VR and AR technology still have much untapped potential in Neuroscience especially in approaches where the advantages of physical and virtual layers can manifest and complement each other.

List of abbreviations

2D	Two-dimensional
3D	Three-dimensional
AD	Alzheimer's disease
ADHD	Attention deficit-hyperactivity disorder
aMCI	Amnestic mild cognitive impairment
AR	Augmented reality
ATS	Affective tutoring system
BCI	Brain–computer interface
BVP	Blood volume pressure
CNV	Contingent negative variety
CT	Computed tomography
DICOM	Digital imaging and communications in medicine
EDA	Electrodermal activity
EEG	Electroencephalography
EMG	Electromyography
ERP	Event-related potential
HCI	Human–computer interaction
HMD	Head-mounted display
HR	Heart rate
ICT	Information and communication technology
IoT	Internet of Things
ITS	Intelligent tutoring system
MCI	Mild cognitive impairment
MMSE	Mini-mental state examination
mTBI	Mild traumatic brain injury
MR	Mixed reality
MRI	Magnetic resonance imaging

MRA	Magnetic resonance angiography
NF	Neurofeedback
NLI	Neurological level of injury
OF	Optic flow
ROF	Radial optic flow
RTAAs	Real-time affective analytics
rtfMRI	Real-time functional magnetic resonance imaging
rTMS	Repetitive transcranial magnetic stimulation
SCI	Spinal cord injury
SEP	Somatosensory-evoked potentials
SNR	Signal-to-noise ratio
SSVEP	Steady-state visual-evoked potential
TBI	Traumatic brain injury
VEP	Visual-evoked potential
VL	Virtual lab
VR	Virtual reality
VRE	Virtual reality environment

References

[1] Faller, Josef, Brendan Z. Allison, Clemens Brunner, *et al.* 2017. "A Feasibility Study on SSVEP-Based Interaction with Motivating and Immersive Virtual and Augmented Reality," January. http://arxiv.org/abs/1701.03981.

[2] Navarro, K.F. 2004. "Wearable, Wireless Brain Computer Interfaces in Augmented Reality Environments." In *International Conference on Information Technology: Coding and Computing, 2004. Proceedings. ITCC 2004.*, 643-647 Vol.2. IEEE.

[3] Bassolino, M., M. Franza, J. Bello Ruiz, *et al.* 2018. "Non-Invasive Brain Stimulation of Motor Cortex Induces Embodiment When Integrated with Virtual Reality Feedback." *European Journal of Neuroscience* 47 (7): 790–99.

[4] Bolognini, Nadia, Cristina Russo, and Giuseppe Vallar. 2015. "Crossmodal Illusions in Neurorehabilitation." *Frontiers in Behavioral Neuroscience* 9: 212. https://doi.org/10.3389/fnbeh.2015.00212.

[5] Vignemont, Frédérique de. 2011. "Embodiment, Ownership and Disownership." *Consciousness and Cognition* 20 (1): 82–93.

[6] Gatta, Francesco Della, Francesca Garbarini, Guglielmo Puglisi, Antonella Leonetti, Annamaria Berti, and Paola Borroni. 2016. "Decreased Motor Cortex Excitability Mirrors Own Hand Disembodiment during the Rubber Hand Illusion." *ELife* 5: e14972. https://doi.org/10.7554/eLife.14972.

[7] Christ, Oliver, and Miriam Reiner. 2014. "Perspectives and Possible Applications of the Rubber Hand and Virtual Hand Illusion in Non-Invasive Rehabilitation: Technological Improvements and Their Consequences." *Neuroscience & Biobehavioral Reviews* 44: 33–44.

[8] Kaldoudi, Eleni, Stathis Konstantinidis, and Panagiotis D. Bamidis. 2010. "Web 2.0 Approaches for Active, Collaborative Learning in Medicine and Health." In *Ubiquitous Health and Medical Informatics*, 127–49. Hershey, PA: IGI Global.

[9] Dunleavy, Matt, Chris Dede, and Rebecca Mitchell. 2009. "Affordances and Limitations of Immersive Participatory Augmented Reality Simulations for Teaching and Learning." *Journal of Science Education and Technology* 18 (1): 7–22.

[10] Dede, Chris. 2009. "Immersive Interfaces for Engagement and Learning." *Science (New York, N.Y.)* 323 (5910): 66–69.

[11] Scheckler, Rebecca K. 2003. "Virtual Labs: A Substitute for Traditional Labs?" *The International Journal of Developmental Biology* 47 (2–3): 231–36.

[12] Astaras, Alexander, Nikolaos Moustakas, Alkinoos Athanasiou, and Aristides Gogoussis. 2013. "Towards Brain-Computer Interface Control of a 6-Degree-of-Freedom Robotic Arm Using Dry EEG Electrodes." *Advances in Human-Computer Interaction*. https://doi.org/10.1155/2013/641074.

[13] Guger, Christoph, Natalie Mrachacz-Kersting, and Brendan Z. Allison, eds. 2019. *Brain-Computer Interface Research*. SpringerBriefs in Electrical and Computer Engineering. Cham: Springer International Publishing.

[14] Alchalabi, Bilal, and Jocelyn Faubert. 2019. "A Comparison between BCI Simulation and Neurofeedback for Forward/Backward Navigation in Virtual Reality." *Computational Intelligence and Neuroscience* 2019: 2503431.

[15] Nierula, Birgit, Bernhard Spanlang, Matteo Martini, Mireia Borrell, Vadim V Nikulin, and Maria V Sanchez-Vives. 2019. "Agency and Responsibility over Virtual Movements Controlled through Different Paradigms of Brain-Computer Interface." *The Journal of Physiology*, October. https://doi.org/10.1113/JP278167.

[16] Škola, Filip, Simona Tinková, and Fotis Liarokapis. 2019. "Progressive Training for Motor Imagery Brain-Computer Interfaces Using Gamification and Virtual Reality Embodiment." *Frontiers in Human Neuroscience* 13: 329.

[17] Choi, Kang-min, Seonghun Park, and Chang-Hwan Im. 2019. "Comparison of Visual Stimuli for Steady-State Visual Evoked Potential-Based Brain-Computer Interfaces in Virtual Reality Environment in Terms of Classification Accuracy and Visual Comfort." *Computational Intelligence and Neuroscience* 2019: 1–7.

[18] Vourvopoulos, Athanasios, Octavio Marin Pardo, Stéphanie Lefebvre, *et al* 2019. "Effects of a Brain-Computer Interface With Virtual Reality (VR) Neurofeedback: A Pilot Study in Chronic Stroke Patients." *Frontiers in Human Neuroscience* 13: 210.

[19] Vourvopoulos, Athanasios, Carolina Jorge, Rodolfo Abreu, Patrícia Figueiredo, Jean-Claude Fernandes, and Sergi Bermúdez I Badia. 2019. "Efficacy and Brain Imaging Correlates of an Immersive Motor Imagery BCI-Driven VR System for Upper Limb Motor Rehabilitation: A Clinical Case Report." *Frontiers in Human Neuroscience* 13: 244.

[20] Pfurtscheller, Gert, Robert Leeb, Claudia Keinrath, *et al* 2006. "Walking from Thought." *Brain Research* 1071 (1): 145–52.

[21] Cruz-Neira, Carolina, Daniel J. Sandin, and Thomas A. DeFanti. 1993. "Surround-Screen Projection-Based Virtual Reality." In *Proceedings of the 20th Annual Conference on Computer Graphics and Interactive Techniques – SIGGRAPH '93*, 135–42. New York, New York, USA: ACM Press. https://doi.org/10.1145/166117.166134.

[22] Chmura, Jennifer, Joshua Rosing, Steven Collazos, and Shikha J. Goodwin. 2017. "Classification of Movement and Inhibition Using a Hybrid BCI." *Frontiers in Neurorobotics* 11: 38. https://doi.org/10.3389/fnbot.2017.00038.

[23] Brouwer, Anne-Marie, Jasper van der Waa, and Hans Stokking. 2018. "BCI to Potentially Enhance Streaming Images to a VR Headset by Predicting Head Rotation." *Frontiers in Human Neuroscience* 12: 420. https://doi.org/10.3389/fnhum.2018.00420.

[24] Coogan, Christopher G, and Bin He. 2018. "Brain-Computer Interface Control in a Virtual Reality Environment and Applications for the Internet of Things." *IEEE Access: Practical Innovations, Open Solutions* 6: 10840–49. https://doi.org/10.1109/ACCESS.2018.2809453.

[25] Bohil, Corey J., Alicea Bradly, and Frank A. Biocca. 2011. "Virtual Reality in Neuroscience Research and Therapy." *Nature Reviews Neuroscience* 12 (12): 752–62.

[26] Salisbury, David B, Marie Dahdah, Simon Driver, Thomas D Parsons, and Kathleen M Richter. 2016. "Virtual Reality and Brain Computer Interface in Neurorehabilitation." *Proceedings (Baylor University. Medical Center)* 29 (2): 124–27.

[27] Zanier, Elisa R., Tommaso Zoerle, Daniele Di Lernia, and Giuseppe Riva. 2018. "Virtual Reality for Traumatic Brain Injury." *Frontiers in Neurology* 9: 345.

[28] Rábago, Christopher A., and Jason M. Wilken. 2011. "Application of a Mild Traumatic Brain Injury Rehabilitation Program in a Virtual Realty Environment." *Journal of Neurologic Physical Therapy* 35 (4): 185–93.

[29] Sessoms, Pinata H., Kim R. Gottshall, John-David Collins, Amanda E. Markham, Kathrine A. Service, and Seth A. Reini. 2015. "Improvements in Gait Speed and Weight Shift of Persons With Traumatic Brain Injury and Vestibular Dysfunction Using a Virtual Reality Computer-Assisted Rehabilitation Environment." *Military Medicine* 180 (3S): 143–49.

[30] Johnson, N. N., J. Carey, B. J. Edelman, *et al.* 2018. "Combined rTMS and Virtual Reality Brain-Computer Interface Training for Motor Recovery after Stroke." *Journal of Neural Engineering* 15 (1): 016009.

[31] Perez-Marcos, Daniel, Odile Chevalley, Thomas Schmidlin, *et al.* 2017. "Increasing Upper Limb Training Intensity in Chronic Stroke Using Embodied Virtual Reality: A Pilot Study." *Journal of NeuroEngineering and Rehabilitation* 14 (1): 119.

[32] Donati, Ana R. C., Solaiman Shokur, Edgard Morya, *et al.* 2016. "Long-Term Training with a Brain-Machine Interface-Based Gait Protocol Induces Partial Neurological Recovery in Paraplegic Patients." *Scientific Reports* 6 (1): 30383.

[33] Shokur, Solaiman, Simone Gallo, Renan C. Moioli, *et al.* 2016. "Assimilation of Virtual Legs and Perception of Floor Texture by Complete Paraplegic Patients Receiving Artificial Tactile Feedback." *Scientific Reports* 6 (1): 32293.

[34] Wood, Guilherme, Silvia Erika Kober, Matthias Witte, and Christa Neuper. 2014. "On the Need to Better Specify the Concept of "control" in Brain-Computer-Interfaces/Neurofeedback Research." *Frontiers in Systems Neuroscience* 8: 171. https://doi.org/10.3389/fnsys.2014.00171.

[35] Kleih, S.C., F. Nijboer, S. Halder, and A. Kübler. 2010. "Motivation Modulates the P300 Amplitude during Brain–Computer Interface Use." *Clinical Neurophysiology* 121 (7): 1023–31.

[36] Pandria, Niki, Dimitris Spachos, and Panagiotis D. Bamidis. 2016. "Evaluation of Neurofeedback on ADHD Using Mobile Health Technologies." In , 1292–96.

[37] Strehl, Ute, Sarah M. Birkle, Sonja Wörz, and Boris Kotchoubey. 2014. "Sustained Reduction of Seizures in Patients with Intractable Epilepsy after Self-Regulation Training of Slow Cortical Potentials – 10 Years After." *Frontiers in Human Neuroscience* 8: 604.

[38] Pandria, Niki, Alkinoos Athanasiou, Nikos Terzopoulos, *et al.* 2018. "Exploring the Neuroplastic Effects of Biofeedback Training on Smokers." *Behavioural Neurology* 2018: 1–19.

[39] Lorenzetti, Valentina, Bruno Melo, Rodrigo Basílio, *et al.* 2018. "Emotion Regulation Using Virtual Environments and Real-Time FMRI Neurofeedback." *Frontiers in Neurology* 9: 390. https://doi.org/10.3389/fneur.2018.00390.

[40] Fernández-Caballero, Antonio, Elena Navarro, Patricia Fernández-Sotos, *et al.* 2017. "Human-Avatar Symbiosis for the Treatment of Auditory Verbal Hallucinations in Schizophrenia through Virtual/Augmented Reality and Brain-Computer Interfaces." *Frontiers in Neuroinformatics* 11: 64. https://doi.org/10.3389/fninf.2017.00064.

[41] Mikhail, Mirriam, Karim Mithani, and George M. Ibrahim. 2019. "Presurgical and Intraoperative Augmented Reality in Neuro-Oncologic Surgery: Clinical Experiences and Limitations." *World Neurosurgery* 128: 268–76.

[42] Contreras López, William Omar, Paula Alejandra Navarro, and Santiago Crispin. 2019. "Intraoperative Clinical Application of Augmented Reality

in Neurosurgery: A Systematic Review." *Clinical Neurology and Neurosurgery* 177: 6–11.

[43] Vávra, P., J. Roman, P. Zonča, *et al.* 2017. "Recent Development of Augmented Reality in Surgery: A Review." *Journal of Healthcare Engineering* 2017: 1–9. https://doi.org/10.1155/2017/4574172.

[44] Zhalmukhamedov, Emil, and Timur U Urakov. 2019. "Current and Future Use of Virtual and Augmented Reality in Neurosurgery: A Literature Review." *Indonesian Journal of Neurosurgery* 2 (2): 50–54.

[45] Besharati Tabrizi, Leila, and Mehran Mahvash. 2015. "Augmented Reality–Guided Neurosurgery: Accuracy and Intraoperative Application of an Image Projection Technique." *Journal of Neurosurgery* 123 (1): 206–11.

[46] Urakov, Timur M., Michael Y. Wang, and Allan D. Levi. 2019. "Workflow Caveats in Augmented Reality–Assisted Pedicle Instrumentation: Cadaver Lab." *World Neurosurgery* 126: e1449–55.

[47] Taswell, S Koby, Teja Veeramacheneni, and Carl Taswell. 2017. "BrainWatch Software for Interactive Exploration of Brain Scans in 3D Virtual Reality Systems." *Conference Proceedings: ... Annual International Conference of the IEEE Engineering in Medicine and Biology Society. IEEE Engineering in Medicine and Biology Society. Annual Conference* 2017: 3704–7.

[48] Hendricks, Benjamin K, Jerome Hartman, and Aaron A Cohen-Gadol. 2018. "Cerebrovascular Operative Anatomy: An Immersive 3D and Virtual Reality Description." *Operative Neurosurgery (Hagerstown, Md.)* 15 (6): 613–23.

[49] Perin, Alessandro, Tommaso Francesco Galbiati, Enrico Gambatesa, *et al.* 2018. "Filling the Gap between the OR and Virtual Simulation: A European Study on a Basic Neurosurgical Procedure." *Acta Neurochirurgica* 160 (11): 2087–97.

[50] Winkler-Schwartz, Alexander, Ibrahim Marwa, Khalid Bajunaid, *et al.* 2019. "A Comparison of Visual Rating Scales and Simulated Virtual Reality Metrics in Neurosurgical Training: A Generalizability Theory Study." *World Neurosurgery* 127: e230–35.

[51] Clark, Anna D., Damiano G. Barone, Nicholas Candy, *et al.* 2017. "The Effect of 3-Dimensional Simulation on Neurosurgical Skill Acquisition and Surgical Performance: A Review of the Literature." *Journal of Surgical Education* 74 (5): 828–36.

[52] Antoniou, Panagiotis E., Lazaros Ioannidis, and Panagiotis D. Bamidis. 2016. "OSCase: Data Schemes, Architecture and Implementation Details of Virtual Patient Repurposing in Multi User Virtual Environments." *EAI Endorsed Transactions on Future Intelligent Educational Environments* 2 (6): 151523.

[53] Antoniou, Panagiotis E., Christina A. Athanasopoulou, Eleni Dafli, and Panagiotis D. Bamidis. 2014. "Exploring Design Requirements for Repurposing Dental Virtual Patients from the Web to Second Life: A Focus Group Study." *Journal of Medical Internet Research* 16 (6): e151.

[54] Antoniou, Panagiotis E., Eleni Dafli, George Arfaras, and Panagiotis D. Bamidis. 2017. "Versatile Mixed Reality Medical Educational Spaces;

Requirement Analysis from Expert Users." In *Personal and Ubiquitous Computing*, 21:1015–24.

[55] Scheckler, Rebecca K. 2003. "Virtual Labs: A Substitute for Traditional Labs?" *The International Journal of Developmental Biology* 47 (2–3): 231–36.

[56] Jong, Ton de, Linn Marcia C, and Zacharia Zacharias C. 2013. "Physical and Virtual Laboratories in Science and Engineering Education." *Science (New York, N.Y.)* 340 (6130): 305–8.

[57] Weller, Jennifer M. 2004. "Simulation in Undergraduate Medical Education: Bridging the Gap between Theory and Practice." *Medical Education* 38 (1): 32–38.

[58] Weller, Jennifer M. 2004. "Simulation in Undergraduate Medical Education: Bridging the Gap between Theory and Practice." *Medical Education* 38 (1): 32–38.

[59] "Microsoft Hololens Product Page." n.d. Accessed October 30, 2019. https://www.microsoft.com/en-us/hololens.

[60] Chiu, Jennifer L., Crystal J. DeJaegher, and Jie Chao. 2015. "The Effects of Augmented Virtual Science Laboratories on Middle School Students' Understanding of Gas Properties." Computers & Education 85: 59–73.

[61] Dede, Chris. 2009. "Immersive Interfaces for Engagement and Learning." *Science (New York, N.Y.)* 323 (5910): 66–69.

[62] Wu, Hsin-Kai, Silvia Wen-Yu Lee, Hsin-Yi Chang, and Jyh-Chong Liang. 2013. "Current Status, Opportunities and Challenges of Augmented Reality in Education." *Computers & Education* 62: 41–49.

[63] Klopfer, Eric, and Kurt Squire. 2008. "Environmental Detectives—the Development of an Augmented Reality Platform for Environmental Simulations." *Educational Technology Research and Development* 56 (2): 203–28.

[64] Dunleavy, Matt, Chris Dede, and Rebecca Mitchell. 2009. "Affordances and Limitations of Immersive Participatory Augmented Reality Simulations for Teaching and Learning." *Journal of Science Education and Technology* 18 (1): 7–22.

[65] Olympiou, Georgios, and Zacharias C. Zacharia. 2012. "Blending Physical and Virtual Manipulatives: An Effort to Improve Students' Conceptual Understanding through Science Laboratory Experimentation." *Science Education* 96 (1): 21–47.

[66] Pfurtscheller, G, D Flotzinger, M Pregenzer, J R Wolpaw, and D McFarland. n.d. "EEG-Based Brain Computer Interface (BCI). Search for Optimal Electrode Positions and Frequency Components." *Medical Progress through Technology* 21 (3): 111–21.

[67] Farwell, L. A., and E. Donchin. 1988. "Talking off the Top of Your Head: Toward a Mental Prosthesis Utilizing Event-Related Brain Potentials." *Electroencephalography and Clinical Neurophysiology* 70 (6): 510–23.

[68] Bayliss, J. D., and D. H. Ballard. 2000. "A Virtual Reality Testbed for Brain-Computer Interface Research." *IEEE Transactions on Rehabilitation Engineering* 8 (2): 188–90.

[69] Farwell, L. A., and E. Donchin. 1988. "Talking off the Top of Your Head: Toward a Mental Prosthesis Utilizing Event-Related Brain Potentials." *Electroencephalography and Clinical Neurophysiology* 70 (6): 510–23.

[70] Sutton, S., M. Braren, J. Zubin, and E. R. John. 1965. "Evoked-Potential Correlates of Stimulus Uncertainty." *Science (New York, N.Y.)* 150 (3700): 1187–88.

[71] Chapman, R. M., and H. R. Bragdon. 1964. "Evoked Responses To Numerical And Non-Numerical Visual Stimuli While Problem Solving." *Nature* 203: 1155–57.

[72] Polich, J. 1998. "P300 Clinical Utility and Control of Variability." *Journal of Clinical Neurophysiology: Official Publication of the American Electroencephalographic Society* 15 (1): 14–33.

[73] Pavani, Francesco, Charles Spence, and Jon Driver. 2000. "Visual Capture of Touch: Out-of-the-Body Experiences With Rubber Gloves." *Psychological Science* 11 (5): 353–59.

[74] Azizian, Allen, Antonio L. Freitas, Mohammad A. Parvaz, and Nancy K. Squires. 2006. "Beware Misleading Cues: Perceptual Similarity Modulates the N2/P3 Complex." *Psychophysiology* 43 (3): 253–60.

[75] Sambo, Chiara F, Helge Gillmeister, and Bettina Forster. 2009. "Viewing the Body Modulates Neural Mechanisms Underlying Sustained Spatial Attention in Touch." *The European Journal of Neuroscience* 30 (1): 143–50.

[76] Harjunen, Ville J, Imtiaj Ahmed, Giulio Jacucci, Niklas Ravaja, and Michiel M. Spapé. 2017. "Manipulating Bodily Presence Affects Cross-Modal Spatial Attention: A Virtual-Reality-Based ERP Study." *Frontiers in Human Neuroscience* 11: 79. https://doi.org/10.3389/fnhum.2017.00079.

[77] Eimer, M. 2001. "Crossmodal Links in Spatial Attention between Vision, Audition, and Touch: Evidence from Event-Related Brain Potentials." *Neuropsychologia* 39 (12): 1292–1303.

[78] Rupp, Greg, Chris Berka, Amir Meghdadi, *et al.* 2019. "EEG Acquisition During the VR Administration of Resting State, Attention, and Image Recognition Tasks: A Feasibility Study." In, 250–58.

[79] Rupp, Greg, Chris Berka, Amir Meghdadi, *et al.* 2019. "EEG Acquisition During the VR Administration of Resting State, Attention, and Image Recognition Tasks: A Feasibility Study." In, 250–58.

[80] Albers, Mark W., Grover C. Gilmore, Jeffrey Kaye, *et al.* 2015. "At the Interface of Sensory and Motor Dysfunctions and Alzheimer's Disease." *Alzheimer's & Dementia: The Journal of the Alzheimer's Association* 11 (1): 70–98.

[81] Yamasaki, Takao, Yoshinobu Goto, Yasumasa Ohyagi, *et al.* 2012. "Selective Impairment of Optic Flow Perception in Amnestic Mild Cognitive Impairment: Evidence from Event-Related Potentials." *Journal of Alzheimer's Disease: JAD* 28 (3): 695–708.

[82] Waninger, Shani, Chris Berka, Amir Meghdadi, *et al.* 2018. "Event-Related Potentials during Sustained Attention and Memory Tasks: Utility as Biomarkers for Mild Cognitive Impairment." *Alzheimer's & Dementia (Amsterdam, Netherlands)* 10: 452–60. https://doi.org/10.1016/j.dadm.2018.05.007.

[83] Kim, Nam-Gyoon. 2015. "Perceiving Collision Impacts in Alzheimer's Disease: The Effect of Retinal Eccentricity on Optic Flow Deficits." *Frontiers in Aging Neuroscience* 7: 218. https://doi.org/10.3389/fnagi.2015.00218.

[84] Chou, Y.-H., R. C. Wagenaar, E. Saltzman, *et al.* 2009. "Effects of Optic Flow Speed and Lateral Flow Asymmetry on Locomotion in Younger and Older Adults: A Virtual Reality Study." *The Journals of Gerontology Series B: Psychological Sciences and Social Sciences* 64B (2): 222–31.

[85] Picard, R. W., S. Papert, W. Bender, *et al.* 2004. "Affective Learning—A Manifesto." *BT Technology Journal* 22 (4): 253–69.

[86] Stein, Nancy L., and Linda J. Levine. 1991. "Making Sense Out of Emotion: The Representation and Use of Goal-Structured Knowledge." In *Memories, Thoughts, and Emotions Essays in Honor of George Mandler*, edited by William Kessen, Andrew Ortony, and Fergus Craik, 1st ed. New York: Psychology Press.

[87] Goleman, Daniel. 1995. *Emotional Intelligence.* New York: Bantam Books, Inc.

[88] Stipek, Deborah J. 2001. *Motivation to Learn: Integrating Theory and Practice.* 4th ed. Boston, MA: Pearson.

[89] Tobias, Sigmund. 1994. "Interest, Prior Knowledge, and Learning." *Review of Educational Research* 64 (1): 37.

[90] Miserandino, Marianne. 1996. "Children Who Do Well in School: Individual Differences in Perceived Competence and Autonomy in above-Average Children." *Journal of Educational Psychology* 88 (2): 203–14.

[91] Chang, Chuan-Yu, Jun-Ying Zheng, and Chi-Jane Wang. 2010. "Based on Support Vector Regression for Emotion Recognition Using Physiological Signals." In *The 2010 International Joint Conference on Neural Networks (IJCNN)*, 1–7. IEEE.

[92] Niu, Xiaowei, Liwan Chen, and Qiang Chen. 2011. "Research on Genetic Algorithm Based on Emotion Recognition Using Physiological Signals." In *2011 International Conference on Computational Problem-Solving (ICCP)*, 614–18. IEEE.

[93] Takahashi, K. 2004. "Remarks on SVM-Based Emotion Recognition from Multi-Modal Bio-Potential Signals." In *RO-MAN 2004. 13th IEEE International Workshop on Robot and Human Interactive Communication (IEEE Catalog No.04TH8759)*, 95–100. IEEE.

[94] Kalimeri, Kyriaki, and Charalampos Saitis. 2016. "Exploring Multimodal Biosignal Features for Stress Detection during Indoor Mobility." In *Proceedings of the 18th ACM International Conference on Multimodal Interaction – ICMI 2016*, 53–60. New York, New York, USA: ACM Press.

[95] Frantzidis, C. A., C. Bratsas, M. A. Klados, *et al.* 2010. "On the Classification of Emotional Biosignals Evoked While Viewing Affective Pictures: An Integrated Data-Mining-Based Approach for Healthcare Applications." *IEEE Transactions on Information Technology in Biomedicine* 14 (2): 309–18.

[96] Ollander, Simon, Christelle Godin, Aurelie Campagne, and Sylvie Charbonnier. 2016. "A Comparison of Wearable and Stationary Sensors for Stress Detection." In *2016 IEEE International Conference on Systems, Man, and Cybernetics (SMC)*, 004362–66. IEEE.

[97] "Empatica E4 Product Page." n.d. Accessed October 30, 2019. https://www.empatica.com/en-eu/research/e4/.

[98] "Nexus-10 Product Page." n.d. Accessed October 30, 2019. https://www.mindmedia.com/en/products/nexus-10-mkii/.

[99] Karrasch, Mira, Matti Laine, Pekka Rapinoja, and Christina M Krause. 2004. "Effects of Normal Aging on Event-Related Desynchronization/Synchronization during a Memory Task in Humans." *Neuroscience Letters* 366 (1): 18–23.

[100] Klimesch, W. 1999. "EEG Alpha and Theta Oscillations Reflect Cognitive and Memory Performance: A Review and Analysis." *Brain Research. Brain Research Reviews* 29 (2–3): 169–95.

[101] Başar-Eroglu, C., E. Başar, T. Demiralp, and M. Schürmann. 1992. "P300-Response: Possible Psychophysiological Correlates in Delta and Theta Frequency Channels. A Review." *International Journal of Psychophysiology: Official Journal of the International Organization of Psychophysiology* 13 (2): 161–79.

[102] Gruzelier, John. 2009. "A Theory of Alpha/Theta Neurofeedback, Creative Performance Enhancement, Long Distance Functional Connectivity and Psychological Integration." *Cognitive Processing* 10 (Suppl 1:) S101-9.

[103] Kubota, Y., W. Sato, M. Toichi, *et al.* 2001. "Frontal Midline Theta Rhythm Is Correlated with Cardiac Autonomic Activities during the Performance of an Attention Demanding Meditation Procedure." *Brain Research. Cognitive Brain Research* 11 (2): 281–87.

[104] Lagopoulos, Jim, Jian Xu, Inge Rasmussen, *et al.* 2009. "Increased Theta and Alpha EEG Activity during Nondirective Meditation." *Journal of Alternative and Complementary Medicine (New York, N.Y.)* 15 (11): 1187–92.

[105] Missonnier, P., M.-P. Deiber, G. Gold, *et al.* 2006. "Frontal Theta Event-Related Synchronization: Comparison of Directed Attention and Working Memory Load Effects." *Journal of Neural Transmission (Vienna, Austria: 1996)* 113 (10): 1477–86.

[106] Grunwald, M., T. Weiss, W. Krause, *et al.* 2001. "Theta Power in the EEG of Humans during Ongoing Processing in a Haptic Object Recognition Task." *Brain Research. Cognitive Brain Research* 11 (1): 33–37.

[107] Jensen, Ole, and Claudia D. Tesche. 2002. "Frontal Theta Activity in Humans Increases with Memory Load in a Working Memory Task." *The European Journal of Neuroscience* 15 (8): 1395–99.

[108] Gevins, A. S., G. M. Zeitlin, J. C. Doyle, R. E. Schaffer, and E. Callaway. 1979. "EEG Patterns during 'cognitive' Tasks. II. Analysis of Controlled Tasks." *Electroencephalography and Clinical Neurophysiology* 47 (6): 704–10.

[109] Gevins, A. S., G. M. Zeitlin, C. D. Yingling, *et al.* 1979. "EEG Patterns during 'cognitive' Tasks. I. Methodology and Analysis of Complex Behaviors." *Electroencephalography and Clinical Neurophysiology* 47 (6): 693–703.

[110] Niedermeyer, Ernst, and F. H. Lopes da Silva, eds. 2005. Electroencephalography: Basic Principles, *Clinical Applications, and Related Fields*. 5th ed. Baltimore: Lippincott Williams & Wilkins.

[111] Drachen, Anders, Lennart E. Nacke, Georgios Yannakakis, and Anja Lee Pedersen. 2010. "Correlation between Heart Rate, Electrodermal Activity and Player Experience in First-Person Shooter Games." In *Proceedings of the 5th ACM SIGGRAPH Symposium on Video Games – Sandbox '10*, 49–54. New York, New York, USA: ACM Press.

[112] Fowles, D. C. 1980. "The Three Arousal Model: Implications of Gray's Two-Factor Learning Theory for Heart Rate, Electrodermal Activity, and Psychopathy." *Psychophysiology* 17 (2): 87–104.

[113] Charalambous, Efrosini, Sean Hanna, and Alan Penn. 2017. "Visibility Analysis, Spatial Experience and EEG Recordings in Virtual Reality Environments: The Experience of 'knowing Where One Is' and Isovist Properties as a Means to Assess the Related Brain Activity." *Proceedings – 11th International Space Syntax Symposium, SSS 2017*, 128.1–128.12.

[114] Lin, Chin-Teng, I-Fang Chung, Li-Wei Ko, Yu-Chieh Chen, Sheng-Fu Liang, and Jeng-Ren Duann. 2007. "EEG-Based Assessment of Driver Cognitive Responses in a Dynamic Virtual-Reality Driving Environment." *IEEE Transactions on Bio-Medical Engineering* 54 (7): 1349–52.

[115] Antoniou, Panagiotis E., Dimitris Spachos, Panagiotis Kartsidis, Evdokimos I. Konstantinidis, and Panagiotis D. Bamidis. 2017. "Towards Classroom Affective Analytics. Validating an Affective State Self-Reporting Tool for the Medical Classroom." *MedEdPublish* 6 (3). https://doi.org/10.15694/mep.2017.000134.

[116] Jong, Ton de, Marcia C. Linn, and Zacharias C. Zacharia. 2013. "Physical and Virtual Laboratories in Science and Engineering Education." *Science (New York, N.Y.)* 340 (6130): 305–8.

[117] VanLehn, Kurt. 2011. "The Relative Effectiveness of Human Tutoring, Intelligent Tutoring Systems, and Other Tutoring Systems." *Educational Psychologist* 46 (4): 197–221.

[118] Mohanan, Ramya, Catherine Stringfellow, and Dittika Gupta. 2017. "An Emotionally Intelligent Tutoring System." In *2017 Computing Conference*, 1099–1107. IEEE.

[119] Sarrafzadeh, Abdolhossein, Samuel Alexander, Farhad Dadgostar, Chao Fan, and Abbas Bigdeli. 2008. "'How Do You Know That I Don't Understand?' A Look at the Future of Intelligent Tutoring Systems." *Computers in Human Behavior* 24 (4): 1342–63.

[120] Athanasiou, Alkinoos, Torstein R. Meling, Alexandros Brotis, *et al.* 2020. "3D Printing in Neurosurgery." In *3D Printing: Application in Medical*

Surgery Volume 2, edited by Vassilios Papadopoulos, Vassilios Tsioukas, Jasjit Suri, and George Tsoulfas. Elsevier B.V.

[121] Pampaloni, Niccolò Paolo, Michele Giugliano, Denis Scaini, Laura Ballerini, and Rossana Rauti. 2019. "Advances in Nano Neuroscience: From Nanomaterials to Nanotools." *Frontiers in Neuroscience* 12. https://doi.org/10.3389/fnins.2018.00953.

[122] Athanasiou, Alkinoos, George Arfaras, Niki Pandria, *et al.* 2017. "Wireless Brain-Robot Interface: User Perception and Performance Assessment of Spinal Cord Injury Patients." *Wireless Communication and Mobile Computing* 2017: 2986423. https://doi.org/10.1155/2017/2986423.

Chapter 5
EEG-based biometric systems

Jardel das C. Rodrigues[1], *Pedro P. Rebouças Filho*[1], *Robertas Damaševičius*[2], and *Victor Hugo C. de Albuquerque*[3,4]

The neuroimaging techniques allow the neuroscientists to understand how the brain develops and controls our perception of the world, our interactions with it and how human brain systems operate [1,2]. The noninvasive brain techniques such electroencephalogram (EEG), transcranial magnetic stimulation (TMS), functional magnetic resonance imaging (fMRI), magnetoencephalogram (MEG), positron emission tomography (PET), and near-infrared spectroscopy (NIRS) allow researchers and clinicians to observe and analyze brain activities in resting state or while the subjects perform perceptual, motor, and/or cognitive tasks [3]. Among those, PET, MEG and fMRI imaging are technically onerous and expensive. These techniques depend on blood flow, have extended time constants and consequently are less susceptible to fast communication. In contrast, the EEG can work in most conditions and demand inexpensive and simple equipment [4–6]. The functional imaging is widely employed to diagnose neurodegenerative disorders, especially Parkinson's disease and Alzheimer's disease [7–10]. Besides that, these tools are also crucial for rehabilitation research [2].

EEG is a noninvasive technique utilized to estimate the neuronal electrical activity in the human brain. The brain signals recorded by EEG can be influenced by factors such as mental state and stress of the subjects. The EEG recordings are obtained using electrodes placed on the scalp [11]. Two of the most used systems to position electrodes for monitoring the EEG signals are the 10–20 International Federation System, proposed by Jasper (1958), and 10-10 International Federation System (an extension of the 10-20 system). The latter later was later extended to 10-5 system [12]. Figure 5.1 gives an illustration of the placement of scalp electrodes.

[1]Laboratório de Processamento de Imagens, Sinais e Computação Aplicada, Instituto Federal de Educação, Ciência e Tecnologia do Ceará, Fortaleza, CE, Brazil
[2]Department of Software Engineering, Kaunas University of Technology, Kaunas, Lithuania
[3]Laboratory of Bioinformatics (LABIO), University of Fortaleza (UNIFOR), Ceará, Brazil
[4]Data Science Director at the Superintendency for Research and Public Safety Strategy of Ceará State (SUPESP/CE), Brazil

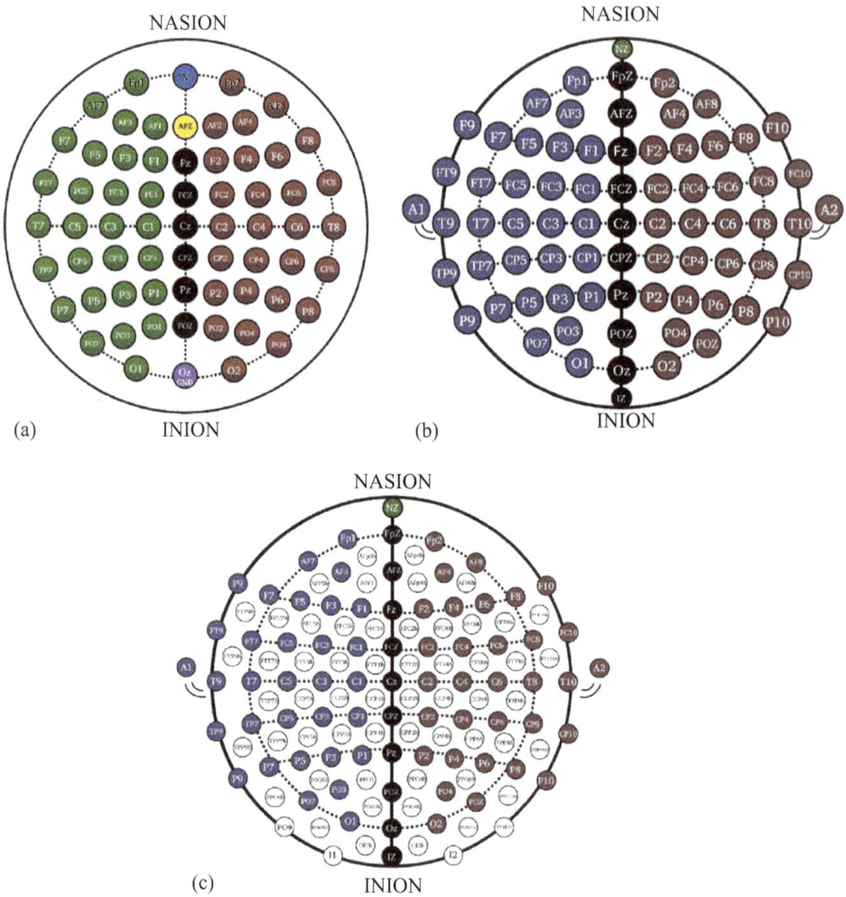

Figure 5.1 Electrodes position on the scalp according to (a) 10-10 International System, (b) 10-20 International System and (c) 10-5 International System. The numbers refer to the electrodes distance along specific contours over the scalp, e.g., in the 10-20 system, the electrodes are placed at distances of 10% and 20%. In general, the 10-20 International System is the prevailing system used in biometric applications. The 10-10 and 10-5 International Systems are suitable for high-density EEG applications. However, high-density EEG have the disadvantage to be time-consuming and are most suitable for epilepsy-related researches, for example

The biometric systems are used to identify or to authenticate subjects based on their characteristics, such as physical or behavioral attributes. These systems are often employed for security purposes. Biometrics are also being employed to mitigate the use of passwords, which are targets for attacks [13]. The traditional

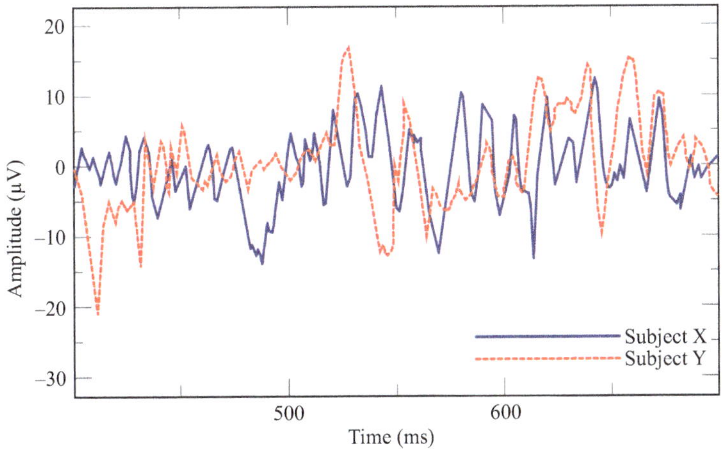

Figure 5.2 EEG signals of two random subjects: X and Y. The EEG signal is measured in μV and represents the difference of voltages between a pair of electrodes. Since EEG signals are in the order of approximately 100 μV, an amplifier module must be used

biometric systems include voice, gait, signature, fingerprint, palm vein, face, retina or iris, hand geometry, and so forth. However, these characteristics are vulnerable to numerous attacks [14]. Based on the assumption that EEG signals are exclusive for each person which are related to the subject's genetic traits, the EEG uniqueness makes the biometric hard to steal and replicate. The main difficulty of using EEG signals is the setup for the recordings [15]. The EEG signals are also contaminated by noise and endogenous or exogenous artifacts due to cerebral activity. In addition to the events generated by stimuli in the reading of EEG signals, there is a background noise, related to heartbeat, muscle pursuit and eye movements [16]. However, various researchers are developing techniques to mitigate the influence of the noise and spurious artifacts [17,18] and to improve the acquisition of EEG signals at the same time that this task is facilitated [19]. An example of EEG signals of two subjects is shown in Figure 5.2.

The authentication is the act to confirm or to deny an identity asserted by a person, the identification, in turn, attempts to recognize a subject from a combination of subjects based on the identification claimed [20]. The conventional person authentication systems are based on passwords inputted by any peripheral device. The use of biometrics can improve the security in this scenario. Therefore, since the use of EEG signals provides more safety, as previously mentioned, its use must be taken into account in this field. So, the block of intruders, anti-spoofing and other real-world applications can take advantage of the EEG benefits.

Wearable devices can obtain biosignals such as electrocardiogram (ECG), EEG, blood pressure, and heart rate and can be use used for remote healthcare. Besides that, urgent signals must be predicted because many diseases need a suitable treatment to avoid patient's death [21]. Portable EEG devices are already used in brain–computer interface (BCI) applications, also referred to brain–machine interface (BMI) [22], medical assistance systems [23], anxiety detection [24], driver vigilance [25], and diagnosis of epilepsy [26]. BCI is a dominant communication paradigm, which assumes that a person does not use his/her usual brain output passages (muscles and nerves) for sending messages to the outside. The messages in an BCI system are encoded in EEG activity [4]. Initially, BCI systems were developed aiming biomedical applications, mainly in rehabilitation cases or in applications where it is instantaneously difficult to move and the response time is crucial. Several other applications can use BCI functionalities, such as smart environments, neuro-marketing, advertisement, entertainment, and others [27].

5.1 Survey scope

Due to the security issues involving network technologies, ubiquitous computing, health applications and smart environments, EEG shows a high potential to address the vulnerabilities problems. This motivated us to produce this work and to study the feasibility of using brain signals in these environments to serve as a high-level reference model to increase the security levels. Therefore, the characteristics of the EEG signals are well suited in this scenario because of the uniqueness, permanence, collectability, performance, acceptability, and circumvention properties [28].

This is the most thorough survey aimed to security using authentication/identification based on EEG signals. The main difference between this work and the works of other researchers is the focus on the applicability of EEG covering the entire landscape from technologies, methodologies, to results and datasets. It was elaborated a systematic review of the literature about recent EEG biometric systems. Thus, the principal techniques used in the state-of-the-art were explored, providing a technical background to help security professionals and researchers to develop and improve such systems.

5.2 Contributions

The main contributions of this review pertinent to the current research in the area can be summarized as follows:

- This survey provides a detailed analysis of the state-of-the-art methods associated to EEG-based subject authentication/identification. The main feature extraction and classification techniques developed by authors are discussed aiming to guide researchers and developers in the design of security systems;
- The most employed datasets of EEG signals are also presented to facilitate the dissemination of useful data for the development of new methodologies;
- Description of commercially available brain signal capture devices are provided;

- The significant functional and nonfunctional requirements that researchers should follow to implement an EEG-based security authentication system are reported;
- Future perspectives based on the emerging evolution of brain signal processing techniques are also discussed.

The rest of the chapter is organized as follows. Section 5.1 details the works surveyed in the state-of-the-art. Section 5.3 exhibits the most employed datasets that contain EEG recordings used in authentication/identification tasks as well as the main devices used for brain signals acquisition commercially available. Section 5.4 identifies the main threats existing in EEG-based security systems and summarizes the performance metrics adopted to evaluate EEG-based biometric systems. Section 5.5 characterizes the essential requirements for an EEG-based biometric system, advantages and disadvantages brought by using EEG signals for security area. Futures perspectives and open issues are discussed in Section 5.6. Finally, Section 5.7 presents the lessons learned about the developed theme and concludes the article.

5.3 EEG-based person authentication and identification systems

Three main keywords were used to identify the selected works in this survey: EEG, authentication, and identification. Thus, it was mounted a large dataset comprised of the state-of-the-art methodologies. A feasibility study for the application of EEG signals aimed for security can be performed after a detailed analysis of them. The primary objective of this section is provide to researchers a collection of works and their respective methods of features extraction and classification for the EEG signals. These techniques can then be adjusted and used for wide variety of biometric applications. A total of 61 works were analyzed from the last six years of research (2013–2019). The surveyed works are briefly explained below and were categorized according to the classification techniques used to classify EEG signals which presented the best performance.

5.3.1 Artificial neural networks, convolutional neural networks and extensions

An user authentication system in a robot environment using EEG signals was proposed by Kumari and Vaish [29]. The EEG signals were acquired while the human was interacting with the robot. The preprocessing step was performed to enhance the signal-to-noise ratio (SNR). For this purpose, a Butterworth fifth-order filter was applied. Daubechies Wavelet Transform decomposed the EEG signals into five levels. Then, the features were computed for all sub-bands. The authors used the mean, energy, and standard deviation as features. The classification was performed using a learning vector quantization (LVQ) neural network. Six users participated in this study and the authors used the visual paradigm stimulus. A total of six sessions were conducted.

Kumari and Vaish [30] used information-theoretic (IT) measures for individual identification. The preprocessing of the signal consists of the application of a Butterworth filter. The data were analyzed using a technique, namely empirical mode decomposition (EMD). The methodology decomposed the signals into a sum of elements, each with slowly verifying phase and amplitude. Once the signal has reproduced in amplitude and phase mode, they could be analyzed using Hilbert–Huang transform (HHT). Thus, the EMD was used to decompose the EEG signals in a set of IMF. The authors have discarded low-frequency IMF which are ineffective to obtain essential information. The power-harmonic-ratio (PHR) was calculated for all IMFs using the ensemble empirical mode decomposition (EEMD) to facilitate the identification of higher power IMF with a large PHR. The IT method was used to obtain the features to perform classification task. The entropy and mutual information were used to form the features vector. The fuzzy learning vector quantization (FLVQ) is a hybrid learning technique that employs a fuzzy theory. This extension of the LVQ was used for classification.

An EEG-based biometric security framework in four different scenarios for user identification and authentication was presented by Gui *et al.* [20]. In the first step, an ensemble averaging technique was applied to the raw signals to reduce noise. A low-pass filter at 60 Hz was performed to eliminate the noise after the ensemble averaging process. A wavelet packet decomposition (WPD) was employed at the preprocessed data and the mean, standard deviation and entropy were determined to form the feature set. In their study, the authors used multi-layer perceptron (MLP) for EEG pattern classification. In the first scenario, the goal was to identify each one of all subjects accurately. In the second scenario, the identification of the subjects was performed using a side-by-side scheme, which generates various smaller size training subsets formed by unique combinations of distinct subjects. The third scenario aims to estimate the performance of how precisely the framework can distinguish one subject from others. Finally, the fourth scenario intended to reproduce the situation that the system may concede the access for a small group of users.

In a posterior work proposed by Gui *et al.* [31], an identification method based on the decision-level fusion of multichannel EEG signals utilizing an improved radial basis function (RBF) neural network was approached. The authors also explored the effects and performance of selecting diverse sorts of visual stimulation. At preprocessing, the EEG signals were ensemble averaged to remove noise and keep the event-related potential (ERP). The Euclidean distance was used to evaluate the similarity degree between a specific EEG signal and the reference signal for each channel. The RBF was used for training. Four EEG electrode channels using four types of visual stimuli from subjects were analyzed in this work.

A convolutional neural network (CNN) is an artificial neural network (ANN) which consists of an MLP and owns a particular topology that contains multiple hidden layers. The CNN can be used both for feature extraction and classification.

The proposed CNN topology by Ma *et al.* [32] has two convolutional layers, two pooling layers, and a single fully connected layer. A detailed description of the adopted strategy can be found at [33]. The convolutional layer convolves the input data matrix and produces the filtered data map. The filtered data maps were then forwarded to the average-pooling layer, which decomposes the input-filtered data maps into sets. The filtered data maps were downsampled and the feature maps were obtained. The second convolutional layer used the feature maps as input data. At the final step, the features acquired by the second average-pooling layer were carried to the fully connected layer.

Mu *et al.* [34] presented a feature extraction paradigm based on fuzzy entropy. Also, the Fisher distance was used to obtain the feature set. A self and nonself-photos exhibited at random sequence were used as visual stimulus. A total of 50 photos were used in the experiment. The preprocessing was performed using Scan 4.3 software (Neuroscan) to remove electrooculographic. After preprocessing, the fuzzy entropy of the EEG signals was estimated. The Fisher distance method was applied to examine the features of EEG signals from different channels. A back-propagation ANN was utilized to classify the processed signals.

The permanence issue of EEG signals over an extended period for authentication task was explored by Das *et al.* [35]. Two distinct stimulation protocols were used to elicit the brain signals. A series of nontarget stimuli and target stimuli were presented to subjects and the appearance of the nontarget stimuli is higher than the target stimuli. In the first protocol, namely "Geometric Protocol", eight geometric shapes were chosen as visual stimuli, a circle image was admitted as a target, while other pictures were considered as nontarget. The second protocol, namely "Letters & Numbers Protocol", comprises a total of 62 pictures, which of 26 pictures were formed by uppercase and lowercase letters. The numbers were intended to be the target stimuli and subjects were inquired to focus on them when they were shown. The letters were used as nontarget stimuli and the subjects were required to neglect them. The obtained EEG signals were initially preprocessed to enhance the SNR. A common average referencing (CAR) filter was applied to mitigate artifacts related to inconsistent references. After the preprocessing level, templates were created by signal averaging for all subjects. The created templates were compared during the verification phase using either discrete wavelet transform (DWT) and symbolic aggregate approximation approaches.

An incremental version of the fuzzy-rough nearest neighbor (FRNN) technique, namely, Incremental fuzzy-rough nearest neighbor (IncFRNN) was used by Liew *et al.* [36]. The FRNN model combines the advantages of the fuzzy and rough sets and the nearest neighbor algorithm. The FRNN algorithm replaces the Euclidean distance with fuzzy-rough membership for object similarity measure in the k-nearest neighbor (k-NN) algorithm. The WPD, Hjorth parameter, coherence, mutual information, cross-correlation, and mean of amplitude were used to acquire the feature set. A classifier based on 10-fold cross-validation was implemented. All subjects were modeled as a binary problem. The IncFRNN model showed best results than FRNN model in terms of area under curve (AUC).

Three ANN performing the classification task in three different domains: time, frequency, and time and frequency was explored in the work proposed by Hasan *et al.* [37]. In this work, a method for EEG identification based on the small intra- and large inter-individual discrepancies of the features was employed. The filtering process was performed using a bandpass filter with ranges between 0.5 and 48 Hz for the raw signal obtained from three electrodes. A smoothing function was also used. The spectral data analysis was performed using fast Fourier transform (FFT) and power spectrum density (PSD). The mean, correlation, skewness, median, maximum value, kurtosis, and standard deviation of the EEG signals were utilized to estimate, predict, and to make decisions from the dataset. The experimental data were obtained using the results from intra-subject and inter-subject features comparison.

Tangkraingkij *et al.* [38] studied a suitable number of hidden neurons of an ANN to perform personal authentication. The authors extracted the delta waves to define the number of hidden neurons in the layers. Firstly, the independent component analysis was performed to separate the adequate signal from the collected signal. The SOBIRO algorithm was then used to isolate the proper signal from the raw signal. Time domain was turned into the frequency domain by using FFT. The delta wave range (0–4 Hz) showed a suitable capacity to identify the most significant amount of EEG data, so this band was selected for the identification. The best number of neurons in the hidden layer was ranged between 5 and 26 neurons.

Finally, the last analyzed study that employs the use of neural networks was the method proposed by Mao *et al.* [39]. This strategy implements a CNN method for identity prediction using the EEG signals. The authors used EEG data obtained from the BCIT Experiment Baseline Driving, namely XB Driving, which checks driving induced fatigue. A total of 100 subjects composes the collection. The preprocessing pipeline [40] was used on four datasets including bandpass filtering, re-referencing, and baseline removal. The data were separated into validation, training, and testing sets. The sixth-order autoregressive coefficients and PSD were extracted from all channels. The CNN architecture explored by the authors comprises convolutional layers succeeded by fully connected layers. Multiple filters or kernels were convolved with the vectorized EEG epochs in each convolutional layer. These filters intended to catch distinctive local spatial features. The result of the convolution layer from one kernel was denominated feature map. The fully connected layers were responsible for combining all the feature maps of the last convolution layer. The experiments showed results in support vector machine (SVM) trained on raw signals, CNN trained on raw signals, SVM trained on PSD of all channels, and SVM trained on AR coefficients of all channels.

5.3.2 Cross correlation

Armstrong *et al.* [41] developed a framework that utilizes the averaged ERP and N400 component for biometric purposes instead of conventional EEG since the ERP is less susceptible to background noise. Only EEG signals obtained from a single active sensor were analyzed. Two benchmarks were employed as a classifier, an SVM with a linear discriminant based on cross-correlation and two neural networks, divergent autoincoder (DIVA) and naive discriminant learning (NDL).

A feasibility study employing topological plots of P300 as biometric in the identification of subjects was presented by Yu *et al.* [42]. The analysis tool of BCI2000 was used to create topological schemes for each session of an individual. The temporal and spatial locations of the most probable response were extracted as features to distinguish the enrolled subjects. At preprocessing stage, the correlation was normalized and the cross-correlation was obtained. A 2D cross-correlation function used a template-matching algorithm to compare the correlation values of topological plots to accept or to reject the subject's claim of one's identity.

An EEG-based lightweight biometric method using ERP in addition with a face identification modality, to generate a multimodal authentication system, was developed by Wang *et al.* [43]. The face modality operates by scanning the images from an HD video camera, using probability density function to estimate the similarity between the classes. Principal component analysis (PCA) was applied to convert the images for training into "Eigenfaces" which were used to calculate the matching score. PCA is a dimensionality-reduction method which projects the data through an orthogonal transform into a space in which the components of the data are uncorrelated. This method is widely employed on large datasets. In the ERP modality, the Oddball audiovisual paradigm was used to evoke the P300 and elicit the ERP biometric. Three simple schemes, max, min, and sum, were applied in combination with the scores obtained from eigenface cross-correlation similarities and signature distances. The information was combined at matching score level. Thus, the results showed that ERPs could be coupled with another biometric scheme, reducing the system vulnerability and improving performance.

Harshit *et al.* [44] proposed a two-phase authentication framework. The acquisition of raw data was made using an EMOTIV Epoc headset. In the signal processing stage, a baseline and a bandpass filtering were used to remove direct current (DC)-drifts in the signal and to decompose the multichannel EEG into four EEG subbands, respectively. The power spectrum analysis was used to compute the power in the EEG samples after segmentation. Alpha and beta bands were used in the visual stimuli process and the gamma was used in audio stimuli, these stimuli were used for matching purposes, compounding a total of ten stimuli used in online authentication. The stored templates were paired with the obtained test feature vectors. Five healthy subjects participated of the experiment and 14 channels formed the template vector.

A biometric person identification system using PSD of gamma band (30–50 Hz) comprised of two phases was developed by Thomas and Vinod [45]. At the first phase, namely enrollment, the EEG signals were collected and 19 EEG channels were used for feature extraction. The elimination of the effects of power line interference was performed by using a notch filter at 60 Hz. For each person, the brain patterns related with resting eyes-closed and resting eyes-open conditions were estimated. The EEG signals were bandpass filtered and a first-order infinite impulse response bandpass filter was also used. The power of the bands was evaluated by applying Welch's periodogram method of PSD. The absolute value of PSD was used as a feature and the concatenation of PSD values composed the feature set for each channel. At the authentication stage, test feature vectors were constructed using the PSD features in the selected band utilizing the periodogram for each channel. The cross-correlation value between vectors was estimated and returned the strength of the linear relationship.

The hand synergies and their neural representations was object of study promoted by Patel *et al.* [13]. The subjects were instructed to perform six grasps tasks found in daily living activities. The selected objects, such as a screwdriver, water bottle or CD, were grasped with 25 repetitions in a complete 150 grasping tasks per subject. The subjects used ten sensors that acquired data from the interphalangeal and metacarpophalangeal joints of four fingers. The kinematic data were filtered utilizing a third-order Butterworth filter (3 Hz cutoff frequency) for each pair of electrodes. After this, the magnitude squared coherence was computed. Lately, the spectral density was estimated for each frequency using FFT with a Hanning window. The principal components were determined by using the singular value decomposition. The final template was considered the "biometric template" and was formed by the combination of the EEG data and hand kinematic data.

5.3.3 L1/L2 and cosine distance

The classification of brain signals and face image using cosine and Euclidean similarities was proposed by Al-Hudhud *et al.* [46]. The authors developed a web-based multimodal biometric authentication. After administrator authorization, the user inputs her/his name. The next step is the enrollment phase, where the EEG signals were captured when an image, selected by the user at the start of the experiment, appears. In the login phase, the user input his/her name and he/she will be taken to EEG log page. The login phase is identical to enrollment phase, but the image memorization was not required. The preprocessing was performed subtracting the means of each channel from each value of the channels after application of FFT. The extracted features were composed by signals speed (frequency), PSD, magnitude, SNR, variance, mean, standard deviation, and zero crossing rate from each channel. During the enrollment and login phases, a face recognition approach using PCA was also performed. In the end, at the fusion level, the system gets the decision score from EEG and face recognition modalities.

The first step of the proposed system by Fraschini *et al.* [47] using eigenvector centrality (EC) for subjects classification consists of applying a bandpass filter on raw EEG signals to analyze the frequency-specific function of the brain network. The second step was achieved by calculating the pair-wise statistical association between EEG time series using a functional connectivity estimator, namely phase lag index (PLI). A functional network was described as a weighted graph in the third step of the method, where an edge represented each functional connection and a node in the graph represented each channel. In this case, the value of PLI was adopted as the strength of the association. The final step describes the functional brain organization. The authors concentrated on the centrality measure, EC, to quantify the relevance of every node in the graph. The EC finds the importance of a node by regarding the quality of its connections.

Maiorana *et al.* [48] proposed several cryptosystems which mix the information during the recognition stage. Biometric cryptosystems are used to satisfy the specifications of renewability, security, and performance. The encryption process was based on previous work proposed by Maiorana *et al.* [49]. A binary message was initially randomly created, then turbo encoded into an n-bit string and modulated into a codeword composed by a determined number of symbols from a constellation in the complex domain. The needed helper data template was achieved by binding the average of the information elicited from each portion of the first biometrics (frame) with the codeword through an operator.

Maiorana *et al.* [50] also analyzed the recognition performance obtainable when using a new characterization of EEG data. The suggested EEG description was retrieved by projecting the extracted spectral characteristics into a subspace. This subspace retains most of the original discriminative information. The authors used PCA to define an orthogonal basis for describing EEG spectral information and indicated with the term Eigenbrains (EB) the eigenvectors obtained during PCA perform. A selection process was conducted to define the most discriminative features. Then, an linear discriminant analysis (LDA) was applied to it. In preprocessing phase, a CAR filter was used to exclude any undesired information. The tests were applied to four different feature extraction strategies: PSD, PSD + LDA, EB, and EB + LDA. The nearest neighbors strategy and Mahalanobis distance were used for classification purposes.

In a third work proposed by Maiorana *et al.* [28], the permanence of EEG distinctive features across time within the biometric framework was analyzed. Besides that, different analysis using a longitudinal private database from fifty subjects was also performed. Three distinct acquisition sections were used to collect the EEG signals with a different time interval between them. Both REC and REO protocols have been studied. Three different sets of features were considered: AR modeling, PSD features, and spectral coherence. In the recognition stage, three different classifiers were adopted, namely Manhattan, Euclidean, and cosine distances, to evaluate the similarity of a given recognition sample.

An optimal set of EEG features employing mutual information between intersubject and intra-subject trials was selected by Lee *et al.* [51]. The EEG data were obtained from a wireless EEG device composed of 21 dry electrodes. However, only 16 electrodes were used. During the experiment, the subjects were in REC condition and the EEG data were registered for 2 min. Each user participated in experiment five or seven times on separate days. The complete research was performed in 40 days. The EEG signals were filtered by applying a bandpass finite impulse response (FIR) filter ranged between 2 and 50 Hz. A total of 7 features were calculated, including median, Shannon entropy, three power ratio, sample entropy (SE), and permutation entropy (PE). In the end, the classification was performed through Bhattacharyya distance-based classifier. The authors also employed the leave-one-out cross-validation methodology.

Bansod *et al.* [52] presented an approach using HHT for personal identification. The amplitude and frequency provided by HHT were used for the classification task. Thirteen features were used for proposed methodology, namely, EEGRawValue, EEGRawValueVolts, attention, meditation, blinkStrength, delta, beta low (13–17 Hz), theta (4–7 Hz), gamma middle (41–50 Hz), gamma low (31–40 Hz), alpha low (8–9 Hz), and alpha high (10–12 Hz). The fusion of alpha high and alpha low features was also used and resulted in a better performance. In the classification task, the Manhattan distance was used for pattern matching and, after, the decision task was performed.

A method which combines a feature and match-score levels fusion rather than an individual approach was proposed by Garau *et al.* [53]. The authors employed both La Rocca–Campisi and Fraschini–Marcialis algorithms. The first was based on the spectral coherence while the second algorithm was based on functional connectivity measurement. The authors used a multi-modal biometric methodology based on specific traits to subjects identification since the unimodal approach suffers several problems, such as non-universality, low performance, lack of uniqueness, and vulnerability to spoofing attacks. The combination of the data from the employed algorithms could be executed on each level of the system. Thus, were considered four levels of fusion: decision, feature extraction, sensor, and matching score.

Finally, Kang *et al.* [54] proposed a subject authentication algorithm, which extracts three varieties of EEG features, specifically, nonlinear, spectral, and network features. These features were used as effective biomarkers of genetic or clinic characteristics, for example. The signals were downsampled to decrease the computational requirements. Then, the raw data were bandpass filtered using an FIR filter to enhance the SNR. An initial test was conducted to examine particular brain areas in which suitable EEG features could be obtained for self-identification. The EOG interference induced by eye movements and the unique properties of intensified alpha activity in the rec condition was analyzed to improve the authentication. Thus, the spectral power of all frequency bands using the fast FFT was analyzed to estimate the influence of these two factors. The EEG features were generated based on brain network properties exhibited by the REC condition and nonlinear characteristics: SE, PE, and the maximum Lyapunov exponents. The authors eliminated 22 EEG channels located on the occipital and frontal regions. The connections in the brain network were induced by phase synchronization between EEG signals using the PLI and mean phase coherence (MPC). A classifier based on the distance was employed to perform the classification.

5.3.4 Random forest

The use of EEG signals to construct a subject identification system using REC and REO scenarios of data recording was investigated by Kaur and Singh [14]. Daubechies-4 wavelet transform was employed to decompose the EEG data into five levels. Only the gamma band was used to obtain statistical features from brain waves. These features were mean, standard deviation, and the root mean square.

Lately, a PCA was applied to reduce the dimensionality of feature space. Two classifiers were employed to perform the classification: an SVM and a random forest (RF).

5.3.5 SVM, support vector data description, and extensions

A paradigm for EEG-based subject authentication using nonself-face and self-face pictures as stimulus was studied by Yeom *et al.* [55]. The proposed paradigm validates the subject's identity using the similarity of the collected biometric with those stored in the dataset. According to the authors, the proposed methodology creates specific brain patterns in response to presented pictures used to stimulate brain activity. The authors also proposed a selection process for the choice of the time interval and for the choice of the adequate number of channels using a pointwise biserial correlation coefficient. The one-tailed *t*-test significant test was also employed.

Nguyen *et al.* [56] examined a subject verification method activity-dependent using EEG data during the execution of motor imagery tasks. All subjects performed identical tasks in the experiment, specifically motor imagery of both hands. The best motor imagery task for each subject was selected to distinguish the subjects. The signals from three electrodes were utilized to extract the features, resulting in a feature set of 99 features. Support vector data description (SVDD) and a Gaussian mixture model (GMM) were used for modeling persons.

Liu *et al.* [57] compared the potentials of EEG signals in relaxed and visual evoked tasks as biometric authentication features. The authors performed re-reference and downsampling operations in preprocessing step to improve SNR. The classification task used spectrum of the time-domain (TPS), AR models, phase-locking value (PLV), and power spectrum of frequency-domain (FPS) as features. SVM classified these feature vectors and the machine-recursive feature elimination (SVM-RFE) was used to obtain the prominent channels. In the first task, the subjects were seated on an chair in REC condition. In the second task, 260 images from Revised Snodgrass and Vanderwart object pictures set [58] were shown to the subjects. The subjects were required to identify and to remember the pictures. One-second duration measures after each stimulus were kept for a detailed analysis.

Pham *et al.* developed two works involving EEG biometrics. In the first work [59], a multi-factor system for user authentication using EEG containing the gender and the age as information. In the enrollment phase, the user firstly provides this information. Next, the user performed some tasks, e.g., imagining moving a specific body part or relax and EEG signals related to these tasks were recorded. In the authentication phase, the mental tasks were used as a component of the credentials. The spectral power was computed for each channel. The relative power in a determined frequency band, divided by the total power in all frequencies, jointly with the total power, were utilized as features. Also, 11th-order AR model was responsible for generating 11 AR coefficients. Three Hjorth parameters, complexity, mobility and activity, were selected for each channel. All these features were employed to train the person model. In the verification stage, when a subject desires to access the system, the EEG signal of the user was recorded by repeating the tasks executed in the enrollment phase. The EEG signals were processed identically as in the enrollment. The modeling technique consists of using the small sphere two large margins support vector data description (SS2LM-SVDD) [60]. The basic idea of this approach is to create an optimal hypersphere in feature space to incorporate EEG signals of a single subject and remove the EEG signals of others subjects.

In the second work [61], the authors analyzed the applicability of brain signals for EEG authentication systems in addition to effects of the emotions on such systems. The 2-D adopted model of emotions presented axes of arousal in vertical and axes of valence in horizontal. The valence varies from unpleasant to pleasant while the arousal varies from calm to excited. In summary, the proposed system has two stages: enrollment and verification. At the first stage, the EEG signals were recorded while the subject performed some task. Only the EEG data associated to the calm state was elected and employed to implement the training model for this subject. In the verification phase, the user repeated the tasks as in enrollment phase. But, the EEG signals associated with the emotional states of stress, excited, and calm were registered and processed. The features were extracted via PSD estimation and AR models. The authors used the SVDD algorithm for classification task.

The impacts of epilepsy on EEG-based identification systems was investigated by Phung *et al.* [62]. The epilepsy is a neurological dysfunction characterized by unexpected and the recurrent seizures. There have been no papers about the consequences of epilepsy on the biometric systems until the publication of this work. It was observed that are notable discrepancies between normal and epileptic EEG. The level of complexity for normal EEG signals are higher than epileptic signals. Besides that, non-epileptic EEG signals are more irregular and more complicated than the epileptic signals and entropy is higher in non-epileptic activity. The AR models and the approximate entropy (ApEn) were used for features extraction. The features were obtained and classified independently for three sub-bands, i.e. alpha, beta, and gamma. The features were used to train a linear SVM. During the classification, one-third was used for testing and two-thirds were used for training using cross-validation.

Davis *et al.* [63] evaluated a biometric identification system using a linear SVM. Consumer digital video library video clips [64] and pictures from LIVE Image Quality Database were presented to all subjects. All EEG signals were bandpass filtered and the line noise at 60 Hz was removed. ICA was also applied to the data to eliminate some undesired objects, including muscle movements and eye blinks. Feature vectors consisting of 20 cepstral coefficients were produced for EEG signals. Two scenarios were analyzed: subjects classification using EEG signals from subjects seeing video clips that were utilized in the training of the linear SVM and subjects classification utilizing EEG signals from subjects watching video clips that were not utilized in training phase. The authors also explored the impacts on the classification of using single-channel versus combined channels.

A methodology that consists of obtaining the EEG signals for the motor imagery tasks of body parts abstraction movement was presented by Nieves and Manian [15]. An adequate number of trials of EEG signals for these tasks were collected and the spectrogram was generated for each signal. The spectral content was obtained using the short-time Fourier transform (STFT). The mean, variance, and skewness were used as features after determining the energy distribution for each task. The sixty best features were selected based on the sum of the distance of the features and the standard deviation. An SVM and an ANN were employed to perform the classification task.

An advanced machine-learning method to perform human identification with a low-cost EEG device was used by Bashar *et al.* [65]. A bandpass eighth-order FIR filter was applied to reduce noise and artifacts in the EEG signal. Four extraction methods were used in this experiment. The local morphology of the EEG signals at multiple scales was extracted using a STFT and comprised the first method. The second extraction method was a multiscale wavelet packet statistics (WPS) into four levels of decomposition. For this method, the entropy, mean values, and standard deviation of each coefficient were computed. Finally, the multiscale wavelet packet energy (WPES) and the alpha-beta statistics (ABS) were used to extract the wavelet energy and to extract the brainwave features, respectively, and comprised the third and fourth methods. For classification purposes, an SVM classifier with an error-correcting output code (ECOC) multiclass model was used.

Keshishzadeh *et al.* [66] established an improved authentication system using the PhysioNet EEG motor movement/imagery (EEGMMI) dataset. The first two tasks of all 109 EEG records obtained from the subjects were used in the experiment and only six channels of the 64 recorded channels were selected according to the literature. At the preprocessing level, a 14th-order low-pass Butterworth filter filtered the EEG signals. Three AR coefficients with the smallest p-values were chosen to form the feature set. These features were selected after performing the student's t-test to reduce the dimensionality. The SVM has been employed for classification task using a one-to-one matching process comparison. To implement a single classifier, the authors employed a technique based on the selection of reference sweeps as in [67]. The genuine and impostor subjects were prepared using the reference sweeps. Feature classification was performed using different proportions (10% to 90%) of sweeps in training and testing stages. The one-step normalization of variance and mean provided an improvement on the performance independently of the training data percent. Thus, by utilizing only 10% of the data for training, a satisfactory performance was obtained. Besides that, the REC protocol provided better results than the REO protocol.

The extracted features of the EEG signals and the channels which provided stable results despite the subjects emotional states was investigated by Vahid and Arbabi [68]. At the end of the presentation of 40 one-minute emotional videos, the levels of valence and arousal were stated with self-assessment manikins (SAM). The valence and arousal planes were classified into five regions which were declared high arousal low valence (HALV), high arousal high valence (HAHV), low arousal high valence (LAHV), low arousal low valence (LALV), and baseline. The subjects with a minimum of two valence-arousal values in each emotional region were selected. For each of the 32 channels, 54 features were obtained and classified into two main groups: temporal/statistical features and frequency features. The Haar wavelet transform was used to discover the most discriminative frequency bands for person identification. For reducing feature dimension and also increasing the stability, the authors computed the variance, third central moment and the mean of all wavelet coefficients, resulting in twelve features. Finally, a sequential floating forward selection (SFFS) was employed as wrapper method and a *t*-test was used for filtering. An SVM with RBF kernel employing 10-fold cross validation performed the classification task.

The EEG feature selection, extraction, and classification for personal authentication was the focus of study of Kang *et al.* [69]. Firstly, the recorded EEG signals were bandpass filtered with an FIR filter. The PSD using FFT was obtained for each segmented epoch and the maximum value of the positive Lyapunov exponents were calculated. The authors also obtained the Shannon entropy and median frequency of the PSD. Six EEG feature candidates for personal authentication based on the calculated spectral power and Lyapunov exponents were calculated: the ratio of alpha to theta power, the ratio of alpha to beta power, the ratio of theta to beta power, the entropy of PSD, the median frequency across the total frequency range (2–50 Hz), and the maximum of the positive values of Lyapunov exponents. The feature vector has 96 feature dimension. Different trials performed on different days were used to define the optimal features that minimized the differences in all EEG epochs. The criteria index derived from three types of variances was also calculated aiming to select some good feature candidates among the extracted features.

Nguyen *et al.* [70] examined the application of EEG lossy compression for an EEG-based subject authentication system. The proposed system comprised two phases, namely enrollment and verification. In the enrollment phase, EEG data corresponding to a specific task from a user was captured, for example, moving the left arm. The EEG signals were preprocessed and the features were extracted to make the user models. In the verification phase, the subject repeated the same tasks as enrollment stage and the EEG signals were obtained. Similar to the enrollment stage, the EEG signal was processed to obtain the features and these features were used for authentication. The features were obtained using AR and power spectral analysis. To perform the classification task, an SVM was employed.

A linear SVM to classify 20 individuals was used by Khalil *et al.* [71]. The single electrode (Cz) data were acquired from the database for emotion analysis using physiological signals (DEAP) dataset. A total of 40 trials were used in the experiment in which 30 trials were used for training and ten trials were considered for testing. The PSD maximum, mean, FFT maximum, FFT mean, gamma band power, delta band power, theta band power, alpha band power, and beta band power were extracted as features for classification.

The last work surveyed in which the SVM outperformed the others classifiers was proposed by Nakamura *et al.* [72]. The authors developed a readily deployable EEG-based biometric system using a generic in-ear EEG sensor. The unobtrusiveness, robustness, and user-friendliness are some of the advantages of the in-ear EEG sensing in biometric applications. The in-ear EEG sensor was fabricated with two conductive flexible electrodes and a viscoelastic memory-foam substrate. The acquired EEG signals were filtered with a fourth-order Butterworth filter. After the preprocessing, PSD and the coefficients of an AR model were used as features. Three classification algorithms using two approaches (parametric and nonparametric) were employed. In the parametric approach, an LDA and an SVM were used for classification, while in the nonparametric method, the minimum cosine distance was applied.

5.3.6 Bayes classifier

An investigation of six publicly available datasets following a sequential experimentation approach to explore the EEG activity for subject identification was performed by DelPozo-Banos *et al.* [73]. The experimentation method was divided into three blocks. During the experimentation step, the configuration of the STFT that increases the discriminant information was defined. Some issues such as Hamming window length, percentage values for window overlap and frequency range were explored. After this step, different time and frequency representations were evaluated. Some tests were performed to describe some of the characteristics of the discriminant information, specifically: spatial and frequency distributions, as well as the uniqueness and the permanence of patterns. The *z*-normalized PSD coefficients were used as input for classification. For cross-validation, the authors combined a Monte Carlo technique with stratified k-folds.

5.3.7 k-Nearest neighbors

Yang and Deravi [74] presented an EEG-based biometric identification system applying HHT [75]. The WPD was used to filter the signal. Four-time domain signals were constructed carrying the frequency segments of frequency bands between the range of 0–128 Hz. The EEMD was used to process each reconstructed band to produce 12 empirical modes. Then, the instantaneous frequency (InsFreq) and instantaneous amplitude (InsAmp) functions were estimated to extract the features. The features were produced by calculating the sum of the instantaneous frequencies of the InsFreq and the mean of InsAmp for each window. A k-NN algorithm was used to classify the InsFreq-based features and an LDA was employed for the InsAmp features.

The method proposed by Wang and Najafizadeh [76] used different bands and was based on the spatial distribution of the spectral power to extract subjects EEG-based signatures. In the first step, artifact removal was done by ICA. Thus, the undesired information such as eye movements interference and background noise were removed. Wavelet Morlet transform was applied to construct the feature vectors. The temporal domain was divided into segments and the average spectral power was computed. The classification task was performed using the k-NN algorithm. Two testing scenarios were considered aiming to explore the invariance of the obtained features for classification. The first scenario handles with intra-subject testing within a session and the second scenario performs intra-subject testing across sessions using the training and testing data obtained from distinct sessions. PCA was applied to diminish the dimension of the feature set and the quantity of recording channels.

The k-NN and a BP feed-forward algorithm were employed in the studied scenarios of Rahman and Gavrilova [77] for classification of the EEG frequencies bands of overt mental stimuli brain signals. The proposed methodology was intended to obtain a proper band or bands combination for identification. The DWT was used to decompose the bands and to obtain the features of each frequency band. Daubechies-4 provided six levels of wavelet transform detail coefficients, generating the alpha, beta, and theta frequencies. The study also considered the combination of frequency and time domain features, such as energy, scale variance, power, roll off, variance, approximate entropy, zero crossing, mean absolute value, and modified mean absolute value.

5.3.8 Gaussian mixture model

Wang and Su [78] proposed a removing method for time-frequency EEG artifacts and compared its performance with EEMD, commonly employed in the state-of-the-art. A subject-adapted threshold was computed to eliminate undesired artifacts, followed by an stationary wavelet transform (SWT) filter, to eliminate the interference. The comparison aspects analyzed were artifacts removal, discriminative features, accuracy, and computation efficiency. The experiments were conducted using a single-channel EEG signal. The authors used AR model parameters of 49th order and PSD to compare the feature discrimination power between EEMD-ICA and the proposed method.

On the other hand, Ward and Obeid [79] verified the separability of subject's EEG signals via clustering methods and data-driven metrics. The removed subject-specific information was used to compose an algorithm that can estimate the correlation between EEG signals despite of the subject state. Primarily, the signals were bandpass filtered using a fifth-order Butterworth filter and Fourier transformed. The data were clustered via GMMS aiming to create unique models in each frequency band for each subject. Then, these models were averaged mutually over the frequency bands to form an aggregate GMM for the considered subject. Euclidean distances between the model centroids were calculated for the aggregate GMMs. In the final step, the distance between each aggregate GMM for all epochs were calculated to allow the identification.

Altahat *et al.* [80] explored the reduction of the electrodes number used in the EEG signal acquisition. In the preprocessing stage, the data for each channel were divided into six parts and segmented into one segment. No filtering was applied to remove EEG artifacts. The feature extraction was performed using the PSD in the range of 4–52 Hz (8 Hz frequency resolution) to decrease the size of the feature vector. Although 106 subjects were used in the experiment, only 50 individuals were used for channel selection to avoid overfitting. The means of the feature vectors distribution distances were calculated by using the average Mahalanobis distance was computed to form a stable channel subset. A sequential forward selection algorithm was executed on stability value of the channels to determine the best channel subset. For the recognition task, a subject authentication framework based on GMM classifier was built. It was suggested the use of eight EEG channels in biometrics instead to use all EEG channel set.

5.3.9 Linear/quadratic classifiers

Rocca *et al.* [81] analyzed parametric time-frequency maps of the captured EEG data. At the preprocessing stage, the raw signals are downsampled applying a decimation factor to the collected data. The obtained signals were segmented into overlapped frames. Next, the DC component was removed from each EEG segment and a z-score normalization was applied to each signal. The wavelet decomposition of EEG signals was achieved by employing complex Morlet waveform for all subjects individually and for each channel. A "bump" modeling was then used to obtain a parametric description of the maximum amount of energy contained in the wavelet maps. The bump modeling is a technique for modeling a time-frequency map, with the aim of describing a map with a limited number of basic functions. This technique aims to reduce a large number of parameters that describe the time-frequency map. Then, discriminant features were extracted by analyzing the obtained parametric maps.

In a posteriori paper, Rocca *et al.* [82] studied the EEG signals to extract biometric traits for identification purposes. The authors tested distinct acquisition protocols using sets of three and five electrodes as well as the proper positioning to obtain stable and repeatable features. At the preprocessing level, a decimation factor was used in the raw data after an anti-aliasing fir filtering. The alpha, beta, delta and theta subbands and their combinations have been considered in their experiments. A CAR filter was utilized to analyze the potential at each location. After the preprocessing stage, each collected signal was modeled as an AR stochastic process. The classification task was performed using a linear transformation of the features vector and the transformation matrix was obtained using minimum mean square error (MMSE). Two simple tasks were conducted to evaluate the repeatability and stability of the EEG signals, performing 1-min REC and REO. Information from nine subjects in REO and REC protocols was used in the research.

An EEG biometric identification system based on real cepstrums (RCeps) coefficients and linear discriminant classifier (LDC) was developed by DelPozo-Banos *et al.* [83]. The preprocessing was performed by applying a normalization to remove idiosyncrasies of the recordings. The EEG signals were also filtered using an fir filter. EEG channels with high noise were identified and discarded using the time kurtosis. The authors also proposed to use RCeps as alternative features. Thus, short time cepstral analysis (STCA) was applied on the signal to obtain the discriminant information contained in the spectral shape of the signal. The LDC was selected for classification of cepstral coefficients. The first stage of the experimental methodology was to identify the lowest optimal number of cepstral coefficients, while in the second stage the system was evaluated using the complete PhysioNet EEGMMI dataset.

Mao *et al.* [84] proposed a closed-set biometric identification system. The identification is defined as closed-set if the person is known to be in the dataset. The authors extracted the EEG signals from an individual channel placed on the scalp. The subjects were in REC condition. A low-pass filter at 40 Hz was applied to exclude higher frequency artifacts followed by a seventh-order Haar wavelet transform used to identify and eliminate EOG using soft thresholding. A total of 24 features for each epoch was extracted from 11 subjects. Thus, eight features from each rhythm were extracted. The statistical features were computed by the average amplitude and the variance of the signals. The AR models composed the frequency domain features using Akaike's information criterion [85]. The authors also used a z-score normalization and the classification task was performed using a Local Probability Center (LPCA) [86], KNN, SVM, LDA, ANN, and NBC.

Yang *et al.* [87] performed an investigation of the impact of the mental activity on the EEG-based biometric system. Four movement/imagery tasks were used as a stimulus and were performed in both REC and REO conditions. Different scalp regions were investigated to evaluate the best performance. Nine electrodes grouped in three different scalp regions were picked for investigation. The first protocol (P1) has the purpose of investigating the effect of pairing electrodes regions and task types on system performance. The second protocol (P2) objectives to examine the effect of distinctive motor/imagery tasks for stages of testing and training. The last protocol (P3) investigates if data from different kinds of tasks may be coupled to train the model to reach a reliable performance.

Chen *et al.* [88] suggested an EEG authentication system based on the ERP elicited by a rapid serial visual presentation (RSVP). Different symbols are serially presented one-by-one and at the same location in the display in the RSVP paradigm. The RSVP grants presenting a substantial amount of stimuli in a short interval of time. Notwithstanding the rapid stimulus exhibition, the RSVP evokes prominent ERPs. In the registration phase, the subject chooses a password, which was presented in an arbitrary order concomitantly with 22 non-password symbols in RSVP. Afterward, the EEG data were estimated and the model was determined. In the login stage, all symbols were showed in RSVP form. The result of the classifier accepts or refuse the login request. The ERP components were classified using a shrinkage LDA.

The proposal made by Yu *et al.* [89] consists of a P300-BCI-based authentication system. Once the same sequence of stimuli is presented to both client and imposers, only the client would be able to distinguish the low-probability target from the high-probability nontarget stimulus. These stimuli were pictures of persons and the low-probability targets were pictures of client's known people. The client's P300 wave can be detected from watching a sequence of stimuli in Oddball paradigm and used as data to authorize the system. At preprocessing stage, filtering using 6th Butterworth bandpass filter was used (1.0–2.0 Hz cutoff frequency). A decimation of order 15 was used to decrease the size of the feature set and the winsorization was performed to eliminate the effects of the outliers in EEG data for improving the signal quality. The P300 classification models were trained using Fischer's LDA. The P300-BCI-based authentication system was performed in two different methods. In the first method, the classifier outputs from only one block of P300 BCI were used to accept or to reject the claim. In the second method, the classifier outputs were averaged separately for all target and nontarget EEG epochs in five blocks of a single trial.

Jayarathne *et al.* [90] developed an authentication system that uses a selection of four-digits as the PIN number. Three main phases, namely neutralization, visual stimulation and thinking stimulation, perform the brain stimulation in a single trial. For visual stimuli, a four-digit number was displayed on a computer display. The subjects were required to think about the showed number in REC condition. For the neutralization period, calm music was played to neutralize the subjects. The pre-processing step consists of neutral and visual stimulation removal, extraction of data in the specific frequency band and separation of classes. Alpha and beta frequency bands were used for analysis. The common spatial pattern (CSP) was used for feature extraction and LDA was used to perform the classification. The data for all subjects were divided into two segments for classification. Among those, five trials were used for classification and ten trials were used for training.

The feasibility of an affordable wireless EEG device for subject identification was demonstrated by Koike-Akino *et al.* [91]. The Oddball was used to evoke the P300. The subjects were notified to answer the target stimuli which appear occasionally and to neglect the nontarget stimuli. The EEG signals were obtained from 25 healthy adults. Among these, four subjects participated in a single session and 21 subjects experimented two sessions. The experimental protocol focused on a card counting task and each epoch has a duration of 800 ms. The ocular artifacts were removed using blind source separation canonical correlation analysis (BSS-CCA). The dimensionality reduction was performed using PCA and partial least squares (PLS) techniques. A total of eight machine-learning algorithms were responsible for classification: LDA, quadratic discriminant analysis (QDA), NBC, logistic regression (LR), decision tree (DT), SVM, k-NN, and deep neural network (DNN). Among those, the QDA outperformed the others classifiers in terms of accuracy.

5.3.10 Classifiers not defined

The individual identification based on visual evoked potential (VEP) using the features extracted utilizing four algorithms was explored by Bai *et al.* [92]. Besides, the authors evaluated the most discriminative features of the EEG signals and the reduction of the channels to turn the system most reliable. The ICA was performed to remove EOG signals. The re-reference and downsampling were also applied to enhance the SNR. Four types of algorithms were used to feature extraction, including AR modeling, TPS, and PLV. SVM was responsible for performing the classification task. The features extraction and channels optimization to remove redundant information were performed using three algorithms, namely genetic algorithm , SVM-RFE, and fisher discriminant ratio (FDR).

Sundararajan *et al.* [93] proposed a wavelet-based generic framework using a single-trail brain-wave response in a biometric system. Some properties of the importance of biometric template security were raised, such as revocability, securable, diversity, and performance. Different wavelets were pointed to remove the noise and to obtain the ERPs from the EEG, although it was observed that quadratic B-Spline wavelet is best suited for this kind of application. The critical requirement to be served by the wavelet function is that it should resemble the ERPs to be extracted from the EEG. Independent of the wavelet selected, three steps must be performed to extract the expected ERP from the EEG signals. The first step is the wavelet decomposition, followed by nonlinear thresholding (second step), in which a soft threshold is preferred due to circumventing the presences of spikes in the resolution. The final step is the inverse wavelet reconstruction. In the enrollment phase, the signal parameters were collected into a biometric template and then saved in an encrypted format. During the authentication stage, the individuals were subjected to comparable classifiers that they were presented during the enrollment phase and the EEG signals were obtained before noise removing. For classification, personal one-class classifiers (OCCs) were used because they usually have a better performance than ANN or GAs.

A two-level authentication method that uses an embedded system was designed by Soni *et al.* [94]. Three main blocks comprised the proposed system. The first block, namely BCI block, is responsible for connecting the human brain to the computer and uses brain sensors consisted of dry electrodes. The captured raw data were converted into packets and transmitted to the second block using wireless Bluetooth transmission. The second block executes data extraction from obtained raw data. This way, the level differentiator unit (LDU) gets the brain signals in real-time and tests if the generated pattern matches with the predefined pattern. The last block, namely checker module (CM), gets the data from the LDU and is responsible for providing two-level authentication. An ARM processor composed the CM with an LCD and a touch panel. The first lock and the second lock of the authentication system open for a six times eye blink and an attention level of 90, respectively. MATLAB® read the transmitted packets and display the values of attention signals and, if the condition is satisfied, a message is displayed. Finally, the CM asks for a passkey to finish the authentication task.

5.3.11 *Final considerations*

ANNs are being used over time to classify the EEG signals with a good precision. The CNN, in turn, provide a more efficient and refined strategy to increase the accuracy of the experiments. Besides that, in addition to improving the recognition rate, the ANN and CNN grant data compression, noise reduction and increase the processing and transmission of the EEG signals, which is essential in the environment where authentication is required, such as smart homes, for example. The development of a more flexible platform and a novel model of interaction using this modality of biometry can guide the researchers.

The SVM is one of the most used machine-learning technique for biometric security in existing literature, providing reasonable access control, malware detection schemes to protect users data and secure offloading. This classifier also plays an essential role in EEG signals classification task for biometric and also for medical purposes in which high accuracy rates are obtained using this classifier for subjects identification and diagnosis of pathologies. Thus, the SVM is a powerful and flexible tool that can be used efficiently for both applications. Like the SVM, the Bayesian classifier also is widely used in biometric application.

In addition to the tools as mentioned earlier, the other machine-learning techniques, such as k-NN, RF, GMM, among others are being used recently to solve issues related to security, communication efficiency, source discovery, and indexing [95,96] and also security for authentication when speaking about biometrics involving brain signals, genuinely raised in the this section.

So many others works can be found in the literature. However, many of them are an extension of the ones presented in this section and do not show significant differences. It can be observed that majority of the works use a conventional methodology as depicted in the Figure 5.3. Bandpass and bandstop filters, such as Butterworth and notch filters, are widely employed to remove noise and spurious

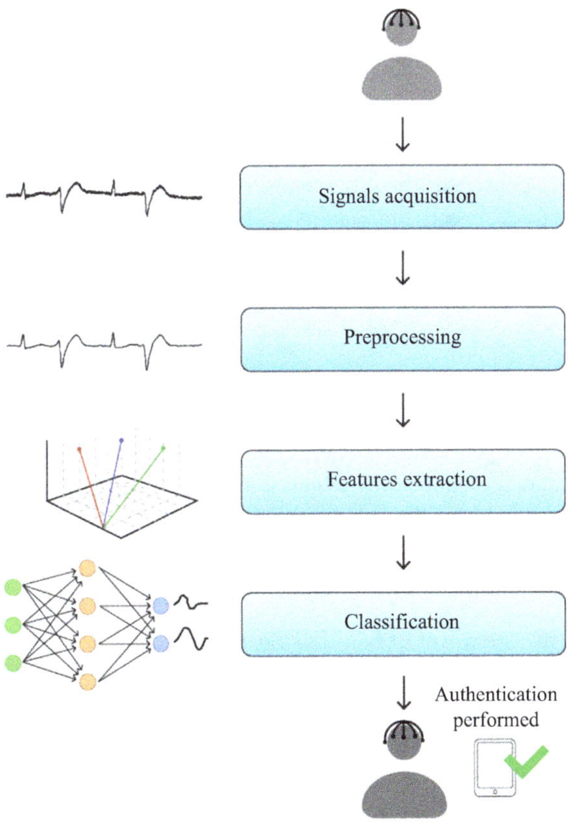

Figure 5.3 Methodology commonly employed by the state-of-the-art authors to classify EEG signals

artifacts. The features were mostly extracted using wavelet decomposition, AR components, and PSD estimation, while SVMs, ANN, k-NN, and L1 and L2 distances are the classifiers most used by authors. Table 5.1 shows the performance measurements of the surveyed works. Besides, Table 5.2 summarizes the techniques for feature extraction and classification. The security issues are discussed in a further section of this chapter.

5.4 Paradigms for signals acquisition

5.4.1 REO and REC

These protocols are widely used in EEG signals acquisition. In general, the subjects are conveniently seated in a reclining seat with both forearms in a relaxed position. The external sounds are mitigated and the illumination is under control.

The variability in PSD is evident among these two protocols along the literature. The difference relies on brain activity, where the activity in some regions, such as bilateral middle occipital gyrus, cerebellum, and frontoparietal cortex, for example, is higher in the REO protocol. Besides that, the distribution of the entropy itself is wider in the time domain and the spectral entropy is narrower for both REO and REC protocol. The visual information inflicts more prominent temporal disorder in time-domain entropy measures, shifting the distribution to the right from REC to REO [97].

There is no consensus in the literature as to which modality presents the highest distinction between individuals in biometric the authentication systems. The efficiency of authentication algorithms varies in a wide range of parameters, such as the number of electrodes, acquisition system, time, and sampling rate, among others.

5.4.2 ERP

The ERP is a potential measured in the brain in response to an event and can be categorized into endogenous components, which is related to subject action, such as VEP, and exogenous components, which is related to external stimulus characteristics and are recorded over the first 150 ms following the eliciting event, e.g., P300 [98].

5.4.2.1 VEP

The VEP is the measurement of the EEG electrical activity response in the occipital cortex after visual stimulus presentation. The VEP is merged with the natural EEG signal and is small in amplitude. Thus, it needs to be amplified and separated using signal-averaging techniques. Besides that, the cerebral cortex, optic radiation and nerve, chiasma and eye affects the VEP, reducing the signal, if there is any damage along the path. The minimum VEP analysis time is 250 ms and the traditional VEP signal used in biometric applications consists of a negative peak (N75), followed by a positive peak (P100), followed by another negative peak (N135).

Table 5.1 Performance measurements of the state-of-the-art works

Authors	Performance (best scenario)
Yeom *et al.* [55]	Acc = 86.1%; FAR = 13.9%; FRR = 13.9%
Nguyen *et al.* [56]	ERR = 0.0409%
Rocca *et al.* [81]	CRR = 95.74%
Rocca *et al.* [82]	CRR = 100%
Liu *et al.* [57]	CRR = 97.25%
Gui *et al.* [20]	CRR = 99.87%
Pham *et al.* [59]	Acc = 97.1%
Yang *et al.* [74]	CRR = 99.50%
Phung *et al.* [62]	Acc = 95.1%
Bai *et al.* [92]	Acc = 96.25%
Wang and Su [78]	–
Davis *et al.* [63]	Acc = ~100%
DelPozo-Banos *et al.* [73]	–
DelPozo-Banos *et al.* [83]	Acc = 97.18%
Gui *et al.* [31]	Acc = 91.62%
Kumari and Vaish [29]	Spc = 97.16%; Sen = 90.33%; Bal. Acc = ~92%
Kumari and Vaish [30]	CRR = ~97.23%–99.65%
Armstrong *et al.* [41]	Acc = 97.6%
Al-Hudhud *et al.* [46]	TAR = 90%; FAR = 0%
Pham *et al.* [61]	FRR = ~100; ERR = ~0.2
Ward and Obeid [79]	–
Fraschini *et al.* [47]	EER = 0.044
Maiorana *et al.* [48]	FRR = 9.46%; FAR = 6.43%
Maiorana *et al.* [50]	CRR = 99%
Sundararajan *et al.* [93]	–
Mao *et al.* [84]	Acc = 97%
Altahat *et al.* [80]	HTER = 12.08%
Ma *et al.* [32]	Acc = 88%
Maiorana *et al.* [28]	CRR = 90.8%
Lee *et al.* [51]	Acc = 95%
Yang *et al.* [87]	CRR = 99%; EER = 4.5%
Yu *et al.* [42]	HTER = 0; FPR = 0.0101
Mu *et al.* [34]	Acc = 87.3%; FAR = 5.5%; FRR = 5.6%
Nieves and Manian [15]	Acc = ~98%
Chen *et al.* [88]	TAR = 3.33e−05; Acc = 87.8% (5.1)
Das *et al.* [35]	Acc = 98.8%
Liew *et al.* [36]	Acc = 95.08%; AUC = 0.8843
Wang *et al.* [43]	Acc = ~90%
Yu *et al.* [89]	Acc = 83.1%; Sen = 57.0%; Spe = 89.7%; FRR = 0.325
Soni *et al.* [94]	–
Jayarathne *et al.* [90]	Acc = 96.97%
Harshit *et al.* [44]	FAR = 13.91%; FRR = 26.66%; Acc = 79.73%
Bashar *et al.* [65]	Acc = 94.44%
Bansod *et al.* [52]	Acc = 61%
Keshishzadeh *et al.* [66]	Acc = 97.43%; Spe = 97.71; Sen = 94.24%; EER = 2.57%
Hasan *et al.* [37]	MSE = 6.7232e−08; Acc = 100%

(Continues)

Table 5.1 (*Continued*)

Authors	Performance (best scenario)
Vahid and Arbabi [68]	CRR = 88%–99%; EER = 5%–35%.
Garau *et al.* [53]	FAR = 1%; FRR = 1.76%; EER = 1.42%
Wang and Najafizadeh [76]	CRR = 92.58%
Koike-Akino *et al.* [91]	Acc = 96.7%
Rahman and Gavrilova [77]	Acc = 84.4%
Kang *et al.* [69]	Acc = 94.9%
Thomas and Vinod [45]	EER = 0.0196; AUC = 0.002
Kaur and Singh [14]	Acc = 98.16%
Patel *et al.* [13]	EER = 7.5%
Tangkraingkij *et al.* [38]	Acc = 95.51% (6.32)
Nguyen *et al.* [70]	EER = ~0.1%
Khalil *et al.* [71]	Precision = 0.77; Acc = 0.96; Sen = 0.75; Spe = 0.99; AUC = 0.99
Mao *et al.* [39]	Acc = 97%
Nakamura *et al.* [72]	HTER = 5.5%; Acc = 99.0%
Kang *et al.* [54]	EER = 0.73%; FAR = 1.10%; CRR = 98.93%

5.4.2.2 P300

Differently from VEP, P300 is a component of the ERP related to decision making which evoke a positive peak around 300 ms after a decision taking. Frequently, P300 is also identified when the oddball paradigm is used. In this paradigm, the subject is asked to identify a target stimulus from a nontarget one. The subject needs to respond the oddball (rare) stimulus that is mixed with a sequence of common stimuli. For example, the subjects may be instructed to count the frequency of a specific card within a set of cards. The P300 are more prominent at frontral, central, and parietal regions of the brain.

5.4.2.3 N400

The N400 corresponds to a negative peak in the central parietal region around 400 ms post-stimulus in response to an extensive variety of potentially meaningfully stimuli. The N400 component is related to access attempt to the natural semantic network by the language comprehension system [41]. The potentials elicited in familiar subjects differs from potentials recorded from unfamiliar subjects. Also, different objects elicit different N400 profiles in the same familiar group of subjects, making this ERP suitable for biometric applications.

5.4.3 RSVP

In the RSVP paradigm, different symbols are displayed serially in the same location of the screen. The RSVP paradigm allows to exhibit a large amount of stimuli in a short period. RSVP is a variant of the oddball paradigm. The advantage of this

Table 5.2 Main techniques for EEG-based identification/authentication systems

Authors	Stimulus	# of ch.	# of sub.	Features	Classifier
Yeom *et al.* [55]	Visual (self- and nonself-face images)	18	10	Temporal, dynamic, AR	Nonlinear SVM with Gaussian function
Nguyen *et al.* [56]	Motor imagery	3	9	AR, PSD	SVDD, GMM
Rocca *et al.* [81]	REC	3	36	Wavelet Morlet	LDA (Mahalanobis distance)
Rocca *et al.* [82]	REC/REO	3, 5	9	AR	Linear classifier
Liu *et al.* [57]	REC, Visual [58]	32	20	AR, TPS, FPS, PLV	SVM-RFE
Gui *et al.* [20]	Visual (text)	1	32	Wavelet	MLP
Pham *et al.* [59]	REC/REO	8	40	AR, Hjorth parameters, PSD	SS2LM-SVDD
Yang and Deravi [74]	Visual, motor/imagery	1	118	Wavelet Daubechies 4, HHT (EMD)	LDA, k-NN
Phung *et al.* [62]	REC/REO	23	40	AR, ApEn	Linear SVM
Bai *et al.* [92]	Visual [58]	32, 37	20	AR, TPS, FPS, PLV	GA, FDR, RFE-SVM
Wang and Su [78]	REC	1	36	AR, PSD (raw, EMD-ICA, and SWT)	–
Davis *et al.* [63]	Visual (video)	128	20	Cepstral coefficients	Linear SVM
DelPozo-Banos *et al.* [73]	REC/REO, motor/imagery task, visual, auditory, intellectual tasks, elicited emotions, multiple-tasks	6, 16, 18	100	PSD	Bayes classifier
DelPozo-Banos *et al.* [83]	Motor/imagery tasks	16	100	Cepstral coefficients	LDC
Gui *et al.* [31]	Visual (words)	1, 4	37	Raw	RBF network, frequency-based RBF network, probability-based RBF network
Kumari and Vaish [29]	Visual (number password)	1, 4, 7, 11, 14	6	Wavelet Daubechies	LVQ neural network

(Continues)

Table 5.2 (Continued)

Authors	Stimulus	# of ch.	# of sub.	Features	Classifier
Kumari and Vaish [30]	Mental task	6	7	HHT (EMD), information theoretic-measures	FNLQV
Armstrong *et al.* [41]	Visual (words)	1	45	Raw	SVM, LDC (cross correlation), DIVA, NDL
Al-Hudhud *et al.* [46]	Visual, mental task	3, 4	30	Statistical features	Cosine, Euclidean similarities
Pham *et al.* [61]	Visual (video), emotional state	8	32	AR, PSD	SVDD
Ward *et al.* [79]	REC	64	50	PSE	GMM (Euclidean, Mahalanobis)
Fraschini *et al.* [47]	REC/REO	64	109	EC	Euclidean distance
Maiorana *et al.* [48]	REC	19	40	AR	L1, L2 distances
Maiorana *et al.* [50]	REC	19	60	PSD, LDA, EB	L1, L2, cosine, Mahalanobis distances
Sundararajan *et al.* [93]	–	–	–	–	–
Mao *et al.* [84]	REC	1	11	Statistical, Hjorth, AR	LPC
Altahat *et al.* [80]	REC/REO, motor task, mental task	8	106	PSD	GMM
Ma *et al.* [32]	REC/REO	64	10	CNN	CNN
Maiorana *et al.* [28]	REC/REO	19	50	AR, PSD, COH	L1, L2, cosine distance
Lee *et al.* [51]	REC	16	7	PSD	Bhattacharyya distance
Yang *et al.* [87]	Motor/imagery task	9	108	Wavelet Daubechies 4	LDA
Yu *et al.* [42]	Visual, mental task	8	24	Topological plot	Correlation
Mu *et al.* [34]	Visual (self- and nonself-face images)	2	10	Fuzzy entropy	ANN
Nieves and Manian [15]	Motor imagery	2, 4, 6	10	PSE	SVM (polynomial kernel), ANN

Chen *et al.* [88]	Visual (RSVP)	16, 28	29	Topographical maps	RLDA
Das *et al.* [35]	Visual (geometric shapes images)	17	40	CNN	CNN
Liew *et al.* [36]	Visual (quiet and noise environment)	8	37	Wavelet, coherence, cross-correlation, Hjorth parameter, mutual information, and mean of amplitude	IncFRNN
Wang *et al.* [43]	Visual, auditory (oddball)	19	6	raw	Cross correlation
Yu *et al.* [89]	Visual (oddball)	8	4	raw	LDA
Soni *et al.* [94]	Eye blink	1	–	–	–
Jayarathne *et al.* [90]	Visual, mental task (EC)	14	12	CSP	LDA
Harshit *et al.* [44]	Visual, auditory (self, familiar, unfamiliar)	14	5	PSE	Correlation
Bashar *et al.* [65]	REC/REO	5	9	MSD, WPS, WPES and ABS	SVM (ECOC)
Bansond *et al.* [52]	Mental task (letter composition)	1	40	Statistical	Manhattan distance
Keshishzadeh *et al.* [66]	REC/REO	6	104	AR	SVM
Hasan *et al.* [37]	EC mental task	3	3	PSD and FFT	ANN
Vahid and Arbabi [68]	Visual (emotional states)	32	32	Statistical and temporal, PSD	SVM (RBF)
Garau *et al.* [53]	REC/REO	56, 64	109	EC, COH	Mahalanobis distance
Wang and Najafizadeh [76]	REC	48, 128	4	Wavelet Morlet (average spectral power)	k-NN
Koike-Akino *et al.* [91]	Visual (counting cards task)	14	25	PCA (raw)	LDA, QDA, NB, DT, k-NN, SVM, LR, DNN
Rahman and Gravilova [77]	Visual (overt mental stimuli)	16	10	AR, Wavelet Daubechies 4	ANN, k-NN
Kang *et al.* [69]	REC	16	7	PSD (Lyapunov exponents)	linear SVM
Thomas and Vinod [45]	REC/REO	19	109	PSD	Cross-correlation

(Continues)

Table 5.2 (Continued)

Authors	Stimulus	# of ch.	# of sub.	Features	Classifier
Kaur and Singh [14]	REC/REO	64	109	PCA, raw	SVM and RF
Patel *et al.* [13]	Motor task (hand synergies)	18	10	FFT (magnitude squared coherence)	Correlation
Tangkraingkij *et al.* [38]	Resting state	4	30	FFT	ANN
Nguyen *et al.* [70]	REC/REO, motor imagery	3, 5	9, 40	AR, PSD features compressed with Wavelet biorthogonal	SVM
Khalil *et al.* [71]		1	20	PSD, FFT	linear SVM
Mao *et al.* [39]	Driving induced fatigue	64	100	Raw, AR, PSD	LDA, SVM, CNN
Nakamura *et al.* [72]	REC	1	15	PSD, AR	Cosine distance, LDA, SVM
Kang *et al.* [54]	REC/REO	34	109	PSD, nonlinear (Lyapunov exponents, SE, and PE), network (MPC, PLI)	Euclidean distance, threshold comparison

paradigm is because the perceptual error caused by attentional blink and/or repetition blindness does not occur between two stimuli flashes. In this case, the accuracy is not impaired [99].

5.4.4 Motor movement/motor imagery

This signal acquisition paradigm consists of subjects performing motor movement or motor imagery of a given action, such as hand movements, foot movements, and tongue movements, according to a cue. The motor imagery paradigm is also employed in BCI applications which plays an important role in rehabilitation or neurorehabilitation training, smart homes, military, and other fields. The main idea behind this paradigm is the use of the discriminative information in frequency bands caused by event-related synchronization and event-related desynchronization to identify the type of the motor movement [100].

5.4.5 Steady-state evoked potentials (SSEVP)

The SSEVP are responses to visual stimulation at frequency range of 3.5–75 Hz [101]. The SSEVP are evoked at the same, or multiple, frequency of the visual stimulus in the visual cortex. Auditory and somatosensory stimulation also evoke these potentials in different frequency ranges. SSEVP is one of the most researched paradigm because it enables an rapid and precise command input, have an excellent SNR and is relatively immune to artifacts. Also, different individuals evokes different potentials, which makes this acquisition paradigm more attractive for biometric applications.

5.5 Datasets and devices

Although the acquisition of the EEG signals is becoming increasingly simple and various technologies are being developed to facilitate the capture process, many authors still prefer to use existing datasets. In general, there is no EEG dataset publicly available for biometric purposes. Many of them are used for studies of neurological disorders, such as epilepsy, for example, while others are derived from healthy people [102]. Therefore, these datasets are widely used in biometric identification/authentication systems [103]. In this way, the works can be compared between themselves and a fairer performance metrics can be obtained. Besides that, the authors do not need to worry about the signal capture, subject's preparation, sessions organization, among others.

In this section, it was described the most utilized datasets in EEG-based biometric systems. The main characteristics of these datasets, such as the number of subjects in the experiment and their respective state conditions, trials, stimulus, and the number of electrodes are also provided. Some datasets, in turn, are hard to be tracked [104] and some of them are proprietary. Table 5.3 summarizes the main characteristics of the reported datasets.

5.5.1 UCI EEG dataset

This dataset, also known as Zhang Dataset or KDD Dataset [105], was originated from an extensive study to examine the correlation of EEG signals of subjects with

Table 5.3 Summary of the most employed dataset in the state-of-the-art and their main characteristics

Dataset	# of subjects	Sampling rate	Protocol	# of electrodes	
UCI EEG	122	256 Hz	Visual	64	
Graz-BCI	2003A	3	250 Hz	Motor imagery	64
	2003B	3	125 Hz	Motor imagery	64
	2008A	9	250 Hz	Motor imagery	22
	2008B	9	250 Hz	Motor imagery	3
Australian EEG		40	167 Hz	REC/REO	23
Poulos		79	128 Hz	REC	2
Keirn		7	250 Hz	REC/REO, mental tasks	6
BCI CSU	Biosemi ActiveTwo	14	1,024 Hz	Mental task	32
	g.Tec g.GAMMAsys	13	256 Hz	Mental task	8
	Neuropulse mindset	13	512 Hz	Mental task	19
PhysioNet EEGMMI		109	160 Hz	REC/REO, mental tasks	64
Yeom		10	250 Hz	Visual	18
Miyamoto		23	128 Hz	REC	2
DEAP		32	512 Hz	REO	32
Ullsperger		5	200 Hz	Mental task	61
Mu		10	1,000 Hz	Visual	30
Cho		52	512 Hz	Motor imagery, non-task related	64
PhysioUnicaDB		22	1,024 Hz	REC/REO	64
Shin		29	1,000 Hz	Motor imagery, mental tasks	30

genetic predisposition to alcoholism. The study aimed to investigate cognitive impairments in nonamnesic alcoholic patients. The authors also intended to check if exists any variation between control patients and alcoholic patients in visual memory potential (VMP) [106]. The subjects were divided into a control group and alcoholic group. The control group consists of 48 right-handed male subjects with no particular and family history of drug or/and alcohol misuse or any history of psychiatric disease or neurological disorder. The subjects were selected from hospital employees. The alcoholic group consists of 77 male alcoholics with the mean age of 35.83 $\pm$ 5.33 years ranged between 22.3 and 49.8 years. Clinical psychiatrists assessed all subjects and the first diagnosis of alcohol dependence or misuse was obtained by the psychiatrist at the Addictive Disease Hospital in Brooklyn following DSM-III criteria. All alcoholic patients were hospitalized at least 30 days on a restricted room before engaging in the ERP investigations.

The presented stimulus was performed using the Snodgrass and Vanderwart [58] pictures dataset. The images of this set represent different concrete objects with a natural designation. All images were paired into non-matching and matching conditions. In the first condition, the second stimuli (S2) was repeated as first stimuli (S1). While in the second, the S2 was followed by an image entirely different from S1, even regarding its semantic category. A total of 40 trials were performed. On half of the trials, S1 was distinct from S2 and, on the other half, S1 was equal to S2. No stimulation was repeated. The stimuli were shown on a blank background at the center of a monitor with the same configuration for all participants. The subjects were required to click a mouse button in one hand if the S1 matched S2 and to click a mouse button in the other hand if the S1 differs from S2. The subjects were seated in a reclining chair and the computer screen was located 1 m distant from vision.

A 61-lead electrode cap (ECI, Electrocap International) was utilized in the experiment in according to 10-20 International System. A nose electrode was responsible for grounding the subjects and the impedance of the electrodes was always below 5 kΩ. The brain signals were amplified using an EPA-2 amplifier bandpass (bandpass of 0.02 and 50 Hz) and sampled with a sampling rate of 256 Hz during a period of 190 ms of pre-stimulation baseline and 1,440 ms after each stimulus exhibition. The considered ERPs without artifacts were sampled during three conditions of stimulation: presentation of S1, S1 matching with S2, and S1 non-matching with S2. The choice of EEG segments was limited to those trials in which the user performed right behavioral answers in both match or non-match trials. The dataset can be downloaded at [107].

5.5.2 Graz-BCI dataset

The Graz-BCI dataset was provided by Department of Medical Informatics of the Institute for Biomedical Engineering at University of Technology Graz (Gert Pfurtscheller, Alois Scholögl). This dataset was used in different works and also is employed in BCI competitions [108].

Dataset 2003A: This dataset is comprised of three subjects in resting state condition. The task for each subject was to perform motor imagery of both hands, tongue or foot movements according to a cue. The research consisted of various runs with 40 trials each with the following methodology:

- 0–2 s: silence.
- 2–3 s: audio stimuli indicating the start of the trial simultaneously with a cross "+" presentation at a display.
- 3–7 s: presentation of an arrow to up, down, right or left and the subjects were required to imagine a tongue, foot, right hand, or left hand movements, respectively.

The data sampling rate was of 250 Hz and a notch filter between 1 and 50 Hz was responsible for filtering the data.

Dataset 2003B: The data were obtained from three subjects with a similar acquisition protocol of Graz Dataset 2003A. The experiment contains three sessions for each subject. Each session includes of 4 to 9 runs. In this dataset, the participants performed motor imagery classes of right and left hands. The EEG data were obtained utilizing three bipolar channels at a sampling rate of 125 Hz and notch filtered between 0.5 and 30 Hz.

Dataset 2008A: This dataset comprises of EEG data from nine subjects. Again, the cue-based BCI paradigm of tongue, foot, left hand and right hand, motor imagery tasks, was performed. Two sessions on separate days were performed for all subjects. All sessions were composed of six runs separated by small intervals. One run consisted of 48 trials which 12 trials were performed for each of the four classes, forming a total of 288 trials per session. At the start of each session, a 5 min recording was conducted to evaluate the EOG influence. A total of 22 silver chloride (AgCl) electrodes were utilized to record the raw data. The EEG signals were recorded with right mastoid serving as ground the left mastoid serving as a reference. The signals were sampled at a 250 Hz and bandpass filtered between 0.5 and 100 Hz. A 50 Hz notch filter was used to remove the line noise. Additionally, three monopolar EOG channels were registered and also sampled with 250 Hz and bandpass filtered between 0.5 and 100 Hz, following a notch filtering.

Dataset 2008B: This dataset was obtained from a study published in [109]. Nine subjects participated in the experiment. Each subject performed five sessions. The first two sessions contain the training data without feedback. These sessions were recorded on two different days within two weeks. The remaining three sessions were recorded with feedback. Each session comprised six runs with ten trials each and two classes of imagery of left and right hands, class 1 and class 2, respectively. This paradigm generated 120 trials per session (20 trials per run). The recording was split into three blocks as in Dataset 2008A. The artifact block was subdivided into four sections, whereby 15 s of artifacts compose each session with 5 s of resting between them. The subjects were oriented to perform either rolling, eye blinking, left–right or up–down movements. At the start of each task, a low warning sound was played and, in the end, a high warning sound was emitted. Each session was initiated with a recording of around 5 min to measure the EOG influence. The EEG data were recorded using three bipolar channels at 250 Hz sampling rate. The signals were bandpass filtered between 0.5 and 100 Hz and notch filtered at 50 Hz. The EOG was recorded using with three monopolar electrodes using the same amplifier settings.

5.5.3 Australian EEG dataset (AED)

The Australian EEG dataset (AED) consists of EEG recordings recorded at a public hospital over 11 years. The AED project started in 2001 as a collaboration between a team from the University of Newcastle and the John Hunter Hospital, New South Wales, Australia. The EEG signals were recorded using 23 electrodes with bipolar connections and standard 10-20 International System. The signals of the individuals were sampled at 167 Hz for approximately 20 min. All subjects were in REC and REO conditions during the acquisition. The dataset contains recordings of 80 subjects, divided into two groups, epileptic and normal, each one with 40 subjects. This dataset can be found at [110].

5.5.4 Poulos dataset

This dataset was originated from the studies of Poulos *et al.* [111–113]. The main purpose of these works was to examine healthy and pathological cases to constitute a one-to-one match between the genetic information of the individuals and specific relevant features of the EEG signals. A total of 45 EEG recordings were taken for each of 4 individuals, referred to as A, B, C and D. The second subset of data was composed of EEG recordings from 75 different subjects to form an "X" group. The data were acquired utilizing a digital electroencephalograph with PHY-100 Stellate software, with the subjects in rec condition, for 3 min. The voltage difference was recorded between leads CZ and O2 at 128 Hz sampling rate. The signal was bandpass filtered at 1–30 Hz to retain only useful information.

5.5.5 Keirn dataset

Keirn and Aunon [114] recorded EEG data from seven subjects, of which one is female and six male, while the subjects were performing one of four tasks or in resting state for 10 s. The four tasks were: to solve nontrivial multiplication problems, mentally rotate an image of a 3-D complex object, mentally write a letter to a family member or a friend and to visualize numbers being sequentially printed on a blackboard. Each condition was replicated five times within 2 weeks and two sessions were recorded for each subject. The EEG acquisition was performed at 250 Hz using the channels C4, C3, P4, P3, O2, O1, and EOG, concerning A1 and A2. The recordings were also done with a bank of Grass 7P511 amplifiers and bandpass filtered at 0.1 to 100 Hz. This dataset can be found at [115].

5.5.6 BCI CSU dataset

The dataset from Colorado State University was obtained between the years 2011 and 2012 by the Brain-Computer Research Group. Tree different EEG systems were used for each of subject: Biosemi ActiveTwo (32 channels and sample rate at 1,024 Hz), g.Tec g.GAMMAsys (8 channels and sample rate at 256 Hz), and Neuropulse Mindset (19 channels and sample rate at 512 Hz). Two group of subjects composed the dataset: not impaired, which the EEG signals were recorded in the lab and impaired, which signals were recorded in the home of the subjects. A total of seven subjects taken part in the experiment. This dataset can be found at [116].

5.5.7 PhysioNet EEGMMI dataset

This dataset was generated and provided to PhysioBank [117] by Gerwin Schalk and his associates at the BCI R & D Program, Wadsworth Center, New York State Department of Health, Albany. The recordings of 109 subjects using the BCI 2000 system composed of the PhysioNet EEG motor movement/imagery dataset (EEGMMI) and was described in [118]. The data were collected at a sampling rate of 160 Hz in a 64-electrode scheme according to the 10-10 International System. The subjects performed 14 experimental runs. Two runs were performed using a baseline recording of 1 min, which one run was performed with eyes open and one run was performed with eyes closed (REC and REO protocols). Three runs were performed in 2 min, each one the subjects were executing one of the four following tasks:

- Task 1: The subjects open or close their right or left fists according to the target appearing on the screen.
- Task 2: The subjects imagine opening or closing their right or left fists according to the target appearing on the screen.
- Task 3: The subjects open and close both fists if the target appears on the top of the screen. If the target appears on the bottom of the screen, the subjects open and close both feet.
- Task 4: The subjects imagine opening and closing both fists if the target appears on the top of the screen. If the target appears on the bottom of the screen, the subjects imagine opening and closing both feet.

These tasks were performed until the target disappears, then the subjects relax. This dataset can be downloaded at [119].

5.5.8 Yeom dataset

This dataset contains EEG recordings of ten healthy Korean students subjects, including monozygotic twins (ages: 20–29; mean: 26.67). All of the subjects had normal visual acuity and were right-handed. The data were recorded at 250 Hz and bandpass filtered between 0.1 and 100 Hz. A 50 Hz notch filter was employed to eliminate power line noise. In this dataset, the subjects viewed photos of themselves or others as a visual stimulus. In general, two sessions were performed on different days. A neuroscan SynAmps2 and BCI20001 were used for the EEG recording. The data were acquired from 18 scalp electrodes, positioned according to the 10–20 International System. The dataset was constituted of EEG signals received from two sessions performed on separate days. Each session involved two runs separated by small intervals and each run has further consisted of 50 trials. A total of 100 trials were performed for each session. For each trial, 20 face pictures were exhibited, which ten images comprise self-face and the other ten comprises nonself-face photos.

5.5.9 Miyamoto dataset

The EEG recordings of this dataset were obtained from 23 healthy male subjects with $\pm$ 20 years old. Each subject was seated at the rec condition in a soundless room. The EEG signals were recorded during continuous 3 min. A total of ten EEG signals were recorded from each subject and the recordings were performed on the same day. The authors used the nonmedical or commercial equipment. The recordings were performed at 128 Hz sampling rate and filtered between 1 and 24 Hz. The minimum and maximum voltage were of 5 and 80 μVpp, respectively. The frontal electrodes FP1 and FP2 were used for signals acquisition.

5.5.10 DEAP dataset

The Database for Emotion Analysis Using Physiological Signals (DEAP dataset) [120] is a multimodal dataset used to analyze the affective states using the exhibition of music videos. The analyses were performed in two laboratories with restrained illumination. The peripheral physiological data and the EEG signals were acquired using a Biosemi ActiveTwo system. The music videos have between 40 and 60 s of duration and were exhibited on a 17″ screen (1,280 × 1,024, 60 Hz). The video stimuli were presented at 800 × 600 resolution, covering approximately two-thirds of the screen, to minimize eye movements. The sound was played through stereo speakers. The EEG data were recorded at 512 Hz sampling rate utilizing 32 active AgCl electrodes, attached according to the 10-20 International System. The frontal face video was recorded from 22 of the 32 subjects, which half were male and another half female. The participants aged between 19 and 37 (mean: 26.9). The experiment began with a baseline recording of 2 min, during which a fixation cross was presented to the subject in resting state condition. Then, 40 videos were exhibited in 40 trials, each one consisting of these steps:

1. The current trial number was displayed on the screen to notify the subjects about their progress (duration: 2 s).
2. Baseline was recorded (duration: 5 s).
3. Presentation of the music video (duration: 1 min).
4. Self-assessment for valence, arousal, preference, and dominance.

After 20 trials, the subjects got a small interval. The participants conducted a self-assessment of their levels of arousal, valence, preference, and dominance at the end of each trial. This dataset can be downloaded at [121].

5.5.11 Ullsperger dataset

This dataset comprises the EEG recordings from five subjects. The EEG signals were recorded in response to acoustic stimuli which the subjects need to classify them as either matching or non-matching referring to the meaning of a target word. A total of 61 electrodes with a sampling rate of 200 Hz were responsible for recordings. The signal was also notch filtered between 47 and 53 Hz to eliminate line interference. These recordings can be found at [122].

5.5.12 Mu dataset

In the study published by Mu *et al.* [34], the authors recorded EEG signals from ten healthy male Chinese subjects (ages: 19–40; mean 30.5). In the article, self- and nonself-pictures were utilized for visual stimulation. A black screen was exposed for 250 ms to notify the subjects about the start of the experiment. The subjects were instructed to pay attention to the display and to classify the stimuli into either "self" or "nonself". Each picture was randomly presented on the screen for 1 s, succeeded by a black screen with duration of 250 ms. All images have 400 × 400 pixels. The EEG data were recorded using Scan 4.3 software and acquired using Nuamps (Neuroscan) 40-channel EEG amplifier with a reference electrode installed on the right mastoid. The electrodes were positioned following the 10-20 International System. The data were sampled at 1,000 Hz and filtered using 0.05 Hz high-pass, 200 Hz low-pass, and 50 Hz notch filters.

5.5.13 Cho dataset

The dataset consists of recorded motor imagery (MI) BCI EEG and electromyography (EMG) datasets recorded from 52 healthy subjects while performing MI movement of the left and right hands [123]. EEG data were captured using Biosemi ActiveTwo device with 64 AgCl active electrodes positioned according to the 10-10 International System and recorded at 512 Hz sampling rate. They performed 20 trials of real hand movement of EEG and EMG for each subject. The non-task-related data included six types of noise data (head movement, eye blinking, eyeball movement left/right and up/down, jaw clenching, and resting state). For preprocessing, their used fourth-order Butterworth filter for high-pass and bandpass filtering. This dataset can be downloaded at [124].

5.5.14 PhysioUnicaDB dataset

The dataset [125] contains the records of EEG and ECG data acquired from 22 healthy subjects. During the recording, each subject was required to close their eyes, stay awake and to diminish eye movements. Signals were obtained simultaneously, using a 64 channel protocol, digitized with 1,024 Hz sampling frequency, and then resampled to 256 Hz and baseline trend removed.

5.5.15 Shin dataset

In the study [126], the authors collected EEG data from two BCI experiments (left vs. right hand MI and mental arithmetic vs. resting state). EEG was recorded employing a multichannel BrainAmp EEG amplifier (Brain Products GmbH, Gilching, Germany) with 30 electrodes and sampled at 1,000 Hz. At the same time, the EOG, ECG, respiration, and NIRS data were recorded. The experiment included three sessions of left- and right-hand MI (dataset A), mental arithmetics (dataset B) and baseline (rest state) tasks. The task began with 2 s of a visual introduction, followed by 10 s of a task and a resting period given from 15–17 s. Five types of motion artifacts provoked by eye and head movements (dataset C) were also recorded. Raw datasets are accessible for download at [127].

5.5.16 EEG devices

There are many EEG devices available in the market. Table 5.4 summarizes the most employed devices used in the surveyed papers for brain signals acquisition. However, there are other devices than those mentioned in this table. The devices have varying costs, ranging from US$ 100.00 to US$ 1,000.00 on average. The choice of the device is closely related to application needs. Studies using small number of channels showed the effectiveness of low-cost devices for biometric purposes [65,72].

Table 5.4 Overview of the main devices employed for EEG signals acquisition available in the market

Manufacturer	Product	General characteristics	Category
Compumedics Neuroscan	SynAmps RT	64 up to 512 monopolar channels; 4 per Headbox, 16 per system unit bipolar channels; 2 per Headbox, 8 per system high-level channels; bandwidth DC to 3,500 Hz; 20,000 Hz sampling rate	Amplifier
	Neuvo	64 up to 512 monopolar channels; 4 per Headbox, 16 per system unit bipolar channels, 2 per Headbox, 8 per system high-level channel; bandwidth DC to 3,500 Hz; 20,000 Hz sampling rate	Amplifier
	Grael	29, 39, 45, and 60 monopolar channels; 8 per system bipolar channel; bandwidth DC-1,150 Hz; 4,096 Hz sampling rate	Amplifier
	Siesta Wireless	32 monopolar channels; 1,024 Hz sampling rate	Amplifier
	CURRY 8x	Artifact removal, reduction methods (PCA, ICA, Covariance, and subtraction), suitable for BCI, compatible with EEG Lab	Tool
	Profusion EEG recording	Windows-based package, easily set-up and maintained	Tool
	Quik-Cap	32 to 256 channels, International 10-20 electrode placement, 4 integrated bipolar leads for vertical and horizontal EOG (VEOG, HEOG), ECG and EMG; diverse material composition according to model, e.g., Lycra, elastic neoprene, among others	Cap
Brain products GmbH	actiCHamp	32 to 160 channels + 8 AUX channels for extra physiological sensors; 100,000 Hz sampling rate; bandwidth DC up to 7,500 Hz; integration with MATLAB, EEGLAB, OpenVibe, and LSL	Amplifier
	BrainAmp	8 or 16 channels (8 bipolar or 8 bipolar + 8 AUX channels); fifth order Butterworth filter; 5,000 Hz sampling rate	Amplifier
	LiveAmp	8 to 64 channels; 1,000 Hz sampling rate; wireless communication;	Amplifier
	MOVE	32 or 64 analog channels; 250 Hz 3-pole Butterworth filter; 954 samples/s at 2 MBits/s or 508 samples/s at 1 MBits/s	Amplifier/transmitter
	V-Amp	16 EEG ground, reference, monopolar channels; 20,000 Hz sampling rate	Amplifier
	actiCap	8, 16 or 32 channels; dry electrodes	Cap
	EasyCap	32 or 64 channels; Ag/AgCl electrodes; spare electrodes for EOG, ECG recordings	Cap
	LiveCap	Up to 64 channels; Ag/AgCl electrodes;	Cap

(Continues)

Table 5.4 (Continued)

Manufacturer	Product	General characteristics	Category
	R-Net	32, 64, 96 or 128 channels; Ag/AgCl electrodes with saltwater interchangeable sponges; International 10-20 system	Cap
EMOTIV	EPOC+	14 channels; saline base electrodes; 2,048 internal sampling rate; wireless communication; fifth-order Sinc filter	Cap
	Insight	5 channels; semi-dry polymer sensor; wireless communication; fifth-order Sinc filter; 128 Hz sampling rate	Cap
	EPOC Flex	Up to 32 channels; saline or gel sensors; wireless communication	Cap
NeuroSky	MindWave Mobile 2	1 channel; 512 Hz sampling rate; Bluetooth communication	Cap
Neuracle	DSI-24	24 channels; dry electrodes; 300 or 600 Hz sampling rate; wireless communication	Cap
OpenBCI	EEG Electrode Cap	21 channels; Ag/AgCl coated or sintered electrodes;	Cap
Cognionics	Quick	8, 20, 30 channels; dry electrodes; 250/500/1,000/2,000 samples per second; Bluetooth communication	Cap
	Mobile	64, 128 channels; 1,000 samples/s at 64; Active Ag/AgCl electrodes	Cap

5.6 Biometric systems: general characteristics

The main purpose of applying biometrics to user authentication is to enable legitimate users to access data and detect impostors based on their physiological and behavioral characteristics.

With the advances of biometric systems, a large number of applications in different domains can take advantages of its benefits. These applications include surveillance [128], temperature monitoring [129], elderly care [130], health care [131,132], smart home and cities [133,134], safety [135], big data [136], and several others [137].

5.6.1 Performance metrics

Several performance metrics are commonly employed to evaluate EEG-based biometric systems. The choice of appropriate performance metrics is of paramount importance because it evaluates the reliability of the systems, especially in unbalanced databases [36]. This section lists some of the main metrics used in authentication/authorization systems.

- *Accuracy (Acc)*: corresponds to the probability of correctly identifying a subject, either an impostor or legitimate user.

$$Acc = \frac{\text{\# of correctly identified subjects}}{\text{total number of attempts}} \tag{5.1}$$

- *Sensitivity (Sen)*: corresponds to the percentage of legitimate subjects correctly classified as legitimate.

$$Sen = \frac{\#\text{ of true acceptances}}{\#\text{ of true acceptances} + \#\text{ of false rejections}} \tag{5.2}$$

- *Specificity (Spc)*: corresponds to the percentage of impostor subjects correctly classified as impostor.

$$Spc = \frac{\#\text{ of true rejections}}{\#\text{ of true rejections} + \#\text{ of false acceptances}} \tag{5.3}$$

- *False acceptance rate (FAR)*: corresponds to the probability of accepting an impostor.

$$FAR = \frac{\#\text{ of false acceptances}}{\text{total number of attempts}} \tag{5.4}$$

- *False recognition rate (FRR)*: corresponds to the probability of rejecting a legitimate user.

$$FRR = \frac{\#\text{ of false rejections}}{\text{total number of attempts}} \tag{5.5}$$

- *Correct recognition rate (CRR)*: corresponds to the probability of accepting a legitimate user.

$$CRR = \frac{\#\text{ of correct acceptances}}{\text{total number of attempts}} \tag{5.6}$$

- *Half total error rate (HTER)*: corresponds to the average of FRR and FAR.

$$HTER = \frac{\text{FRR} + \text{FAR}}{2} \tag{5.7}$$

- *AUC*: corresponds to the overall classification performance for legitimate subjects and impostors in terms of sensitivity and specificity. AUC indicates whether on average a true positive is ranked higher than a false positives

$$AUC = \int_{x=0}^{1} Sen(Spc^{-1}(x))dx \tag{5.8}$$

5.7 Requirements for security based on EEG authentication

Reducing data volume is crucial for system performance. Since EEG information occupies a lot of space, often in the range of gigabytes, strategies need to be adopted to provide effective storage of the data. The EEG lossy compression to EEG-based user authentication was analyzed in Nguyen *et al.* [70] where various experiments using an SPIHT compression algorithm [138] were performed. In the study performed by Wiechert *et al.* [139], the identification of the subjects was obtained using a headband called Muse [140]. The recordings were made using four EEG signals and a compact representation of the input streams from multiple channels was obtained to stream the data over Bluetooth. These signals were compressed via Golomb Encoding and further quantized to decrease its volume. Specific devices can be designed to obtain, process, and compact the EEG signals. Then, the authentication of the users can be performed using a dedicated module to increase performance. The effects of EEG signals compression techniques can be found at [141,142]. Besides that, the compression also reduce the system power consumption.

Storage location of EEG data is important for the security system design. The device that will store the EEG information needs to be installed in a safe place. External agents may compromise the security of the system if the devices are damaged or even if the EEG data is exposed. In addition to security issues, factors such as system performance related to communication also depend on where nodes are deployed in the network. Thus, the storage of the data must be installed in a way that allows fast user authorization, respecting the network topology and constraints, in case of remote access.

Acquisition of the signals also needs to be done efficiently and easily, as end users do not have technical knowledge about the system. The EEG recordings using wet electrodes are not feasible in this context. However, dry electrodes design would allow EEG to be recorded using a simple cap/hat. Wearable EEG devices are suitable in these platforms because they are versatile to use. The communication with the device designed to process the EEG signals and perform the user authentication must be efficient. Besides that, the acquisition process should not be too long or require a complex preparation. Attempts to access the system by unauthorized users who can circumvent the systems need to be logged.

Encryption mechanisms of EEG data are required to ensure the confidentiality and authenticity of brain signals, in addition to the transmitted data. The data stored on a device, whether the EEG signals or application data, are vulnerable to attacks, compromising the entire system. Lightweight cryptographic algorithms are feasible as they can encrypt stored EEG signals without degrading system performance [143,144]. Obviously, decryption process to obtain the original data would not be a complex task or impose overhead to the system.

Multi-user scenario would be allowed in an EEG-based authentication system. If multiple people of the same family use this authentication system, the platform should allow the use of the devices existing in the network. Different levels of access should also be assigned to each user. In a hypothetical case, children should not have access to home buying systems, for example. After performing authorization, a complete access to the system must be provided to the administrator(s). Furthermore, the fast processing of the EEG signals is imperative for authentication/identification tasks.

Fast-data analytics is primordial in authentication tasks. The manner that systems will response the huge amount of the data generated in the infrastructure will grant the performance and the feasibility of the system. With the use of EEG signals to perform the authentication, there is another concern to researchers related to brain signals processing. The responsible nodes need to collect the EEG signals, start the communication and send the data to authorize the operation in the environment according to the guidelines formulated. The additional overhead generated by the security using the EEG signals in the network needs to be taken into account of the infrastructure in general.

5.7.1 Advantages and disadvantages of EEG biometrics

Applications can use this modality of authentication/identification in a descentralized environment. This modality will able to increase the performance of these tasks, once specific devices and even mobile phones would be able to process the EEG signals for this purpose. Thus, the use of brain signals can be used in domestic fields applications, laboratories, companies and control of machines or brain-to-brain communication.

Some acquisition paradigms are not suitable in the environments in which security must be implemented. It is an unpractical situation for the user to wait for many hours while the brain signals are obtained, processed, and authentication happens. Coupled with this need, the constraints of biometric systems also play a key role in the design of authentication systems based on EEG. The system needs to manage resources, events, codes, communication, energy, and others factors. Besides that, characteristics such as scalability, continuously updates, failure robustness, privacy, automatic interaction with other devices, and high compatibility level must be provided.

The use of EEG signals for authentication requires acquisition of measurable brain characteristics capable of satisfying the requirements of circumvention, collectability, distinctiveness, permanence, and universality [28,145].

Circumvention – so far, there is no information publicly available on any method or technique that would allow the reproduction of brain waves. Recording of EEG requires cooperation of a subject, while any coercion would induce stress that lead to the generation of brain waves differing from the training patterns.

Collectability – EEG recording using medical grade equipment with wet electrodes is inconvenient. The use of dry electrodes is more convenient, without significant reduction of performance [19,51,88,94].

Distinctiveness – personal authentication has been shown to be highly accurate within small groups of subjects, however, large population studies has yet to be performed.

Permanence – the EEG biometric must remain stable over time. However, the stability of EEG-based authentication depends on age, sex, and pathology [146], which underscores the need for increasing the robustness EEG-based security systems.

Universality – the brain of each human produces electrical EEG which can be acquired without any age or health limitations.

5.7.2 Feasibility of EEG signals for security – perspectives

Although some acquisition paradigms cannot be used for authentication/identification of users, the main idea behind these protocols can be modified for these purposes. The signals filtering can be employed in this scenario and the machine-learning or deep learning techniques are already being used efficiently in this domain for data mining and pattern recognition [147].

BCI, sleep disorder diagnosis, epilepsy diagnosis, and other medical application are the main beneficiaries of the EEG wearable systems. Due to this fact, the EEG sensors must be free of wires, besides that, these devices needs to be comfortable, small, discrete, and offering lower power consumption, what is essential in biometric applications. Subdermal electrodes was already used in some studies [148].

Bluetooth and ZigBee communication protocols can be able to transmit EEG signals via wireless. However, these technologies may be interfere to other equipment. In this aspect, Visible Light Communications (VLC) can be able to transmit these signals free from electromagnetic interference with improved security [149].

Since neurologists recognize the need to use wearable EEG recorders, we have a trade-off between signal quality acquisition for real life use cases at the same time that the users need to have comfort and convenience for the biometric use. Besides that, technologies such as MEGs, are not suitable for daily life activities, so EEG is the best choice to be used in the infrastructure. Gtec [150], Mindo [151], Imec [152], and Quasar [153], for example, are good solutions already available and are showing great evolution potentials every year. Notes such as these are already being cited in articles about smart environments for security [154].

Achieving stability and high recognition accuracy requires that the EEG features used for authentication present high stability and repeatability over time and under different external conditions [145]. EEG-based authentication must be flexible enough to allow authentication in case of mental and physiological changes induced, for example, by fatigue, lack of sleep, daily stress, and other negative factors of our daily life. Such flexibility, however, should not interfere with the circumvention requirement so that a subject under coercion is prevented by the system from authentication.

5.8 Discussion, open issues, and directions for future works

A great variability of methodologies related to signals acquisition can be found in the literature. The number and the placement of electrodes can influence the performance of authentication/identification systems, however, outstanding results were also obtained by using just a few channels or a single-channel, as demonstrated in [155–157], where equal error rate (EER) of 11% and CRR ranging from 70% to 87% were obtained for the first and second work, respectively. A variety of stimuli have also been found in the literature. Different authors try to find stimuli that are more appropriate for the classification task, using the developed methodologies. The most commonly used stimuli are: EEG data recorded from the subjects in rec or rec condition, ERPs in EEG signals evoked by specific cognitive tasks, such as VEP in response to visual stimuli, multi-task approach where the subjects performs mathematical operations, mental letters composition, geometric figure rotation, and imagery movements. Most of these works use features extracted from the P300 component [158]. It was possible to observe different sampling frequency rates and recording sessions. All these factors have a direct impact on the volume of data and also influence the performance of the system.

There are five major bands in an EEG signal [93]. The delta band (0–4 Hz) is related to the deep sleep or when the person is unconscious. The theta band (4–8 Hz) that is related during states of quite focus. The alpha band (8–13 Hz) that is found over the parietal-occipital region and indicates relaxed awareness without attention. The beta band (13–30 Hz) is associated with states of increased focused attention and alertness. The gamma band (over 30 Hz) is observed during active information processing [28,90,103]. It has also been observed that there is no specific methodology for the use of a single band. For example, in work proposed by Thomas and Vinod [159], the alpha frequency and delta band power were used in the EEG-based identification system. In work proposed by Chuang *et al.* [160], the alpha and beta frequencies were employed for analysis. Fraschini *et al.* [47], in turn, used the features extracted from the gamma band. However, most studies that employ the resting state paradigm use alpha waves [45]. Therefore, the choice of the frequency band is intrinsically related to the acquisition paradigm, resting state or stimuli response, as well as the set of extracted features, e.g., PSD, wavelet coefficients, among others.

Others means of authentication such as passwords or even biometric approaches are deficient as shown in the Introduction. If a user needs to access a sensitive data, the system must recognize that user as the system administrator and make the data available. Again, n biometric security system based on EEG authentication provides an effective solution to overcome this weakness. Besides that, in a more general context, platforms need to implement different levels of security to allow a user or group of users to have appropriate system access. A solution to this problem may be using multimodal biometrics, which combines EEG authentication with other biometric modalities such as ECG, facial recognition and fingerprinting, to improve accuracy and stability [161–163].

Some practical applications can be cited in a more specific scenario, e.g., smart homes can enjoy the benefits of the EEG-based authentication systems to guarantee the entry into homes, turn on the lights and TV, or perform tasks which require more safety, such as open a safe, for example. Besides that, logging into bank accounts and identifying personal preferences are possible applications which can use EEG signals. The EEG-based systems also can be used for implementing security in which the users have some kind of physical impairment, such as amyotrophic lateral sclerosis, tetra- and paraplegia and vegetative state, if the brain is in normal activity.

5.9 Learned lessons and conclusions

This chapter presented a detailed review about EEG-based authentication/identification systems. In comparison to other neuroimaging techniques, the EEG data bring numerous advantages, such as relatively simple and inexpensive equipment for signals acquisition and the possibility to be used in most environments. A large number of studies found in the state-of-the-art were approached and analyzed in this work. More than 190 papers were analyzed in detail in order to present the main methodologies used by the authors over the last few years. The main datasets and their characteristics were also exposed, such as the number of electrodes, sampling rate, number of subjects, and filtering techniques. The feature extraction and classification techniques of the related works demonstrated high accuracy rates at the same time that the number of the channels is getting reduced and the signals acquisition process is getting less complicated.

The advantages brought by the biometric systems based on brain waves, such as universality, uniqueness, collectability, and permanence, make these systems well suitable in security environments. As shown in this survey, it was concluded that EEG signals are feasible for authentication and identification purposes, since these signals carry unique information for each person. These methods can be adapted considering the requirements exposed in the Section 5.5 and implemented in the most diverse applications, providing or increasing the security in such systems, as learned. Thus, as a result of the advances made in the processing of EEG signals, simultaneously with the evolution acquisition devices, these two technologies will be able to coalesce for a robust and highly reliable authentication.

References

[1] Focus on human brain mapping. Nature Neuroscience. 2017;20:297.

[2] Crosson B, Ford A, McGregor KM, *et al.* Functional imaging and related techniques: an introduction for rehabilitation researchers. Journal of Rehabilitation Research and Development. 2010;47(2):7–34.

[3] Xue G, Chen C, Lu ZL, *et al.* Brain imaging techniques and their applications in decision-making research. Acta Psychologica Sinica. 2010;42 (1):120–137.

[4] Wolpaw JR, Birbaumer N, McFarland DJ, *et al.* Brain-computer interfaces for communication and control. Clinical Neurophysiology. 2018;113 (6):767–791.

[5] Nunes TM, André L V, Lima CAM, *et al.* EEG signal classification for epilepsy diagnosis via optimum path forest - a systematic assessment. Neurocomputing. 2014;136:103–123.

[6] Pereira LAM, Papa JP, Coelho AV, *et al.* Automatic identification of epileptic EEG signals through binary magnetic optimization algorithms. Neural Computing and Applications. 2019;31:1317–1329.

[7] Stoessl AJ. Neuroimaging in the early diagnosis of neurodegenerative disease. Translational Neurodegeneration. 2012;1:5.

[8] Pinheiro PR, Tamanini I, Pinheiro MCD, *et al.* Evaluation of the Alzheimer's disease clinical stages under the optics of hybrid approaches in verbal decision analysis. Telematics and Informatics. 2018;35(4):776–789.

[9] Gupta D, Sundaram S, Ashish Khanna, *et al.* Improved diagnosis of Parkinson's disease using optimized crow search algorithm. Computers & Electrical Engineering. 2018;68:412–424.

[10] Pereira CR, Pereira DR, Rosa GH, *et al.* Handwritten dynamics assessment through convolutional neural networks: an application to Parkinson's disease identification. Artificial Intelligence in Medicine. 2018

[11] Light GA, Williams LE, Minow F, *et al.* Electroencephalography (EEG) and event-related potentials (ERP's) with human participants. Current Protocols in Neuroscience. 2010;6:1–24.

[12] Oostenveld R, and Praamstra P. The five percent electrode system for high-resolution EEG and ERP measurements. Clinical Neurophysiology. 2001;112(4):713–719.

[13] Patel V, Burns M, Chandramouli R, *et al.* Biometrics based on hand synergies and their neural representations. IEEE Access. 2017;5: 13422–13429.

[14] Kaur B, and Singh D. Neuro signals: a future biometric approach towards user identification. In: 2017 7th Int. Conference on Cloud Computing, Data Science Engineering - Confluence; 2017. p. 112–117.

[15] Nieves O, and Manian V. Automatic person authentication using fewer channel EEG motor imagery. In: 2016 World automation Congress (WaC); 2016. p. 1–6.

[16] Campisi P, and Rocca DL. In: Li SZ, Jain AK, editors. EEG Biometrics. Boston, MA: Springer US; 2009. p. 1–11.

[17] Khanom S, and Islam MR. Performance analysis of QR-decomposition RLS and householder sliding window RLS for noise elimination of EEG. In: 2017 IEEE Region 10 Humanitarian Technology Conference (R10-HTC); 2017. p. 594–597.

[18] Li J, Li C, Thakor N, *et al.* a new perspective of noise removal from EEG. In: 2017 8th Int. IEEE/EMBS Conference on Neural Engineering (NER); 2017. p. 501–504.

[19] Srinivasa MG, and Pandian PS. Dry electrodes for bio-potential measurement in wearable systems. In: 2017 2nd IEEE Int. Conference on Recent Trends in Electronics, Information Communication Technology (RTEICT); 2017. p. 270–276.

[20] Gui Q, Jin Z, and Xu W. Exploring EEG-based biometrics for user identification and authentication. In: 2014 IEEE Signal Processing in Medicine and Biology Symposium (SPMB); 2014. p. 1–6.

[21] Shin Y, Lee J, and Kim S. Validity of biosignal processing system based on Haar transform in IoT application. In: 2017 IEEE 17th Int. Conference on Bioinformatics and Bioengineering (BIBE); 2017. p. 206–211.

[22] Ahn JW, Ku Y, Kim DY, *et al.* Wearable in-the-ear EEG system for SSVEP-based brain computer interface. Electronics Letters. 2018;54(7):413–414.

[23] Yao K, Lan S, Xu C, *et al.* A wearable EEG-based control network and emergency medical assistance system. In: 2017 Int. Symposium on antennas and Propagation (ISaP); 2017. p. 1–2.

[24] Chaitanya MN, Jayakkumar S, Chong E, *et al.* A wearable, EEG-based massage headband for anxiety alleviation. In: 2017 39th annual Int.

Conference of the IEEE Engineering in Medicine and Biology Society (EMBC); 2017. p. 3557–3560.
[25] Lin CT, Chuang CH, Huang CS, *et al.* Wireless and wearable EEG system for evaluating driver vigilance. IEEE Transactions on Biomedical Circuits and Systems. 2014;8(2):165–176.
[26] Chen W, Damalerio R, Lim R, *et al.* Simulation analysis of a wearable dry EEG electrodes for epilepsy monitoring. In: 2017 IEEE 19th Electronics Packaging Technology Conference (EPTC); 2017. p. 1–4.
[27] Abdulkader SN, Atia A, and Mostafa M-SM. Brain computer interfacing: applications and challenges. Egyptian Informatics Journal. 2015;16(2): 213–230.
[28] Maiorana E, Rocca DL, and Campisi P. On the permanence of EEG signals for biometric recognition. IEEE Transactions on Information Forensics and Security. 2016;11(1):163–175.
[29] Kumari P, and Vaish A. Brainwave based user identification system: a pilot study in robotics environment. Robotics and Autonomous Systems. 2015;65:15–23.
[30] Kumari P, and Vaish A. Information-theoretic measures on intrinsic mode function for the individual identification using EEG sensors. IEEE Sensors Journal. 2015;15(9):4950–4960.
[31] Gui Q, Jin Z, Xu W, *et al.* Multichannel EEG-based biometric using improved RBF neural networks. In: 2015 IEEE Signal Processing in Medicine and Biology Symposium (SPMB); 2015. p. 1–6.
[32] Ma L, Minett JW, Blu T, *et al.* Resting state EEG-based biometrics for individual identification using convolutional neural networks. In: 2015 37th annual Int. Conference of the IEEE Engineering in Medicine and Biology Society (EMBC); 2015. p. 2848–2851.
[33] Berg Palm R. Prediction as a candidate for learning deep hierarchical models of data. 2012 01.
[34] Mu Z, Hu J, and Min J. EEG-based person authentication using a fuzzy entropy-related approach with two electrodes. Entropy. 2016;18(12):432.
[35] Das R, Piciucco E, Maiorana E, *et al.* Visually evoked potentials for EEG biometric recognition. In: 2016 First Int. Workshop on Sensing, Processing and Learning for Intelligent Machines (SPLINE); 2016. p. 1–5.
[36] Liew SH, Choo YH, Yusoh ZIM, *et al.* Incrementing FRNN model with simple heuristic update for brainwaves person authentication. In: 2016 IEEE EMBS Conference on Biomedical Engineering and Sciences (IECBES); 2016. p. 115–120.
[37] Hasan MM, Sohag MHA, and Ahmad M. EEG biometrics based on small intra-individual and large inter-individual difference of extracted features. In: 2016 2nd Int. Conference on Electrical, Computer Telecommunication Engineering (ICECTE); 2016. p. 1–4.
[38] Tangkraingkij P, Montaphan A, and Nakavisute I. an appropriate number of neurons in a hidden layer for personal authentication using delta brainwave

signals. In: 2017 2nd Int. Conference on Control and Robotics Engineering (ICCRE); 2017. p. 232–236.

[39] Mao Z, Yao WX, and Huang Y. EEG-based biometric identification with deep learning. In: 2017 8th Int. IEEE/EMBS Conference on Neural Engineering (NER); 2017. p. 609–612.

[40] Bigdely-Shamlo N, Mullen T, Kothe C, *et al.* The PREP pipeline: standardized preprocessing for large-scale EEG analysis. Frontiers in Neuroinformatics. 2015;9:16.

[41] Armstrong BC, Ruiz-Blondet MV, Khalifian N, *et al.* Brainprint: assessing the uniqueness, collectability, and permanence of a novel method for ERP biometrics. Neurocomputing. 2015;166:59–67.

[42] Yu YC, Wang S, and Gabel LA. A feasibility study of using event-related potential as a biometrics. In: 2016 38th annual Int. Conference of the IEEE Engineering in Medicine and Biology Society (EMBC); 2016. p. 4547–4550.

[43] Wang M, Abbass HA, and Hu J. Continuous authentication using EEG and face images for trusted autonomous systems. In: 2016 14th annual Conference on Privacy, Security and Trust (PST); 2016. p. 368–375.

[44] Harshit RS, Thomas KP, Smitha KG, *et al.* Online Electroencephalogram (EEG) based biometric authentication using visual and audio stimuli. In: 2016 IEEE EMBS Conference on Biomedical Engineering and Sciences (IECBES); 2016. p. 454–459.

[45] Thomas KP, and Vinod AP. EEG-based biometric authentication using gamma band power during rest state. Circuits, Systems, and Signal Processing. 2017;37(7):277–289.

[46] Al Hudhud G, Alarfag E, Alkahtani S, *et al.* Web-based multimodal biometric authentication application. In: 2015 5th National Symposium on Information Technology: Towards New Smart World (NSITNSW); 2015. p. 1–6.

[47] Fraschini M, Hillebrand A, Demuru M, *et al.* An EEG-based biometric system using eigenvector centrality in resting state brain networks. IEEE Signal Processing Letters. 2015;22(6):666–670.

[48] Maiorana E, Rocca DL, and Campisi P. Cognitive biometric cryptosystems a case study on EEG. In: 2015 Int. Conference on Systems, Signals and Image Processing (IWSSIP); 2015. p. 125–128.

[49] Maiorana E, Blasi D, and Campisi P. Biometric template protection using turbo codes and modulation constellations. In: 2012 IEEE Int. Workshop on Information Forensics and Security (WIFS); 2012. p. 25–30.

[50] Maiorana E, Rocca DL, and Campisi P. EEG-based biometric recognition using EigenBrains. In: 2015 IEEE Int. Conference on Multimedia Expo Workshops (ICMEW); 2015. p. 1–6.

[51] Lee C, Kang JH, and Kim SP. Feature selection using mutual information for EEG-based biometrics. In: 2016 39th Int. Conference on Telecommunications and Signal Processing (TSP); 2016. p. 673–676.

[52] Bansod NS, Dabhade SB, Kazi MM, *et al.* Single electrode brain signal data fusion for security. In: 2016 Int. Conference on Global Trends in Signal

Processing, Information Computing and Communication (ICGTSPICC); 2016. p. 108–112.

[53] Garau M, Fraschini M, Didaci L, *et al.* Experimental results on multi-modal fusion of EEG-based personal verification algorithms. In: 2016 Int. Conference on Biometrics (ICB); 2016. p. 1–6.

[54] Kang JH, Jo YC, and Kim SP. Electroencephalographic feature evaluation for improving personal authentication performance. Neurocomputing. 2018;287:93–101.

[55] Yeom SK, Suk HI, and Lee SW. Person authentication from neural activity of face-specific visual self-representation. Pattern Recognition. 2013;46(4):1159–1169.

[56] Nguyen P, Tran D, Huang X, *et al.* Motor Imagery EEG-Based Person Verification. In: Rojas I, Joya G, Cabestany J, editors. Advances in Computational Intelligence. Berlin, Heidelberg: Springer Berlin Heidelberg; 2013. p. 430–438.

[57] Liu S, Bai Y, Liu J, *et al.* Individual Feature Extraction and Identification on EEG Signals in Relax and Visual Evoked Tasks. In: Biomedical Informatics and Technology. Berlin, Heidelberg: Springer Berlin Heidelberg; 2014. p. 305–318.

[58] Snodgrass GJ, and Vanderwart M. A standardized set of 260 pictures: norms for name agreement, image agreement, familiarity, and visual complexity. Journal of Experimental Psychology: Human Learning and Memory. 1980;6:174–215.

[59] Pham T, Ma W, Tran D, *et al.* Multi-factor EEG-based user authentication. In: 2014 Int. Joint Conference on Neural Networks (IJCNN); 2014. p. 4029–4034.

[60] Le T, Tran D, Ma W, *et al.* An optimal sphere and two large margins approach for novelty detection. In: The 2010 Int. Joint Conference on Neural Networks (IJCNN); 2010. p. 1–6.

[61] Pham T, Ma W, Tran D, *et al.* A study on the stability of EEG signals for user authentication. In: 2015 7th Int. IEEE/EMBS Conference on Neural Engineering (NER); 2015. p. 122–125.

[62] Phung D, Tran D, Ma W, *et al.* Investigating the impacts of epilepsy on EEG-based person identification systems. In: 2014 Int. Joint Conference on Neural Networks (IJCNN); 2014. p. 3644–3648.

[63] Davis P, Creusere CD, and Kroger J. Classification of human viewers using high-resolution EEG with SVM. In: 2014 48th asilomar Conference on Signals, Systems and Computers; 2014. p. 184–188.

[64] Pinson MH. The consumer digital video library [Best of the Web]. IEEE Signal Processing Magazine. 2013;30:172–174.

[65] Bashar MK, Chiaki I, and Yoshida H. Human identification from brain EEG signals using advanced machine learning method EEG-based biometrics. In: 2016 IEEE EMBS Conference on Biomedical Engineering and Sciences (IECBES); 2016. p. 475–479.

[66] Keshishzadeh S, Fallah A, and Rashidi S. Improved EEG based human authentication system on large dataset. In: 2016 24th Iranian Conference on Electrical Engineering (ICEE); 2016. p. 1165–1169.
[67] Rashidi S, Fallah A, and Towhidkhah F. Similarity evaluation of online signatures based on modified dynamic time warping. Applied Artificial Intelligence. 2013;27(7):599–617.
[68] Vahid A, and Arbabi E. Human identification with EEG signals in different emotional states. In: 2016 23rd Iranian Conference on Biomedical Engineering and 2016 1st Int. Iranian Conference on Biomedical Engineering (ICBME); 2016. p. 242–246.
[69] Kang JH, Lee CH, and Kim SP. EEG feature selection and the use of Lyapunov exponents for EEG-based biometrics. In: 2016 IEEE-EMBS Int. Conference on Biomedical and Health Informatics (BHI); 2016. p. 228–231.
[70] Nguyen B, Nguyen D, Ma W, *et al.* Investigating the possibility of applying EEG lossy compression to EEG-based user authentication. In: 2017 Int. Joint Conference on Neural Networks (IJCNN); 2017. p. 79–85.
[71] Khalil R, Arasteh A, and Sarkar AK. EEG based biometrics using emotional stimulation data. In: 2017 IEEE Region 10 Humanitarian Technology Conference (R10-HTC); 2017. p. 246–249.
[72] Nakamura T, Goverdovsky V, and Mandic DP. In-ear EEG biometrics for feasible and readily collectable real-world person authentication. IEEE Transactions on Information Forensics and Security. 2018;13(3):648–661.
[73] DelPozo-Banos M, Travieso CM, Weidemann CT, *et al.* EEG biometric identification: a thorough exploration of the time-frequency domain. Journal of Neural Engineering. 2015;12(5):056019.
[74] Yang S, and Deravi F. Novel HHT-Based features for biometric identification using EEG signals. In: 2014 22nd Int. Conference on Pattern Recognition; 2014. p. 1922–1927.
[75] Huang NE, Shen Z, Long SR, *et al.* The empirical mode decomposition and the Hilbert spectrum for nonlinear and non-stationary time series analysis. Proceedings of the Royal Society of London a: Mathematical, Physical and Engineering Sciences. 1998;454(1971):903–995.
[76] Wang Y, and Najafizadeh L. On the invariance of EEG-based signatures of individuality with application in biometric identification. In: 2016 38th annual Int. Conference of the IEEE Engineering in Medicine and Biology Society (EMBC); 2016. p. 4559–4562.
[77] Rahman MW, and Gavrilova M. Overt mental stimuli of brain signal for person identification. In: 2016 Int. Conference on Cyberworlds (CW); 2016. p. 197–203.
[78] Wang J, and Su F. A new time-frequency method for EEG artifacts removing. In: 2014 IEEE 3rd Int. Conference on Cloud Computing and Intelligence Systems; 2014. p. 341–346.
[79] Ward CR, and Obeid I. Robust EEG. Separability of subject specific records via clustering and data driven metrics. In: 2015 41st annual Northeast Biomedical Engineering Conference (NEBEC); 2015. p. 1–2.

[80] Altahat S, Wagner M, and Marroquin EM. Robust electroencephalogram channel set for person authentication. In: 2015 IEEE Int. Conference on acoustics, Speech and Signal Processing (ICaSSP); 2015. p. 997–1001.

[81] Rocca DL, Campisi P, and Solé-Casals J. EEG based user recognition using BUMP modelling. In: 2013 Int. Conference of the BIOSIG Special Interest Group (BIOSIG); 2013. p. 1–12.

[82] La Rocca D, Campisi P, and Scarano G, Stable EEG. Features for Biometric Recognition in Resting State Conditions. In: Biomedical Engineering Systems and Technologies. Berlin, Heidelberg: Springer Berlin Heidelberg; 2014. p. 313–330.

[83] DelPozo-Banos M, Travieso CM, Ticay-Rivas JR, *et al.* Real cepstrums on electroencephalogram biometric identification. In: 2015 Eighth Int. Conference on Contemporary Computing (IC3); 2015. p. 54–58.

[84] Mao C, Hu B, Wang M, *et al.* EEG-based biometric identification using local probability centers. In: 2015 Int. Joint Conference on Neural Networks (IJCNN); 2015. p. 1–8.

[85] Bozdogan H. Model selection and akaike's Information Criterion (aIC): The general theory and its analytical extensions. Psychometrika. 1987;52(3): 345–370.

[86] Li B, Chen YW, and Chen YQ. The nearest neighbor algorithm of local probability centers. IEEE Transactions on Systems, Man, and Cybernetics, Part B (Cybernetics). 2008;38(1):141–154.

[87] Yang S, Deravi F, and Hoque S. Task sensitivity in EEG biometric recognition. Pattern Analysis and Applications. 2016. Available from: https://doi.org/10.1007/s10044-016-0569-4.

[88] Chen Y, Atnafu AD, Schlattner I, *et al.* A high-security EEG-based login system with RSVP stimuli and dry electrodes. IEEE Transactions on Information Forensics and Security. 2016;11(12):2635–2647.

[89] Yu M, Kaongoen N, and Jo S. P300-BCI-based authentication system. In: 2016 4th Int. Winter Conference on Brain-Computer Interface (BCI); 2016. p. 1–4.

[90] Jayarathne I, Cohen M, and Amarakeerthi S. BrainID: development of an EEG-based biometric authentication system. In: 2016 IEEE 7th annual Information Technology, Electronics and Mobile Communication Conference (IEMCON); 2016. p. 1–6.

[91] Koike-Akino T, Mahajan R, Marks TK, *et al.* High-accuracy user identification using EEG biometrics. In: 2016 38th annual Int. Conference of the IEEE Engineering in Medicine and Biology Society (EMBC); 2016. p. 854–858.

[92] Bai Y, Zhang Z, and Ming D. Feature selection and channel optimization for biometric identification based on visual evoked potentials. In: 2014 19th Int. Conference on Digital Signal Processing; 2014. p. 772–776.

[93] Sundararajan A, Pons A, and Sarwat AI. A generic framework for EEG-based biometric authentication. In: 2015 12th Int. Conference on Information Technology - New Generations; 2015. p. 139–144.

[94] Soni YS, Somani SB, and Shete VV. Biometric user authentication using brain waves. In: 2016 Int. Conference on Inventive Computation Technologies (ICICT). vol. 2; 2016. p. 1–6.

[95] Gomes R, Ahsan M, and Denton A. Random forest classifier in SDN framework for user-based indoor localization. In: 2018 IEEE International Conference on Electro/Information Technology (EIT); 2018. p. 0537–0542.

[96] Hoseinitabatabaei SA, Fathy Y, Barnaghi P, *et al.* A novel indexing method for scalable IoT source lookup. IEEE Internet of Things Journal. 2018;5(3): 2037–2054.

[97] Wei J, Chen T, Li C, *et al.* Eyes-open and eyes-closed resting states with opposite brain activity in sensorimotor and occipital regions: multi-dimensional evidences from machine learning perspective. Frontiers in Human Neuroscience. 2018;12:422.

[98] Wolpaw J, and Wolpaw EW. *Brain-Computer Interfaces: Principles and Practice*. Oxford University Press, USA; 2012. Available from: https://books.google.com.br/books?id=tC2UzuC_WBQC.

[99] Gupta C, Palaniappan R, and Paramesran R. Exploiting the P300 paradigm for cognitive biometrics. International Journal of Cognitive Biometrics. 2012;1:26–38.

[100] Padfield N, Zabalza J, Zhao H, *et al.* EEG-based brain-computer interfaces using motor-imagery: techniques and challenges. Sensors. 2019;19 (6):1423.

[101] Mohammad S, Obaidat IW, and Traore I. Biometric-Based Physical and Cybersecurity Systems. 1st ed. Springer International Publishing; 2019.

[102] Damaševičius R, Maskeliūnas R, Kazanavičius E, *et al.* Combining cryptography with EEG biometrics. Computational Intelligence and Neuroscience. 2018;2018(1867548):11.

[103] Abo Zahhad M, Ahmed SM, and Abbas SN. State-of-the-art methods and future perspectives for personal recognition based on electroencephalogram signals. IET Biometrics. 2015;4(3):179–190.

[104] Pozo-Banos MD, Alonso JB, Ticay-Rivas JR, *et al.* Electroencephalogram subject identification: a review. Expert Systems with Applications. 2014;41 (15):6537–6554.

[105] Zhang XL, Begleiter H, Porjesz B, *et al.* Event related potentials during object recognition tasks. Brain Research Bulletin. 1995;38(6):531–538.

[106] Zhang XL, Begleiter H, Porjesz B, *et al.* Electrophysiological evidence of memory impairment in alcoholic patients. Biological Psychiatry. 1997;42 (12):1157–1171.

[107] UCI EEG Dataset. Accessed: 2019-01-14. https://archive.ics.uci.edu/ml/datasets/eeg+database.

[108] BCI Competition. Accessed: 2019-01-14. http://www.bbci.de/competition/.

[109] Leeb R, Lee F, Keinrath C, *et al.* Correction to "brain–computer communication: motivation, aim, and impact of exploring a virtual apartment".

IEEE Transactions on Neural Systems and Rehabilitation Engineering. 2008;16(1):119.
[110] aED Dataset. Accessed: 2019-01-14. https://demo.ands.org.au/australian-eeg-database-database-aed/69936.
[111] Poulos M, Rangoussi M, Chrissikopoulos V, *et al.* Person identification based on parametric processing of the EEG. In: Electronics, Circuits and Systems, 1999. Proc. of ICECS '99. The 6th IEEE Int. Conference on. vol. 1; 1999. p. 283–286 vol.1.
[112] Poulos M, Rangoussi M, Chrissikopoulos V, *et al.* Parametric person identification from the EEG using computational geometry. In: Electronics, Circuits and Systems, 1999. Proc. of ICECS '99. The 6th IEEE Int. Conference on. vol. 2; 1999. p. 1005–1008 vol.2.
[113] Poulos M, Rangoussi M, and Alexandris N. Neural network based person identification using EEG features. In: 1999 IEEE Int. Conference on acoustics, Speech, and Signal Processing. Proc.. ICaSSP99 (Cat. No.99CH36258). vol. 2; 1999. p. 1117–1120 vol.2.
[114] Keirn ZA, and Aunon JI. A new mode of communication between man and his surroundings. IEEE Transactions on Biomedical Engineering. 1990;37 (12):1209–1214.
[115] Keirn Dataset. Accessed: 2019-01-14. http://www.cs.colostate.edu/eeg/main/data/1989_Keirn_and_aunon.
[116] BCI CSU. Accessed: 2019-01-14. http://www.cs.colostate.edu/eeg/main/data/2011-12_BCI_at_CSU.
[117] Goldberger AL, Amaral LAN, Glass L, *et al.* PhysioBank, PhysioToolkit, and PhysioNet. Circulation. 2000;101(23):e215–e220.
[118] Schalk G, McFarland DJ, Hinterberger T, *et al.* BCI2000: A general-purpose brain-computer interface (BCI) system. IEEE Transactions on Biomedical Engineering. 2004;51(6):1034–1043.
[119] EEG Motor Movement/Imagery Dataset. Accessed: 2019-01-14. https://www.physionet.org/pn4/eegmmidb/.
[120] Koelstra S, Muhl C, Soleymani M, *et al.* DEaP: a database for emotion analysis using physiological signals. IEEE Transactions on Affective Computing. 2012;3(1):18–31.
[121] Database for Emotion analysis using Physiological Signals (DEaP) Dataset. Accessed: 2019-01-14. http://www.eecs.qmul.ac.uk/mmv/ datasets/deap/.
[122] Ullsperger Dataset. Accessed: 2019-01-14. https://sccn.ucsd.edu/wiki/Chapter_02:_STUDY_Creation.
[123] Cho H, Ahn M, Ahn S, *et al.* EEG datasets for motor imagery brain–computer interface. Giga Science. 2017;6(7):1–8.
[124] Cho H, Ahn M, Ahn S, *et al.* Supporting data for "EEG datasets for motor imagery brain computer interface". Giga Science Database; 2017. Available from: https://doi.org/10.5524/100295.
[125] Barra S, Fraschini M, Casanova A, *et al.* PhysioUnicaDB: a dataset of EEG and ECG simultaneously acquired. Pattern Recognition Letters. 2017. Available from: https://doi.org/10.1016/j.patrec.2017.12.017.

[126] Shin J, von Luhmann Alexander, Blankertz B, *et al.* Open access dataset for EEG+NIRS single-trial classification. IEEE Transactions on Neural Systems and Rehabilitation Engineering. 2017;25(10):1735–1745.

[127] Open access dataset for simultaneous EEG and NIRS brain-computer interface (BCI). Accessed: 2019-01-14. http://doc.ml.tu-berlin.de/hBCI.

[128] Hu L, and Ni Q. IoT-driven automated object detection algorithm for urban surveillance systems in smart cities. IEEE Internet of Things Journal. 2017;99:1–1.

[129] Ramesh V, Sankaramahalingam M, Bharathy MSD, *et al.* Remote temperature monitoring and control using IoT. In: 2017 Int. Conference on Computing Methodologies and Communication (ICCMC); 2017. p. 1059–1063.

[130] Al-A Adhab, Altmimi H, Alhawashi M, *et al.* IoT for remote elderly patient care based on Fuzzy logic. In: 2016 Int. Symposium on Networks, Computers and Communications (ISNCC); 2016. p. 1–5.

[131] Raj C, Jain C, and Arif W. HEMAN: Health monitoring and nous: an IoT based e-health care system for remote telemedicine. In: 2017 Int. Conference on Wireless Communications, Signal Processing and Networking (WiSPNET); 2017. p. 2115–2119.

[132] Ahmed Faeq Hussein, Arun Kumar, Burbano-Fernandez M, *et al.* An automated remote cloud-based heart rate variability monitoring system. IEEE Access. 2018; p. 1–1.

[133] Beibei T, and Yi L. Upgraded application of intelligent environment monitoring system in IOT smart home. In: 2015 Sixth Int. Conference on Intelligent Systems Design and Engineering applications (ISDEa); 2015. p. 916–919.

[134] Zanella A, Bui N, Castellani A, *et al.* Internet of things for smart cities. IEEE Internet of Things Journal. 2014;1(1):22–32.

[135] Vishal D, Afaque HS, Bhardawaj H, *et al.* IoT-driven road safety system. In: 2017 Int. Conference on Electrical, Electronics, Communication, Computer, and Optimization Techniques (ICEECCOT); 2017. p. 1–5.

[136] Lakshmanaprabu SK, Shankar K, Khanna A, *et al.* Effective features to classify big data using social Internet of things. IEEE Access. 2018;6:24196–24204.

[137] Al Fuqaha A, Guizani M, Mohammadi M, *et al.* Internet of things: a survey on enabling technologies, protocols, and applications. IEEE Communications Surveys & Tutorials. 2015 Fourthquarter;17(4):2347–2376.

[138] Said A, and Pearlman WA. A new, fast, and efficient image codec based on set partitioning in hierarchical trees. IEEE Transactions on Circuits and Systems for Video Technology. 1996;6(3):243–250.

[139] Wiechert G, Triff M, Liu Z, *et al.* Evolutionary semi-supervised rough categorization of brain signals from a wearable headband. In: 2016 IEEE Congress on Evolutionary Computation (CEC); 2016. p. 3131–3138.

[140] Interaxon, "Protocols: Compressed EEG Packets". Accessed: 2019-01-14. http://developer.choosemuse.com/protocols/bluetooth-packet-structure/compressed- EEG- packets.

[141] Higgins G, Faul S, McEvoy RP, *et al.* EEG compression using JPEG2000: How much loss is too much? In: 2010 annual Int. Conference of the IEEE Engineering in Medicine and Biology; 2010. p. 614–617.
[142] Daou H, and Labeau F. Performance analysis of a 2-D EEG compression algorithm using an automatic seizure detection system. In: 2012 Conference Record of the Forty Sixth asilomar Conference on Signals, Systems and Computers (aSILOMaR); 2012. p. 1632–1636.
[143] Fadele A, Othman M, Abaker Targio Hashema I, *et al.* Internet of things Security: a Survey; 2017. 04;88.
[144] Ahmed I, Saleel AP, Beheshti B, *et al.* Security in the Internet of things (IoT). In: 2017 Fourth HCT Information Technology Trends (ITT); 2017. p. 84–90.
[145] Chan HL, Kuo PC, Cheng CY, *et al.* Challenges and future perspectives on electroencephalogram-based biometrics in person recognition. Frontiers in Neuroinformatics. 2018. Available from: https://doi.org/10.3389/fninf.2018.00066.
[146] Holler Y, Bathke AC, and Uhl A. Age, sex, and pathology effects on stability of electroencephalographic biometric features based on measures of interaction. IEEE Transactions on Information Forensics and Security. 2019;14(2):459–471.
[147] Mohammadi M, Al-Fuqaha A, Sorour S, *et al.* Deep learning for IoT big data and streaming analytics: a survey. IEEE Communications Surveys Tutorials. 2018; p. 1–1.
[148] Martz GU, Hucek C, and Quigg M. Sixty day continuous use of subdermal wire electrodes for EEG monitoring during treatment of status epilepticus. Neurocritical Care. 2009;11(2):223–227.
[149] Aggarwal G, Dai X, Binns R, *et al.* In: Wireless EEG Signal Transmission Using Visible Light Optical Camera Communication. Springer Singapore; 2018. p. 152–161.
[150] Gtec. Accessed: 2018-09-20. http://www.gtec.at.
[151] Mindo. Accessed: 2018-09-20. http://mindo.com.tw.
[152] Patki S, Grundlehner B, Verwegen A, *et al.* Wireless EEG system with real time impedance monitoring and active electrodes. In: 2012 IEEE Biomedical Circuits and Systems Conference (BioCAS); 2012. p. 108–111.
[153] Quasar. Accessed: 2018-09-20. http://www.quasarusa.com.
[154] Singh S, and Mittal K. Internet of everything smart environment. In: 2015 International Conference on Green Computing and Internet of Things (ICGCIoT); 2015. p. 1309–1311.
[155] Nakanishi I, Baba S, and Miyamoto C. EEG based biometric authentication using new spectral features. In: 2009 Int. Symposium on Intelligent Signal Processing and Communication Systems (ISPaCS); 2009. p. 651–654.
[156] Abdullah MK, Subari KS, Loong JLC, *et al.* Analysis of effective channel placement for an EEG-based biometric system. In: 2010 IEEE EMBS Conference on Biomedical Engineering and Sciences (IECBES); 2010. p. 303–306.

[157] Zúquete A, Quintela B, and Cunha JPS. Biometric authentication with electroencephalograms: Evaluation of its suitability using visual evoked potentials. In: Int. Joint Conference on Biomedical Engineering Systems and Technologies. Springer; 2010. p. 290–306.

[158] Donchin E, Spencer KM, and Wijesinghe R. The mental prosthesis: assessing the speed of a P300-based brain-computer interface. IEEE Transactions on Rehabilitation Engineering. 2000;8(2):174–179.

[159] Thomas KP, and Vinod AP. Utilizing individual alpha frequency and delta band power in EEG based biometric recognition. In: 2016 IEEE Int. Conference on Systems, Man, and Cybernetics (SMC); 2016. p. 004787–004791.

[160] Chuang J, Nguyen H, Wang C, *et al.* I think, therefore I am: usability and security of authentication using brainwaves. In: In Proc. of the Workshop on Usable Security, USEC 13; 2013.

[161] Barra S, Casanova A, Fraschini M, *et al.* EEG/ECG signal fusion aimed at biometric recognition. In: New Trends in Image Analysis and Processing – ICIAP 2015 Workshops. Springer International Publishing; 2015. p. 35–42.

[162] Rahman W, and Gavrilova ML. Emerging EEG and Kinect face fusion for biometric identification. In: 2017 IEEE Symposium Series on Computational Intelligence (SSCI). IEEE; 2017.

[163] Saini R, Kaur B, Singh P, *et al.* Don't just sign use brain too: A novel multimodal approach for user identification and verification. Information Sciences. 2018;430–431:163–178.

Chapter 6

The evolution of passive brain–computer interfaces: enhancing the human–machine interaction

Nicolina Sciaraffa[1], Pietro Aricò[1,2,3], Gianluca Borghini[1,2,3], Gianluca Di Flumeri[1,2,3], Antonio Di Florio[2], and Fabio Babiloni[1,2,4]

In the last decade, a real revolution in the field of brain–computer interfaces (BCI) led from the "overt" detection of human intention to the "covert" assessment of the actual human mental states. While the first aspect is the basis of the traditional BCI systems, the latter represents the outcome of the passive BCI applications. In fact, passive BCI derives its outputs from brain activity arising without the purpose of voluntary control, but implicitly related to the human mental state. The necessity of monitoring human mental states driven by safety-critical application has been just the boost to the passive BCIs developing: more in general, passive BCI represents the implicit channel of information that enhances the goal-oriented cooperation of humans and machines as a whole, the so-called human–machine interaction. So far, there have been countless passive BCI applications in a wide range of contexts such as driving, gaming, and surgery. If on the one hand, this has been possible thanks to the development of more and more discrete neurotechnological devices, on the other hand, we must not overlook the significant step forward in the employed algorithms, with the adoption in this field of machine learning and deep learning enhancements. This chapter will retrace not only the major achievements but also the future trends, in terms of technologies, methods, and applications of what concerns the field of passive BCIs. The final aim of the work is to draw a mark on where we are nowadays and the future challenges, in order to make passive BCIs closer to being integrated into day-life applications.

[1]Department of Molecular Medicine, Sapienza University of Rome, Rome, Italy
[2]BrainSigns srl, Rome, Italy
[3]IRCCS Fondazione Santa Lucia, Neuroelectrical Imaging and BCI Lab, Rome, Italy
[4]College of Computer Science and Technology, Hangzhou Dianzi University, Hangzhou, China

6.1 Passive BCI as mind–computer interface

Since 1929, Berger demonstrated that it is possible to record brain activity in a non-invasive way through electrodes placed on the scalp [1], it was immediately clear the great potential that this new instrument could have had in neurophysiology and neuroanatomy [2]. Equally rapid was the awareness that brain signals could have been used as a means of control over external processes. The BCI era had begun. Its origin can be identified in the 1970s when brain signals were used for the first time to create a *brain–computer communication* based on visual-evoked potential, the measurement of the electrical signal produced at the visual cortex in response to visual stimuli [3]. In its classical definition, BCI was considered as "a communication system in which messages or commands do not pass through the brain normal output pathways of peripheral nerves and muscles" [4]. Therefore, a traditional BCI system (i) acquires brain signals preferably with electroencephalography (EEG) or minimally invasive acquisition devices; (ii) analyzes them to obtain features like, for example, the slow cortical potentials, mu rhythm, P300 potentials; (iii) converts them in an output according to the information transfer rates limits [5]. This output can assume different forms because, by means of an external device, it could be used to communicate or to control, for example, a neuroprosthesis or it can be translated directly into actions.

As well as actions and reactions, the *mind* (understood as a set of mental states) can also be considered as a product of the brain [6], therefore, starting from this assumption, it is easy to extend the concept of BCI from intentional (*actions*) to unintentional (*mental states*) brain products. The first to make this distinction was Zander [7], who highlighted the existence of three main types of BCI: the Active, the reactive and the passive BCI. The *Active BCI* is the most traditional of the BCIs, the one created to allow subjects with severe disabilities or reduced mobility, to communicate and control devices belonging to the external environment by means of appropriate processing of signals associated with brain activity. In this case, a subject voluntarily controls his brain activity in terms of slow cortical potentials, mu rhythms, P300 potentials to perform, through a BCI system, an action coherently with the information transfer rates of the BCI technology [8,9]. The second type of BCI is called *Reactive BCI*, and similarly to active BCI is used to control external devices, but in this case, the modulation of brain activity is due to a reaction to an external stimulus. The third type of BCI, the *Passive BCI*, is clearly different from the first two: its aim is not to voluntarily control the external world, but to allow the subject to *unconsciously affects surroundings* by means of his mental states, in what can be defined as a *mind–computer interface*. The most widespread definition of passive BCI says that it allows enriching a human–computer interaction. But what is meant by enriching human–machine interaction and how neurosciences can allow this to be done? This chapter aims to describe these aspects, in particular how a passive BCI system works, in which areas it is essential or will be essential in the near future to enrich human–machine interaction, and finally, how methodological advances and the development of new technologies can promote its dissemination and use.

6.1.1 Passive BCI applications

In line with the BCI evolution, Wolpaw and Wolpaw [10], proposed a new definition of a BCI: it is "a system that measures central nervous system (CNS) activity and converts it into artificial output that replaces, restores, enhances, supplements, or improves natural CNS output and thereby changes the ongoing interactions between the CNS and its external or internal environment". Such definition involves the Passive BCI characteristics. In fact, compared to traditional BCI, the passive BCI is an instrument that allows to monitor a subject mental state, that is not consciously manipulated by the subject itself with the aim to enhance or improve natural neurophysiological output. This is possible because a passive BCI system measures the CNS acquiring different neurophysiological signals and relates these measurements to mental states (such as workload, stress, attention, emotion and several others that could be of interest) computing indexes based on the models. These indexes are given as input of a device to allow it to perceive the subject. This enhances the human–machine interaction because the technology, by interpreting the information provided by the passive BCI about the user's mental state, could understand and adapt itself to the subject condition [11] or provide an indicator of subject condition to be used, for example, to enhance the safety [12]. Already in 1973, Vidal pointed out that other biosignals apart from brain signals should be included in the brain–computer communication as promising for some applications [3]. Therefore, to measure the CNS to date, in addition to the EEG and functional near-infrared (fNIRS) to record the brain signals, neuroscientific researches are based on the use of other types of biosignals such as, electrocardiography (ECG), electrooculography (EOG) and galvanic skin response (GSR) [13].

Understood the potential of passive BCI, several researchers have proposed exploring all the possible applications [14]. Among these researches, five different main passive BCI areas of interest have been identified: *user-state monitoring, training, education and cognitive improvement, gaming and entertainment, evaluation, safety, and security* [15]. All these applications and, more generally, the passive BCI, aim to have an impact on everyday activities and life, from education to work. In this perspective, it is important to be able to take the system out of the laboratory. In fact, there are significant differences between what is studied in a shielded environment such as that of the laboratory, compared to real and noisy environment. Moreover, also the tasks switching from being very controlled to being natural actions are an additional factor that can greatly influence the functioning of the system. In the years that have elapsed since the definition of passive BCI and the Van Erp's article to date, for each of the main areas of interest formerly identified, several studies have been carried out both in laboratory and, with promising results, in realistic contexts.

6.1.1.1 User-state monitoring

The basis for improving the human–machine interaction is the user-state monitoring. This is particularly relevant in safety-critical environments, where the mental state of the user could be used to predict human error or more generally to evaluate human factors [16]. In fact, in such safety-critical environments, users are subject to different stimuli of different nature (just think of all the information is necessary

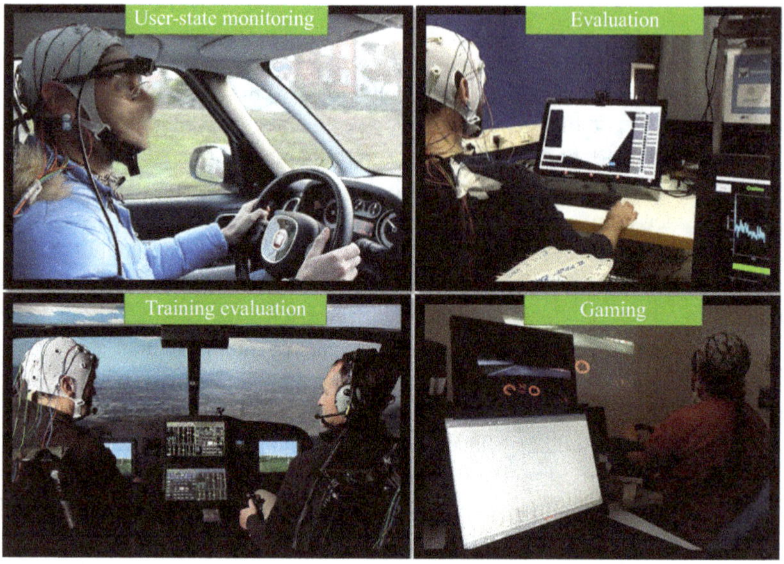

Figure 6.1 Passive BCI applications

to pay attention while driving). Contexts like this require that the attentive resources are continuously divided between different stimuli, which can easily mislead the subject [17]. Several groups have used the passive BCI to monitor the subject while driving [18,19], also to assess the incidence of aids such as cruise control [20], the effects of sleep deprivation [21] and drowsiness [22]. In general, the driving task has been one of the most studied in the field of passive BCI both in the laboratory and during realistic driving [23] and not only of cars (Figure 6.1), but also of boats [24] and planes [25].

6.1.1.2 Training, education, and cognitive improvement

Everything that goes from the acquisition of new knowledge to its consolidation has implications in the brain activity of the subject and then can be examined through the analysis of neurophysiological signals [26]. In general, in fact, the cognitive resources employed by the subject vary in relation to how automatic the action for the subject is [27], inducing in the subject a greater focus of brain activity as well as a functional reorganization of brain areas [28]. Training and evaluation of several relevant professional figures have been positively influenced by the introduction of passive BCI [29], since they allow both to optimize resources and to adapt the learning process to the subject [30] or to define an objective criterion for the personnel selection. In fact, passive BCI-based system has been used to assess professional surgeons performing surgical training task [31], professional air traffic controllers while dealing with realistic activities [32,33], or pilots [34] (Figure 6.1). From the educational point of view, passive BCI has been used to assess the level of attention sustained during regular university lectures [35].

6.1.1.3 Gaming and entertainment

Apart from the possibility of using neurophysiological signals to control a game interface [36], in the field of gaming the passive BCI had a dual purpose (Figure 6.1). It has been used both to adapt game features to the state of the subject [37] and vice versa to maximize the commitment of the subject by progressively increasing the difficulty [38]. The applications in this field are more than ever linked to the introduction of minimally invasive and commercial acquisition systems [39] and have been the cradle of the hybrid "eye-brain computer interface", the EBCI, a method based on a combination of eye-tracking and EEG signal activity, able to differentiate intentional and spontaneous eye movements [40].

6.1.1.4 Evaluation

Because the passive BCI is able to catch emotions and feelings of a subject, it found significant applicability in the neuromarketing and neuroergonomic fields. In the last years, several studies used neurophysiological signals like EEG, GSR, and ECG to show how it is possible to detect hidden signs of the memorization process and pleasantness related to advertisement watching [41], olfactory stimuli [42], and tasting ones [43]. Moreover, passive BCI has been used to evaluate the effects of technologies on the user's mental state. Borghini *et al.* [44] performed a study on professional helicopter pilots performing realistic simulated missions with the aim to compare different avionic technologies according to different experienced workload. More in general, the human–machine interaction experience itself could be evaluated using the passive BCI system (Figure 6.1) [45].

6.1.1.5 Safety and security

In this context, different groups have analyzed the possibility to use brain waves for subjects identification. In particular, it has been proved that it is possible to use just one EEG electrode to analyze the brain waves similarity, that it is not affected by the task the subject is performing [46]. In fact, the identification has been proved also on drivers [47]. The hardware and software implementation of a real security system based on EEG has been used to open a lock [48]. Brain waves have been combined with iris characteristic to enhance the safety of a security system [49].

In addition to these fields of application, it should be noted that in recent years there have been numerous applications during real activities such as artistic experiences [50] and in complex environments like cement, city traffic, and power plant control centers [51].

At this point, it is useful to note that most of these results have been obtained thanks to the use of minimally invasive acquisition systems that are easily adaptable for the use during real activities. However, both the technological and methodological components of the passive BCI system need to be further improved and adapted to allow the real-time monitoring of each mental states during real life.

6.2 Passive BCI system description

As seen so far, the ultimate aim of the passive BCI evolution is to integrate this system into everyday activities. To achieve this goal both technological and methodological advancements are required. The advancements achieved in neurotechnological field

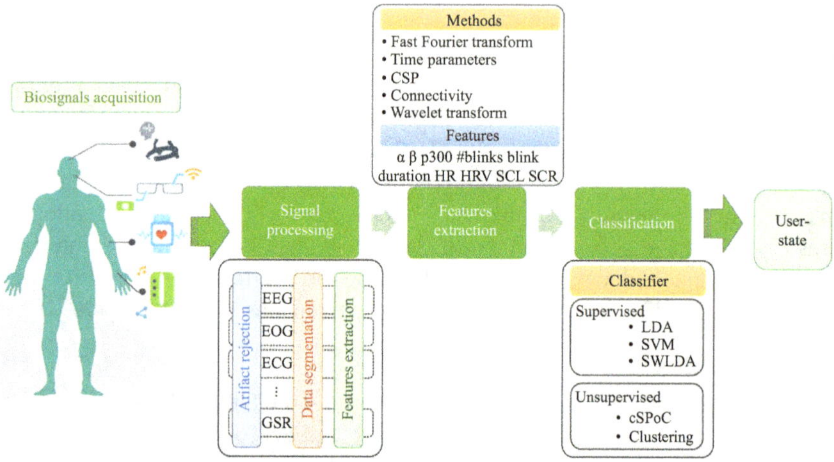

Figure 6.2 Passive BCI system

allow to record reliable user's biosignals with comfortable devices. But to have also a reliable real-time information of the subject's mental state, signal processing methodology needs to be adapted to passive BCI requirements, as well as features extraction and the subsequent classification (Figure 6.2).

6.2.1 New technology for passive BCI

Differently from the traditional concept of BCI, characterized by a static subject, the passive BCI, as a system integrated into the daily activities, is affected by dynamicity of the humans interacting with the machine itself. This is the reason why more in the context of the passive BCI than in traditional BCI, a radical innovation in the biosignals acquisition and processing is required. All the acquisition devices used for passive BCI application have to guarantee high signal quality, the appropriate protection against movements and interferences effects on signals, have to be noninvasive and easy to wear, the comfort must not be compromised over time. In order to ensure these requests, which would make it possible to improve and the large-scale use of passive BCI system, during the last years great advances have been made for what concerning both brain signals and autonomic signals.

For brain signals acquisition, the most trendy solution is to avoid the use of gel in the phase of preparation and, eventually, to replace it with water. On the one hand, easy to apply ring-electrodes whit a sponge soaked in water has been demonstrated to provide reliable EEG [52]. On the other hand, dry electrodes have been proposed along the years, moving from metallic spins electrodes to silicone ones [53] to easily reach the head skin although the hair presence. In this regard, different portable and dry EEG recording systems have been employed in several studies [54,55] in what has been called "the dry revolution" [56]. To avoid the disposition of the EEG electrodes on the scalp, the ears have been involved. In this regard cEEGgrid, consisting in 10 electrodes printed on a flexible adhesive to be placed around the ear [57], or of soft and thin electrodes to be placed in the ear [58] have been developed. Wireless and wearable EEG devices have been adopted during different experiments in real driving [59]. An alternative and futuristic approach to wireless recording of biosignals is the noncontact acquisition by means of Microsoft Kinect: this method has been used for heart rate and breath acquisition [60]. More easily, heart rate and skin conductance could be recorded through wristbands that are already available on the market [61].

For passive BCI applications more than BCI, is necessary to smartly solve the interaction between different measurement devices by means of a multimodal approach [62]. In this regard, von Lühmann proposed a mobile, modular, and multimodal (EEG, fNIRS, ECG) biosignal acquisition architecture (M3BA) [63]. It consists of a microcontroller that allows the integration and customization, Bluetooth connection and 3-D accelerometer. The use of both wet or dry electrodes is possible thanks to high input impedance and common mode rejection ratio. Finally, due to long-lasting duration of biosignals acquisition in the case of use in everyday activities, radio frequency identification (RFID)-based sensor systems have been proposed: passive RFID technology allowing a significant reduction in size for a wearable system thanks to the absence of battery [64].

6.2.2 Signal processing

The aforementioned passive BCI applications have required an extension and adaption of signal processing techniques to real contexts. The datasets recorded in real settings, in fact, are characterized by a huge and dissimilar number of artifacts come from the subject activity, like muscular activity, or environmental noise, like the 50 Hz power-line noise. Also in this case, the general approach is to filter out every part of the signals that do not provide useful information to discriminate, for example, the subject mental state, and, at the same time, it is necessary to not introduce time delay for an online application. In the case of EEG signals many of the most used artifact removal methods (Gratton algorithm, independent component analysis, wavelet-based method) are not directly applicable in realistic passive BCI application due to the necessity of a high number of EEG channels and the necessity to calibrate such methods characterized by high computational complexity. In this regard, several attempts have been made to preserve as much signal as possible respect to the content of the artifacts. For example, in the case of blink artifacts, the REBLINCA algorithm [65] based on regression method has been successfully used to remove blink artifacts during realistic driving [66]. Additional signal preprocessing steps, even though necessary to maximize the signal-to-noise ratio (SNR), would slowdown the process of data analysis distancing the possibility to carry out a real-time analysis. Moreover, noisier realistic data are assumed to have better generalization properties [67].

6.2.3 Features extraction

For its definition, a passive BCI system is characterized by several channels and samples, therefore by a multidimensional features space, with low SNR. So far, both spectral and time domain features have been both successfully used to discriminate different mental states in passive BCI applications. In the frequency domain, a clear correlation between user mental states and EEG rhythms has been proved, like in the case of workload [68,69] or attention [70], and the fast Fourier transform (FFT) is the most commonly applied method among spectral analysis techniques because it is time shift invariant [71]. In the time domain, the most used features have been the event-related potentials like the P300 [72] that allows to know the amount of attentional resources and memory operations, or the event-related negativity that allows to know if a subject is aware that he has just made a mistake [73]. In [74] Frey proved that the combination of temporal and spectral features could provide redundant information, as in this case it has been tested to estimate visual comfort. A time–frequency representation of the neurophysiological signals and an appropriate description of transient events could be obtained by Wavelet transformation [75]. It has been efficiently employed for drowsiness detection [76].

Some authors proposed to use functional connectivity-based features, aiming to investigate alteration in brain networks associated with the different mental states [77]. However, both time domain and connectivity-based features do not allow to obtain real-time information about the mental state variation, because for both is necessary to provide several repetitions to average the noise of EEG [78] and estimate the model parameters [79], respectively. Otherwise, to enhance the powerfulness of spectral features and the single trial analysis, the use of spatial filters has been proposed [80]. Several algorithms based on the common spatial pattern [81] have been developed to extract discriminative oscillatory features, by removing the signal's strong correlation among the original axes, and provide features distributions maximally dissimilar along the new axes maximizing the variance for one class and minimizing the variance for the other class. However, despite the CSP is a high-performance method, it is very sensitive to the quality and quantity of data, then different regularized versions of this feature extraction algorithm have been proposed [82].

6.2.4 Classification techniques

Once obtained the signatures of each mental state in each biosignal, most of the described works used supervised machine learning algorithms to assess the mental state level. The first issue to face with approaching machine learning techniques to implement passive BCI applications is the choice of the model the specific algorithm uses to maximize the differences between mental state classes. For example, linear methods are appropriate when limited data and limited knowledge about the data itself are available and after a proper transformation of the features space [83]. However, in the presence of strong noise or outliers, or if the dimensionality of the features space is too high, linear methods can fail then nonlinear methods are suitable to find potentially complex structure in the data. Regularization methods allow to increase the reliability of the method by manipulating available training data, limiting the influence of outliers or strong noise and the complexity of the discrimination function [84]. Otherwise, kernel-based learning machines like support vector machines [85] allows to apply a linear algorithm in some appropriate nonlinear feature spaces calculated by using specific transformation functions (named kernel functions): all beneficial properties of linear classification are maintained, but at the same time the overall classification is nonlinear in input space, since feature and input space are nonlinearly related. The second issue is related to the amount of data used to train the model itself that could bring to overfitting or underfitting issues. In the former, the model becomes too specific on unique or spurious differences between classes of the training set. The direct consequence is that the model does not perform well over time, and needs repeated calibration sessions [84]. On the contrary, underfitting refers to a technique that can neither model the training data and generalize to new data. Intuitively, underfitting occurs when the model or the algorithm selects too little features to fit the data well enough [86]. There are different techniques that can be used to quantify the model prediction error in order to reduce the under and overfitting issues, such as the cross-validation-based approaches that provide good error estimation with minimal *a priori* assumptions [87]. In addition, linear classifiers are generally more robust and easier to optimize than nonlinear ones, since the latter have limited flexibility (less free parameters to tune) and are, thus, less prone to overfitting [88].

The supervised machine learning algorithms usually used in this context (LDA, SWLDA, SVM, [84]) require to be frequently calibrated to maintain high performance over time. Several efforts are being done in such direction in order to mitigate this issue. For example, Aricò *et al.* proposed a linear regression algorithm, the automatic stop stepwise linear discriminant analysis (asSWLDA) able to evaluate the workload in operational environments without recalibrating the system up to a month [89]. The frequent recalibration represents one of the main limitations for the passive BCIs diffusion, for this reason different attempts have been made toward an unsupervised direction, as in the case of canonical source power co-modulation (cSPoC) analysis which has been used to classify the workload [90]. In this case it does not need any calibration phase, however other attempts try to reduce the calibration time [91]. Anyway, regarding supervised and unsupervised learning, the former is still preferred to the latter since it allows to achieve higher accuracies [92].

Among supervised, the linear methods have usually low computational requirement and provide high passive BCI performance, however, the recent development of deep learning methods for image and speech recognition, pave the way to extend them to EEG signal analysis [93]. Deep learning methods usually require a huge amount of data for an efficient calibration, therefore this is usually done with data from different subjects. Their capability to exploit the underlying patterns learning directly from data allow them to overperform more traditional methods relying on hand-crafted features based on *a priori* knowledge [94]. Different deep learning architectures have been tested on different EEG datasets showing promising results [95].

6.3 Laboratory vs. realistic passive BCI example applications

The considerations made so far have highlighted the open issues still to be resolved in the methodological field to obtain a reliable and real-time evaluation of a mental state. In this regard, this paragraph aims to provide the guidelines of the steps necessary to obtain an index of the mental state of a subject, in particular the mental workload, both during a laboratory and a more realistic task, highlighting the different results in terms of classification performance both in the case of the use of different features extraction methods, and in the case of the use of different classification techniques.

The mental workload has been chosen because its monitoring is of particular interest in safety-critical applications. In general, as cognitive workload increases, maintaining task performance within an acceptable range becomes more difficult because of the use of more cognitive resources than those available [96]. Objective measures of mental workload based on neurophysiological signals could be at the basis of different passive BCI system, used to evaluate the user's state, to allow allocating resources to minimize the errors due to overload. This system could intervene in real time before the operator becomes overloaded while performing safety-critical tasks [97]. In the following, the workload will be assessed using the EEG signals. From the literature in fact, it has been observed that to growing workload has been associated an increasing of frontal activity of the brain in the theta band and a decreasing of parietal activity in the alpha band [98].

The air traffic management (ATM) represents one of the fields where the evaluation of workload has a fundamental role both in training aspects and in operational conditions. In fact, it has been established that it is a high demanding work during which the slightest error could have very ominous effects [99]. Changes in the traffic manipulation, complexity or volume, produced changes in mental resources required and therefore in workload. In this domain, different tasks have been used to investigate workload changes and to create a workload index based on the biosignals, even though most of the results are related to laboratory environment. Hereafter the passive BCI system has been applied on two different ATM tasks with the aim of assessing the workload level: the first based on the LABY microworld that is strictly a laboratory task, the second instead is based on a realistic ATM simulation.

6.3.1 Datasets

6.3.1.1 Laboratory dataset

The LABY microworld is a functional simulation of air traffic control (ATC) that captures the underlying processes involved in ATM with a simplified version of the operational human–machine interface. The LABY microworld is based upon the main task of guiding n planes around a predetermined route, indicated by a green path (Figure 6.3) while avoiding potential conflicts and dealing concurrently with other incoming information. Participants must input numerical values such as heading, flight level, speed, etc., in order to direct flights around the trajectory and

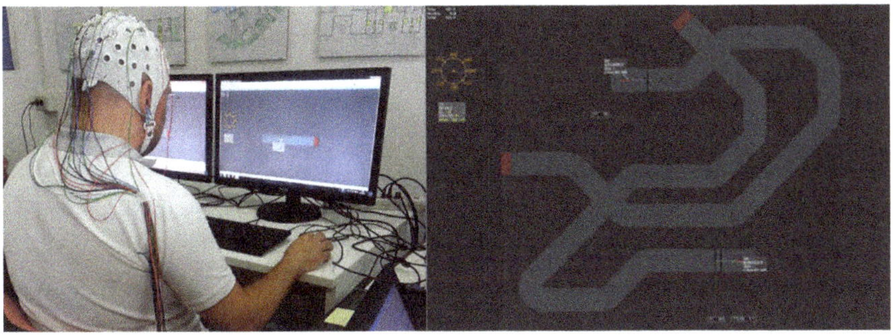

Figure 6.3 Laboratory setup on the left. On the right, an example of the task interface

avoid any obstacles which may occur. Penalties are applied if the aircraft deviates outside the route or if other constraints are not met, and a score is displayed dynamically on the screen. The difficulty of the task, and therefore the workload, is altered according to how many aircrafts the participant must control, the number and type of clearances required over time, and the number/trajectory of other interfering flights; while time pressure is augmented by increasing aircraft speed. Critical events (e.g., conflict with other aircraft), or adverse conditions (e.g., a prohibited area) occur when a guided aircraft enters a particular zone.

By means of LABY have been created different sets of clearances and situations associated with three basic difficulty levels (easy, medium, and hard). The experimental protocol implies the use of 27 min long exercises: three exercises of 3 min each have been grouped to create one exercise of 9 min and repeated three times. Those exercises are created by selecting and ordering randomly three easy, three medium and three hard basic exercises. The process will not use the same exercise twice during the 27 min. A group of 11 professional ATCOs has been enrolled. They were asked to execute the LABY exercises under different conditions, with various difficulty levels.

Their EEG have been recorded by the digital monitoring BEmicro system (EBNeuro system). The 16 (FPz, F3, Fz, F4, AF3, AF4, C3, Cz, C4, P3, Pz, P4, POz, O1, Oz, O2) EEG channels have been collected simultaneously with a sampling frequency of 256 (Hz). All the EEG electrodes have been referenced to both earlobes, and the impedances of the electrodes have been kept below 10 kΩ.

6.3.1.2 Operational dataset

In this case, a realistic ATM protocol has been settled up. The controller position is similar to the ones used in real operational center and in training. The controller working position has two screens, one 30″ to display radar image and a 21″ to interact with the radar image (zoom, move, clearances, and information) and the voice communication between controller and pseudo-pilot uses the same hardware and software like the one used in training (headphones, microphone, and push-to-talk command), very similar to what normally used into operations (Figure 6.4(a)). The

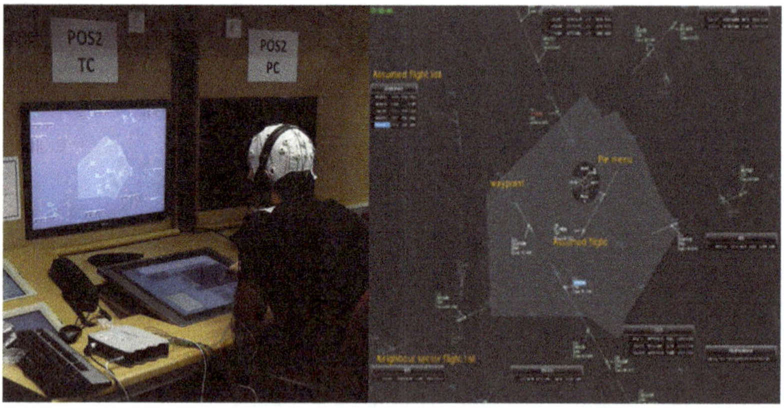

Figure 6.4 Operational setup. On the right, a typical ATM interface

radar picture (Figure 6.4(b)) shows the sector (light grey), routes, waypoints and flights displayed according to their coordination state (white ones are assumed). Information of neighbor flights is displayed in the list.

The experimental task is an ecological ATM scenario of about 45 min, in which different air-traffic conditions take place. For instance, the task could start with an increasing traffic complexity until a hard condition, and then decreasing until a condition similar to the initial one by passing through a medium complexity condition. The variation of the task complexity is necessary to evaluate if the system is able to differentiate the different workload levels. Each scenario lasts globally 45 min, while each session of low (L), medium (M), and high (H) workload level lasts about 15 min. The different levels of traffic, defined by an expert controller and scenario-specialist, vary according to the number of aircrafts, traffic geometry, number of conflicts and subjective assessment of controller's skill. Three sets of traffic samples are considered.

The experimental protocol involves 11 professional ATCOs. ATCOs have been selected in order to have a homogeneous sample in terms of sex, age, and skill level. Sixteen EEG channels (FPz, F3, Fz, F4, AF3, AF4, C3, Cz, C4, P3, Pz, P4, POz, O1, Oz, O2) have been recorded by the digital monitoring BEmicro system (EBNeuro system). with a sampling frequency of 256 Hz. All the EEG electrodes have been referenced to both earlobes, and the impedances of the electrodes have been kept below 10 kΩ.

6.3.2 Methods

6.3.2.1 Signal processing

The EEG signals have been digitally band-pass filtered by a fourth-order Butterworth filter (low-pass filter cut-off frequency: 30 Hz, high-pass filter cut-off frequency: 1 Hz) and the FPz signal has been used to correct eyes-blink artifacts from the EEG data by means of the Reblinca algorithm [65]. It should be underlined that normally in a realistic environment, different sources of artifacts could affect neurophysiological recorded signals, more than in the laboratory environments. For instance, ATCOs normally communicate verbally and perform several movements during their operational activity. Then each trial having amplitude exceeds 100 μV (threshold criteria), or the slope trend higher than 3, or a sample to sample difference higher than 25 μV have been marked as "artifact" and then rejected [100], with the aim to have clean EEG signals from which estimate the brain parameters for the different analyses.

6.3.2.2 Features extraction

Two different features extraction methods have been compared. In the first case the EEG signal has been segmented in periods of 2 s, 0.125 s shifted. After that, for each period, the power spectral density (PSD) by using the FFT has been computed in the EEG frequency bands. The EEG frequency bands were defined accordingly with the Individual Alpha Frequency (IAF) value estimated for each subject. Since the alpha peak is mainly prominent during rest conditions, the participants were asked to keep the eyes closed for a minute before starting with the experiment. Such condition was then used to estimate the IAF value specifically for each subject. In particular, the theta (IAF-6 ÷ IAF-2) and alpha brain activities (IAF-2 ÷ IAF+2) were considered over the EEG frontal and parietal channels in order to assess the workload. In fact, it was widely demonstrated how such EEG bands are the most correlated to mental workload variations. In this regard, a mental workload index (MWI) is defined as:

$$MWI = \frac{PSD_{\theta_F}}{PSD_{\alpha_P}} \tag{6.1}$$

where the PSD_{θ_F} represents the PSD in the theta band estimated over the frontal brain areas and the PSD_{α_P} is the PSD in the alpha band estimated over the parietal brain areas. The MWI in easy and hard condition has then been computed.

In the second case, the common spatial pattern (CSP) has been applied to the EEG signals to obtain the spatial filters. The features have been computed as the logarithm of the variance of the projected signals. The logarithmic variance in frontal theta and parietal alpha band has been also used to define the unsupervised model. The cSPoC algorithm allows to optimize spatial filters maximizing the anticorrelation between theta and alpha rhythms, according to the neurophysiological workload model assumption.

6.3.2.3 Classification

To differentiate easy level from hard level in the considered tasks, the performance of both supervised and unsupervised machine learning techniques has been investigated. In particular, among the supervised have been chosen two traditional classifiers typically used in the context of EEG signals, like the LDA and the SVM, and the asSWLDA that is an updated version of the stepwise regression [89]. From the unsupervised point of view has been compared the use of the theta frontal PSD, theta/alpha PSD (that is the MWI) and the cSPOC [90]. To evaluate the accuracy of the system, values of the area under curve (AUC) of the receiver operating characteristic [101] were calculated from the outputs of the classifiers (for each different refresh rate). The AUC values represent how well the classifier separates two different subtasks, and so how well the classifier could predict the difficulty of the task directly related to the subject's workload level. Three two-way repeated-measures ANOVAs were performed on the AUC values. In the case of PSD features and CSP features the classifiers (asSWLDA, LDA, SVM) and dataset type (LAB and REAL) are the independent variables. In the case of unsupervised analysis, the classification methods (theta/alpha, theta, cSPoC) and dataset type (LAB and REAL) are the independent variables. In addition, Tukey post-hoc tests were performed to test the differences between all the levels.

6.3.3 Results

In Figure 6.5, the accuracy of different passive BCI systems revealed by AUC values has been shown. In the case of PSD features, the ANOVA analysis is not statistically significant ($F(2, 40) = 1.26$, $p = 0.29$), however, the analysis reveals the significant effect of the classifier and the dataset. In particular, considering the classifier factor, the AUC values calculated in the case of LAB dataset are not significantly different using different classifiers, however in the case of REAL dataset the asSWLDA provides performance significantly higher than the LDA

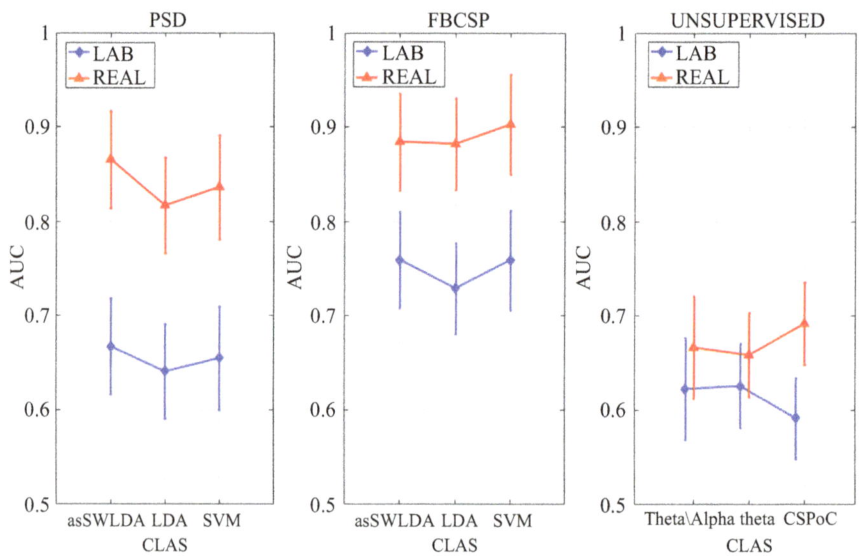

Figure 6.5 Mean values and related standard errors (CI = 0.95) of the AUC values achieved in laboratory (LAB) and operational (REAL) environment using the different classifiers (asSWLDA, LDA, SVM) in the PSD and CSP panel, and theta/alpha, theta, CSPoC in the unsupervised analysis

(p = 0.0004). In the case of CSP features, the ANOVA analysis is statistically significant ($F(2, 40) = 5.61$, $p = 0.007$), therefore for each classifier the performance in the case of LAB dataset is significantly lower than the REAL dataset. The post-hoc test shows that the AUC of the asSWLDA is significantly higher than LDA AUC in the case of LAB dataset ($p = 0.0002$). For the REAL dataset, the SVM AUC is significantly higher than both LDA ($p = 0.01$) and asSWLDA ($p = 0.04$). In the UNSUPERVISED case, the ANOVA analysis is not statistically significant ($F(2, 40) = 1.95$, $p = 0.15$), however it reveals the significant effect of the dataset.

6.3.4 Discussion

This example wanted to highlight which are the typical steps to be carried out in order to obtain an index of a mental state through a passive BCI system. Specifically, the workload has been analyzed, as one of the most versatile and interesting mental states, given its link with the performance of the subjects. Although both datasets aim to analyze the workload in the field of ATM, they are profoundly different as they are based, respectively, on a laboratory ATM task and a realistic ATM task. This diversity affects the passive BCI system ability to distinguish the workload level of a subject while is performing an easy and a difficult task. Supervised (asSWLDA, LDA, SVM) and unsupervised (theta/alpha, theta, CSPoC) classification methods have been analyzed, considering for the supervised also two different methods of features extraction (PSD, CSP). In each of the analyses, therefore independently from the classifiers and features used, it has been obtained that the passive BCI system works better in a realistic environment than in a laboratory environment. This result can be explained in two ways. First, the realistic task has a greater difference between the two workload levels than the laboratory one, because the hard level has been considered very difficult for many of the subjects by subjective measures. Second, as pointed out in the previous paragraph, natural activities can induce misleading features that could help the system in classification, compared to a standard laboratory task. Considering now the supervised analysis, according to the literature the use of the CSP allows to obtain higher performances than the use of the PSD (average increase of the AUC of 10%), reaching 90% of accuracy in the case of the REAL dataset. However, it should be noted that although the CSP can allow a single trial classification [80], for the calculation of the matrix of spatial filters it is necessary to use an appropriate amount of data known in advance. As far as the classifier aspect is concerned, asSWLDA obtained the highest performance in both datasets using the PSD features (0.67 AUC LAB and 0.86 AUC REAL) and significantly higher performance than the LDA using the CSP features (0.76 AUC). The combination of SVM and FBCSP proved to be clearly successful (AUC 0.9) for the REAL dataset, probably thanks to the generalizing capabilities and the overtraining insensitivity typical of the SVM [102]. Finally, as expected from the literature, the use of unsupervised techniques has led to the lowest performance, always on average less than 70% of accuracy.

6.4 Limits, possible solutions, and future trends

This work provides an image of what has become the BCI in the current years, namely the passive BCI, highlighting both the technological and methodological aspects that have contributed to this evolution. Retracing the salient aspects of this chapter in terms of both methodology and technology, it can be seen that the roads of active BCI and passive BCI research are strongly interconnected. During the Sixth International Brain–Computer Interface meeting, one of the main topics for discussion was the analysis of specific signals or technology for BCI advancing. The first point of the discussion was *exploiting the cognitive processes related to the human–machine interaction*, because are assumed to be naturally elicited during the task and then they could be used in two ways: to covertly control an external devices and to know about the user state in order to optimize the user efficiency and BCI experience. The passive BCI facilities, in fact, could be used in parallel to active BCI systems to monitor, for instance, the mental workload. This could be particularly useful, for example, when BCI is used for rehabilitation because this allows to adapt the rehab session and difficulty to the subject, optimizing time and resources. The second point emerged from the BCI meeting was the *necessity to investigate signal analysis techniques and algorithms to improve BCI functions*. Actually, this methodological aspect is superimposable in the active and passive BCI research, therefore every methodological enhancement in one sense could have beneficial effects also on the other. Finally, the third point covered the *hardware developing*. The necessity to limit the invasiveness of acquisition devices in terms of number of electrodes, dimension and usability involves both the active and passive BCI. Accordingly, active and passive BCI also share the same gaps that need to be filled to allow the BCI technology entering everydayness.

To have a system that works in the most diverse conditions, it is necessary to have a unique link between many, ideally all, the possible human mental states. This mapping is missing today, both in the laboratory and, especially, in the real environment. The motivations are essentially two. The first is that mental states are often not independent. That is, a change in workload is often associated with a change in fatigue [103], or there is an undeniable link between stress and attention. In each of these cases, obviously in the laboratory it is easier to discriminate the variation of the single component, i.e., individually workload, attention or stress, because specific tasks are used to elicit that particular component and the boundary conditions in which the activity is carried out are known. Certainly, in a realistic context this independence is lacking [103]. The second motivation, which is a direct consequence of the first, is due to the fact that neurophysiological indicators are too sensitive and not very specific in realistic contexts. For example, a strong alpha-band activity typical of a visual task could hide a variation of parietal alpha due to workload variation. At the same time, an artifact typical of the activity that is taking place could erroneously be selected as features by the system, but it would not be an indicator of the mental state, thus leading to an incorrect classification. In this regard, Fairclough, in his review work on physiological computing systems

[104], stressed such issue to export neurophysiological measures from the laboratory to more realistic applications. In particular, these measures must be tested with a full range of the target population in the field, hence the requirement to establish neurophysiological validity. Neurophysiological measures should be selected on the basis of diagnostic (i.e., the ability of a variable to index target mental state and remained unaffected by related states), sensitivity (i.e., the ability of a variable to respond rapidly to changes in mental state) and reliability (i.e., the consistency of the neurophysiological inference across different individuals and environments). One possible solution to these gaps may be to increase the degrees of freedom, using more neurophysiological signals and fewer constraints for the indexes modeling. This would mean being able to evaluate several mental states simultaneously. Moreover, it is necessary to have an appropriate resolution, which means that it is essential to move from a binary classification to a multi-class classification. In order to obtain accurate indexes it is therefore necessary to provide for an appropriate transfer rate that can allow a real-time evaluation of the mental state.

A further factor that causes a decrease in the usability of the BCI System is the need to perform long and repeated calibrations of the system to allow optimal accuracy of the estimation of the mental state in question. Cross-task calibration does not seem to be a solution to date, and, furthermore, the calibration is affected by the different positions and conditions of the electrodes between one installation and another. In this regard, research has to take several steps forward in improving unsupervised algorithms, that do not require calibration data but still suffer from lower performance with respect to supervised ones. The solution would be to adopt methods that do not require frequent calibration [89] or a semi-supervised approach [105]. In general, it would be very useful to validate existing signal processing chains and classification techniques [106] to find the best algorithm for each investigated mental state and application, however also basic knowledge such as the EEG channels reference has been still on the debate during the last decades [107,108].

The passive BCI evolution is expected to enhance the security of several activities, like for example driving or monitoring, thanks to the fact that knowing the mental state of an operator, the passive BCI system could be able to detect his distraction, fatigue or lack of vigilance avoiding accidents. At the same time, it could be used to understand the causes of human error and performance degradation without the unreliability of subjective measures. The employment of the passive BCI can help in the design of new technologies: for example, by evaluating the workload, it is possible to get an idea of the cognitive resources employed while using a device. Therefore could be promoted the design and the choice of devices and interfaces that allow to use as few cognitive resources as possible. Even more so, the passive BCI outcomes could be used to trigger an Adaptive automation, i.e., the adaptation of the interface of the instrument in relation to the condition of the subject. That is, when the user gets cognitively overloaded, the interface could present less information until the user has more free cognitive resources to better deal with the details. Finally, the passive BCI could be considered a teaching aid, because it allows to adapt training and in general lessons, to student condition monitoring his level of attention or workload.

Since the high number of passive BCI-related works performed in real settings published in the last few years [109], and the big effort that is being done to address the above-mentioned issues, it is possible to infer that passive BCI systems are not too far from being introduced as new consumer product, and in the near future it will not be a science fiction to imagine smart devices that adapt their behavior depending on the user's insight. Finally, the question remaining is: given the growing interest in this area and the technological developments that have allowed the commercialization of many devices used for monitoring biosignals, when will start the commercialization of a passive BCI system? Some years ago estimates of the time to market were made but, in some ways, they have been disregarded [15]. The calibration and EEG technology were identified as the two major issues for marketing, and it is interesting to highlight how such issues have been not totally addressed so far.

References

[1] Berger H. Über das elektrenkephalogramm des menschen. Arch Psychiatr Nervenkr. 1929;87(1):527–70.

[2] Black AH. The operant conditioning of central nervous system electrical activity. Vol. 6. Academic Press: New York, NY, USA; 1972.

[3] Vidal JJ. Toward direct brain-computer communication. Annu Rev Biophys Bioeng. 1973;2(1):157–80.

[4] Wolpaw JR, Birbaumer N, McFarland DJ, Pfurtscheller G, and Vaughan TM. Brain–computer interfaces for communication and control. Clin Neurophysiol. 2002;113(6):767–91.

[5] Shih JJ, Krusienski DJ, and Wolpaw JR. Brain-computer interfaces in medicine. In: Mayo Clinic Proceedings. Elsevier; 2012. p. 268–79.

[6] Barrett LF. The future of psychology: connecting mind to brain. Perspect Psychol Sci. 2009;4(4):326–39.

[7] Zander TO, PhyPA T, and Jatzev S. Detecting affective covert user states with passive brain-computer interfaces. In: 2009 3rd International Conference on Affective Computing and Intelligent Interaction and Workshops, Amsterdam, IEEE; 2009. p. 1–9.

[8] Schettini F, Aloise F, Aricò P, Salinari S, Mattia D, and Cincotti F. Self-calibration algorithm in an asynchronous P300-based brain–computer interface. J Neural Eng. 2014;11(3):35004.

[9] Aloise F, Aricò P, Schettini F, Salinari S, Mattia D, and Cincotti F. Asynchronous gaze-independent event-related potential-based brain–computer interface. Artif Intell Med. 2013;59(2):61–9.

[10] Wolpaw J, and Wolpaw EW. Brain-computer interfaces: principles and practice. OUP USA; 2012.

[11] Aricò P, Borghini G, Di Flumeri G, *et al.* Adaptive automation triggered by EEG-based mental workload index: a passive brain-computer interface

application in realistic air traffic control environment. Front Hum Neurosci. 2016;10:539.
[12] Khan MJ, and Hong K. Passive BCI based on drowsiness detection: an fNIRS study. Biomed Opt Express. 2015;6(10):4063–78.
[13] Aricò P, Reynal M, Di Flumeri G, *et al.* How neurophysiological measures can be used to enhance the evaluation of remote tower solutions. Front Hum Neurosci. 2019;13:303.
[14] Aricò P, Borghini G, Di Flumeri G, Sciaraffa N, Colosimo A, and Babiloni F. Passive BCI in operational environments: insights, recent advances, and future trends. IEEE Trans Biomed Eng. 2017;64(7):1431–6.
[15] Erp JBF Van Lotte F, Tangermann M, *et al.* Brain-computer interfaces: beyond medical applications to cite this version: HAL Id: hal-00688344. 2012.
[16] Arico P, Borghini G, Di Flumeri G, *et al.* Human factors and neurophysiological metrics in air traffic control: a critical review. IEEE Rev Biomed Eng. 2017;10:250–263.
[17] Ma T, Li H, Deng L, *et al.* The hybrid BCI system for movement control by combining motor imagery and moving onset visual evoked potential. J Neural Eng. 2017;14(2):26015.
[18] Di Flumeri G, Borghini G, Aricò P, *et al.* EEG-based mental workload assessment during real driving: a taxonomic tool for neuroergonomics in highly automated environments. In: Neuroergonomics. London: Academic Press; 2018. p. 121–126.
[19] Herff C, Putze F, and Schultz T. Evaluating fNIRS-based workload discrimination in a realistic driving scenario. The First Biannual Neuroadaptive Technology, Conference, Berlin, 2017. p. 69–70.
[20] Brouwer A, Snelting A, Krol L, Jaswa M, and Zander T. Physiological effects of adaptive cruise control behaviour in real driving. In: Proceedings of the 2017 ACM Workshop on An Application-oriented Approach to BCI out of the laboratory. 2017;15–9.
[21] Perrier J, Jongen S, Vuurman E, Bocca ML, and Ramaekers JG. Driving performance and EEG fluctuations during on-the-road driving following sleep deprivation. Biol Psychol [Internet]. 2016;121:1–11.
[22] Liang Y, Horrey WJ, Howard ME, *et al.* Prediction of drowsiness events in night shift workers during morning driving. Accid Anal Prev [Internet]. 2017;(May):0–1.
[23] Maglione A, Borghini G, Aricò P, *et al.* Evaluation of the workload and drowsiness during car driving by using high resolution EEG activity and neurophysiologic indices. In: 2014 36th Annual International Conference of the IEEE Engineering in Medicine and Biology Society. 2014. p. 6238–41.
[24] Miklody D, Uitterhoeve WM, Van Heel D, Klinkenberg K, and Blankertz B. Maritime cognitive workload assessment. In 2017. p. 102–14.
[25] Dehais F, Roy RN, Durantin G, and Gateau T. EEG-engagement index and auditory alarm misperception: an inattentional deafness study in actual flight

condition EEG-engagement index and auditory alarm misperception: an inattentional deafness study in actual flight condition. 2017;(June).

[26] Parasuraman R, and McKinley RA. Using noninvasive brain stimulation to accelerate learning and enhance human performance. Hum Factors. 2014;56 (5):816–24.

[27] Liu Y, and Wickens CD. Mental workload and cognitive task automaticity: an evaluation of subjective and time estimation metrics. Ergonomics. 1994;37(11): 1843–54.

[28] Kelly AMC, and Garavan H. Human functional neuroimaging of brain changes associated with practice. Cereb cortex. 2004;15(8):1089–102.

[29] Borghini G, Aricò P, Di Flumeri G, *et al.* A new perspective for the training assessment: Machine learning-based neurometric for augmented user's evaluation. Front Neurosci. 2017;11:325.

[30] Carelli L, Solca F, Faini A, *et al.* Brain-computer interface for clinical purposes: cognitive assessment and rehabilitation. Biomed Res Int. 2017;2017 (1695290):1–11.

[31] Zander TO, Shetty K, Lorenz R, *et al.* Automated task load detection with electroencephalography: towards passive brain–computer interfacing in robotic surgery. Journal of Medical Robotics Research. 2017;(October).

[32] Borghini G, Aricó P, Di Flumeri G, *et al.* Neurophysiological measures for users' training objective assessment during simulated robot-assisted laparoscopic surgery. In: Engineering in Medicine and Biology Society (EMBC), 2016 IEEE 38th Annual International Conference of the. IEEE; 2016. p. 981–4.

[33] Borghini G, Aricò P, Di Flumeri G, *et al.* EEG-based cognitive control behaviour assessment: an ecological study with professional air traffic controllers. Sci Rep. 2017;7(1):547.

[34] Borghini G, Aricò P, Astolfi L, *et al.* Frontal EEG theta changes assess the training improvements of novices in flight simulation tasks. In: 2013 35th Annual International Conference of the IEEE Engineering in Medicine and Biology Society (EMBC). 2013. p. 6619–22.

[35] Ko L, Komarov O, Hairston WD, and Jung T. Sustained attention in real classroom settings: An EEG Study. 2017;11:1–10.

[36] Martišius I, and Damaševičius R. A prototype SSVEP based real time BCI gaming system. Comput Intell Neurosci. 2016;2016(3861425):1–15.

[37] Krol LR, Zander TO, Jaswa M, *et al.* Online-capable cleaning of highly artefactual eeg data recorded during real driving. In: Proceedings of the 7th Graz Brain-Computer Interface Conference. 2017;254–60.

[38] Ewing KC, Fairclough SH, and Gilleade K. Evaluation of an adaptive game that uses EEG measures validated during the design process as inputs to a biocybernetic loop. Front Hum Neurosci. 2016;10(May):223.

[39] Vecchiato G, Borghini G, Aricò P, *et al.* Investigation of the effect of EEG-BCI on the simultaneous execution of flight simulation and attentional tasks. Med Biol Eng Comput. 2016;54(10):1503–13.

[40] Shishkin SL, Nuzhdin YO, Svirin EP, and Trofimov AG. EEG Negativity in fixations used for gaze-based control: toward converting intentions into actions with an eye-brain-computer interface. 2016;10(November).

[41] Cherubino P, Trettel A, Cartocci G, *et al.* Neuroelectrical indexes for the study of the efficacy of TV advertising stimuli. In: Selected Issues in Experimental Economics. Cham: Springer; 2016. p. 355–71.

[42] Di Flumeri G, Herrero MT, Trettel A, *et al.* EEG frontal asymmetry related to pleasantness of olfactory stimuli in young subjects. In: Selected Issues in Experimental Economics. Cham: Springer; 2016. p. 373–81.

[43] Di Flumeri G, Aricò P, Borghini G, *et al.* EEG-based approach-withdrawal index for the pleasantness evaluation during taste experience in realistic settings. In: Engineering in Medicine and Biology Society (EMBC), 2017 39th Annual International Conference of the IEEE. IEEE; 2017. p. 3228–31.

[44] Borghini G, Aricò P, Di Flumeri G, *et al.* Avionic technology testing by using a cognitive neurometric index: a study with professional helicopter pilots. Conf Proc Annu Int Conf IEEE Eng Med Biol Soc IEEE Eng Med Biol Soc Annu Conf. 2015;2015:6182–5.

[45] Aricò P, Reynal M, Imbert J-P, *et al.* Human-machine interaction assessment by neurophysiological measures: a study on professional air traffic controllers. In: EBMC 2018, 40th International Engineering in Medicine and Biology Conference. 2018.

[46] Chuang J, Nguyen H, Wang C, and Johnson B. I think, therefore I am: usability and security of authentication using brainwaves. 0424422:1–16.

[47] Nakanishi I. Driver authentication using brain waves while route tracing as a mental driver authentication using brain waves while. 2015;(June).

[48] Shah PG, Vastare KC, Huang X, *et al.* Development of a novel EEG wave controlled security system. In: Intelligent Computing and Information Systems (ICICIS), 2015 IEEE Seventh International Conference on. IEEE; 2015. p. 116–20.

[49] Karthikeyan T. Enhancement of multi-modal biometric authentication based on IRIS and brain neuro image coding. 2011;(5):249–56.

[50] Cartocci G, Maglione AG, Modica E, *et al.* The “NeuroDante project”: neurometric measurements of participant’s reaction to literary auditory stimuli from Dante’s “Divina Commedia.” In: International Workshop on Symbiotic Interaction. Springer; 2016. p. 52–64.

[51] Fallahi M, Motamedzade M, Heidarimoghadam R, Soltanian AR, and Miyake S. Effects of mental workload on physiological and subjective responses during traffic density monitoring: a field study. Appl Ergon. 2016;52:95–103.

[52] von Lühmann A., Soekadar S., Müller K-R., and Blankertz B. Headgear for mobile neurotechnology looking into alternatives for Eeg and Nirs Probes. 2017;(April). Available from: https://www.researchgate.net/profile/Alexander_Von_Luehmann/publication/316216040_Headgear_for_mobile_neurotech

nology_looking_into_alternatives_for_EEG_and_NIRS_probes/links/58f64a8d0f7e9b67a34b84c3/headgear-for-mobile-neurotechnology-looking-into-alterna

[53] Yu Y, Chen S, Chang C, Lin C, Hairston WD, and Mrozek RA. New flexible silicone-based EEG dry sensor. 2007.

[54] Amaral CP, Simões MA, Mouga S, Andrade J, and Castelo-Branco M. A novel brain computer interface for classification of social joint attention in autism and comparison of 3 experimental setups: A feasibility study. J Neurosci Methods. 2017;290:105–15.

[55] Vourvopoulos A, Niforatos E, Hlinka M, Skola F, and Liarokapis F. Investigating the effect of user profile during training for BCI-based games. In: 2017 9th International Conference on Virtual Worlds and Games for Serious Applications (VS-Games). IEEE; 2017. p. 117–24.

[56] Di Flumeri G, Aricò P, Borghini G, Sciaraff N, Di Florio A, and Babiloni F. The dry revolution: evaluation of three different EEG dry electrode types in terms of signal spectral features, mental states classification and usability. Sensors. 2019;19(6):1365.

[57] Debener S, Emkes R, De Vos M, and Bleichner M. Unobtrusive ambulatory EEG using a smartphone and flexible printed electrodes around the ear. Sci Rep [Internet]. 2015;5:1–11. Available from: http://dx.doi.org/10.1038/srep16743

[58] Nguyen A, Alqurashi R, and Raghebi Z. In-ear biosignal recording system: a wearable for automatic whole-night sleep staging. 19–24.

[59] Lin CT, Chuang CH, Huang CS, *et al.* Wireless and wearable EEG system for evaluating driver vigilance. IEEE Trans Biomed Circuits Syst. 2014;8(2):165–76.

[60] Procházka A, Schätz M, and Vališ M. Breathing and heart rate analysis. 2016;1–11.

[61] Garbarino M, Lai M, Bender D, Picard RW, and Tognetti S. Empatica E3—A wearable wireless multi-sensor device for real-time computerized biofeedback and data acquisition. In: Wireless Mobile Communication and Healthcare (Mobihealth), 2014 EAI 4th International Conference on. IEEE; 2014. p. 39–42.

[62] Blechert J, Peyk P, Liedlgruber M, and Wilhelm FH. ANSLAB: integrated multichannel peripheral biosignal processing in psychophysiological science. Behav Res Methods [Internet]. 2016;48(4):1528–45.

[63] von Lühmann A, and Müller K. M3BA: New technology for mobile hybrid BCIs M3BA: new technology for mobile hybrid BCIs. 2016;(June):9–10.

[64] Vora S, and Kurzweg TP. Unconventional biosignal sensing with passive RFID tags. Bridg. 2016;112(3):25.

[65] Di Flumeri G, Aricò P, Borghini G, Colosimo A, and Babiloni F. A new regression-based method for the eye blinks artifacts correction in the EEG signal, without using any EOG channel. Vol. 2016, Conference proceedings: ... Annual International Conference of the IEEE Engineering in Medicine and Biology Society. IEEE Engineering in Medicine and Biology Society. Conference. 2016. p. 3187–3190.

[66] Di Flumeri G, Aricò P, Borghini G, *et al.* EEG-based mental workload neurometric to evaluate the impact of different traffic and road conditions in real driving settings. Front Hum Neurosci. 2018;12:509.
[67] Suleiman AR, and Fatehi TA. Features extraction techniques of eeg signal for BCI applications. 2018;(January).
[68] Borghini G, Astolfi L, Vecchiato G, Mattia D, and Babiloni F. Measuring neurophysiological signals in aircraft pilots and car drivers for the assessment of mental workload, fatigue and drowsiness. Vol. 44, Neuroscience and Biobehavioral Reviews. Elsevier Ltd; 2014. p. 58–75.
[69] Mandrick K, Peysakhovich V, Rémy F, Lepron E, and Causse M. Neural and psychophysiological correlates of human performance under stress and high mental workload. Biol Psychol. 2016;121:62–73.
[70] Singh M, and Sharma A. Correlational study of attention task performance and EEG alpha power. Int J Inf Technol Knowl Manag. 2015;8(2):188–96.
[71] Akin M. Comparison of wavelet transform and FFT methods in the analysis of EEG signals. J Med Syst. 2002;26(3):241–7.
[72] Polich J. Updating P300: an integrative theory of P3a and P3b. Clin Neurophysiol. 2007;118(10):2128–48.
[73] Scheffers MK, and Coles MGH. Performance monitoring in a confusing world: error-related brain activity, judgments of response accuracy, and types of errors. J Exp Psychol Hum Percept Perform. 2000;26(1):141.
[74] Frey J, Appriou A, Lotte F, and Hachet M. Classifying EEG signals during stereoscopic visualization to estimate visual comfort. Computational intelligence and neuroscience, 2016:7.
[75] Hazarika N, Chen JZ, Tsoi AC, and Sergejew A. Classification of EEG signals using the wavelet transform. In: Digital Signal Processing Proceedings, 1997 DSP 97, 1997 13th International Conference on. IEEE; 1997. p. 89–92.
[76] Lee B-G, Lee B-L, and Chung W-Y. Mobile healthcare for automatic driving sleep-onset detection using wavelet-based EEG and respiration signals. Sensors. 2014;14(10):17915–36.
[77] Sun Y, Lim J, Meng J, Kwok K, Thakor N, and Bezerianos A. Discriminative analysis of brain functional connectivity patterns for mental fatigue classification. Ann Biomed Eng. 2014;42(10):2084–94.
[78] Donchin E, Spencer KM, and Wijesinghe R. The mental prosthesis: assessing the speed of a P300-based brain-computer interface. IEEE Trans Rehabil Eng. 2000;8(2):174–9.
[79] Antonacci Y, Toppi J, Caschera S, Anzolin A, Mattia D, and Astolfi L. Estimating brain connectivity when few data points are available: Perspectives and limitations. In: Engineering in Medicine and Biology Society (EMBC), 2017 39th Annual International Conference of the IEEE. IEEE; 2017. p. 4351–4.
[80] Blankertz B, Tomioka R, Lemm S, Kawanabe M, and Muller K-R. Optimizing spatial filters for robust EEG single-trial analysis. IEEE Signal Process Mag. 2008;25(1):41–56.

[81] Fukunaga K. Introduction to statistical pattern classification. Academic Press USA; 1990.
[82] Lotte F, and Guan C. Regularizing common spatial patterns to improve BCI designs: unified theory and new algorithms. IEEE Trans Biomed Eng. 2011;58(2):355–62.
[83] Wolpaw JR, Birbaumer N, Heetderks WJ, *et al.* Brain-computer interface technology: a review of the first international meeting. IEEE Trans Rehabil Eng. 2000;8(2):164–73.
[84] Lemm S, Blankertz B, Dickhaus T, and Müller K-R. Introduction to machine learning for brain imaging. Neuroimage. 2011;56(2):387–99.
[85] Mountrakis G, Im J, and Ogole C. Support vector machines in remote sensing: A review. ISPRS J Photogramm Remote Sens. 2011;66(3):247–59.
[86] Fortmann-Roe S. Understanding the bias-variance tradeoff. 2012.
[87] Weiss SM, and Kulikowski CA. Computer systems that learn: classification and prediction methods from statistics, neural nets, machine learning, and expert systems. San Francisco, CA: Morgan Kaufmann Publishers Inc.; 1991.
[88] Vapnik V. The Nature of Statistical Learning Theory. New York, NY: Springer; 2000.
[89] Aricò P, Borghini G, Di Flumeri G, Colosimo A, Pozzi S, and Babiloni F. A passive brain–computer interface application for the mental workload assessment on professional air traffic controllers during realistic air traffic control tasks. Prog Brain Res. 2016;228:295–328.
[90] Schultze-Kraft M, Dähne S, Gugler M, Curio G, and Blankertz B. Unsupervised classification of operator workload from brain signals. J Neural Eng. 2016;13(3):36008.
[91] Wei C-S, Lin Y-P, Wang Y-T, Lin C-T, and Jung T-P. A subject-transfer framework for obviating inter-and intra-subject variability in EEG-based drowsiness detection. NeuroImage. 2018;174:407–19.
[92] Blankertz B, Acqualagna L, Dähne S, *et al.* The Berlin brain-computer interface: progress beyond communication and control. Front Neurosci. 2016;10:530.
[93] Huggins JE, Arbor A, Guger C, Allison B, Anderson CW, and Lotte F. Meeting: defining the future. Vol. 1. 2015. 27–49 p.
[94] Zheng W, and Lu B. Investigating critical frequency bands and channels for EEG-based emotion recognition with deep neural networks. IEEE Transactions on Autonomous Mental Development, 2015;7(3):162–75.
[95] Heilmeyer FA, Schirrmeister RT, and Martin V. A large-scale evaluation framework for EEG deep learning architectures.
[96] Wickens CD. Multiple resources and mental workload. Hum Factors. 2008;50(3):449–55.
[97] Byrne EA, and Parasuraman R. Psychophysiology and adaptive automation. Biol Psychol. 1996;42(3):249–68.
[98] Berka C, Levendowski DJ, Lumicao MN, *et al.* EEG correlates of task engagement and mental workload in vigilance, learning, and memory tasks. Aviat Space Environ Med. 2007;78(5):B231–44.

[99] Metzger U, and Parasuraman R. Automation in future air traffic management: effects of decision aid reliability on controller performance and mental workload. Hum Factors. 2005;47(1):35–49.

[100] Delorme A, and Makeig S. EEGLAB: an open source toolbox for analysis of single-trial EEG dynamics including independent component analysis. J Neurosci Methods. 2004;134(1):9–21.

[101] Bamber D. The area above the ordinal dominance graph and the area below the receiver operating characteristic graph. J Math Psychol. 1975;12 (4):387–415.

[102] Wu W, Gao X, Hong B, and Gao S. Classifying single-trial EEG during motor imagery by iterative spatio-spectral patterns learning (ISSPL). IEEE Trans Biomed Eng. 2008;55(6):1733–43.

[103] Roy RN, Charbonnier S, Campagne A, and Bonnet S. Efficient mental workload estimation using task-independent EEG features. J Neural Eng. 2016;13(2):26019.

[104] Fairclough SH. Fundamentals of physiological computing. Interact Comput. 2009;21(1–2):133–45.

[105] Lotte F, Bougrain L, Cichocki A, *et al.* A review of classification algorithms for EEG-based brain–computer interfaces: a 10 year update. J Neural Eng. 2018;15(3):31005.

[106] Sciaraffa N, Aricò P, Borghini G, Di Flumeri G, Di Florio A, and Babiloni F. On the use of machine learning for EEG-Based workload assessment: algorithms comparison in a realistic task. In: International Symposium on Human Mental Workload: Models and Applications. Springer, Cham; 2019. p. 170–185.

[107] Yao D. Is the surface potential integral of a dipole in a volume conductor always zero? a cloud over the average reference of EEG and ERP. Brain Topogr. 2017;30(2):161–71.

[108] Yao D. A method to standardize a reference of scalp EEG recordings to a point at infinity. Physiol Meas. 2001;22(4):693–711.

[109] Aricò P, Borghini G, Di Flumeri G, Sciaraffa N, and Babiloni F. Passive BCI beyond the lab: current trends and future directions. Physiol Meas. 2018;39(8): 08TR02.

Chapter 7

Neurorobotics: review of underlying technologies, current developments, and future directions

Christos Dimitrousis[1], Sofia Almpani[2], Petros Stefaneas[3], Jan Veneman[4], Kostas Nizamis[5], and Alexander Astaras[1]

7.1 Introduction

Neurorobotics is an interdisciplinary scientific field which focuses on embodied neural systems and spans scientific theory, research, development and clinical medical practice. It involves a variety of science and engineering disciplines, most frequently electronic, mechanical and control engineering (sometimes alternatively described as electrical, automation, computer or automation engineering), informatics, computer science, software engineering and artificial intelligence (AI), while in the faculty of health sciences it is often associated with neuroscience, neurophysiology and neurosurgery.

Due to the element of robotics, apart from involving some type of neural system, neurorobotic devices also need to be able to sense their environment, to possess programmable computational processing power, some form of actuation (movement) as well as a certain degree of autonomy.

While virtual reality (VR) robotics simulations can also be considered part of the neurorobotics field, most neurorobots are required to function in a dynamic and often unpredictable real-world environment. The field also includes work on robotic control devices interfacing directly with the central nervous system, including brain–computer interfaces (BCIs) used to control robots.

The general paradigm found at the heart of the neurorobotics field is that the mind is embodied, while the body is embedded in its environment. The power of

[1]Division of Technology and Science, American College of Thessaloniki, Thessaloniki, Greece
[2]School of Electrical and Computer Engineering, National Technical University of Athens, Athens, Greece
[3]School of Applied Mathematics and Physical Sciences, National Technical University of Athens, Athens, Greece
[4]Hocoma AG, Volketswil, Switzerland
[5]Department of Design, Production and Management, University of Twente, Enschede, The Netherlands

this paradigm stems from the rich input environment in which a natural or artificial "mind" can operate, interact, learn and evolve.

This holds particularly true when robotics is combined with machine learning, an AI field which has seen a dramatic increase in its impact over the past two decades. A robotic body can act as a conduit for a machine learning algorithm, helping evolve a computational "mind" in a sensory-rich, dynamic and unpredictable environment. Such unpredictable, rich sensory data sampled into multi-dimensional data spaces can both provide a challenge and an opportunity for evolving complex machine learning AI entities.

At the same time, such environments cannot generally be navigated by pre-programmed expert system algorithms or other pre-programmed sets of algorithmic rules, due to their unpredictability and infinite variability. They can however be efficiently navigated by algorithms which have evolved in the same or similar environment, under the guidance of an appropriate machine learning scheme.

The field of human–robot interfaces studies the interactive communication between humans and robots. It involves programming robots to respond to control queues obtained from human behavior, sound and movement, or even direct interfacing to the human central nervous system—including BCIs. The latter scenario can be particularly important in cases of medical research aiming to enable or improve control of robotic prosthesis; neural rehabilitation using robotic assistance; physical rehabilitation using robotic assistance; as well as experimentation aiming to improve scientific understanding of the human and animal sensory and neural processing systems. Finally, there is a small field of research and development activities aiming to develop robotic setups and devices using biological tissues, including biological neural control systems.

The present book chapter focuses more on the neuroscience, electronics and AI aspects of neurorobotics. It includes research, development and experimentation taking place in virtual environments, so long as the latter explicitly aim at the development of robotic control devices to be eventually embodied in physical robotics setups [1]. It will also focus on research activity targeting or involving the human central nervous system, so long the sensorimotor loop is somehow involved in robotic control [2].

This work will focus less on research, development and experimentation with general neuro-inspired Informatics algorithms, electro-mechanical robotics, AI algorithms and robotic systems without an explicit biological neural connection, and biological neuron cultures grown on electronics substrates without subsequent embodiment in a robotic setup. Machine learning using artificial neural networks, (ANNs) in order to control a robot, without explicit interfacing to a biological neural network, is considered outside the scope of this chapter (Figure 7.1).

7.2 State of the art: underlying technologies

Robotics always entails sensing, data processing and actuation, while neurorobotics additionally implicates a robotic interface to a neural system, be it biological or in silico. In this section, we consider the technologies which make neurorobotics

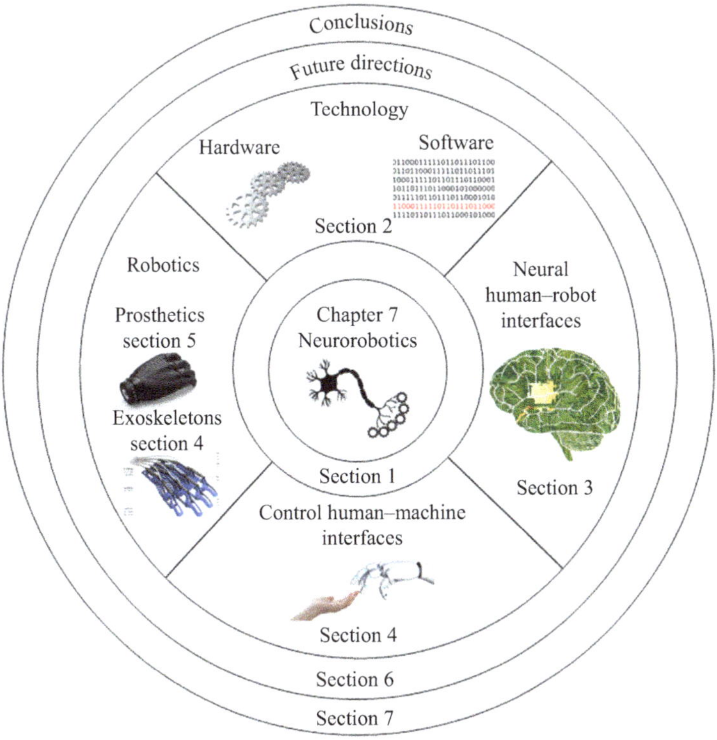

Figure 7.1 Illustrated schematic overview of the contents of this chapter

research and development possible, particularly those driving development in the present and those likely to be shaping the field in the immediate and midterm future. We discuss advances in electronics and software design, electromechanics and neural-silicon interfaces.

7.2.1 Advances in electronics

Since the invention of the transistor in the 1960s, silicon-based electronics has been the technological platform of choice for most designers of computation machines. Partly due to technological limitations and partly due to economic circumstances, the digital machines that have prevailed are almost exclusively based on the Church–Turing hypothesis and the von Neumann architecture, eclipsing alternative forms of computation. Such designs are characterized by a clear separation between hardware and programming code, as well as between programming code and data. They are also designed to implement deterministic algorithms, are notoriously fault-intolerant and ultimately limited by inevitable bottlenecks in data flow through their serial computation CPUs.

Such computation machines were among the first designs to become widely available and managed to overcome—or manoeuvre around—most serious obstacles to achieve fast automated computation. They consequently made it to market before competing computational architectures, such as bio-inspired analog and parallel distributed processing designs. Market-driven investment subsequently ensured that these serial devices became the only prevalent designs, based on digital chips clocked at ever faster frequencies and—mostly in the twenty-first century—multiple CPU cores per chip (a pseudo-parallel design). Analog and parallel distributed computation architectures were largely marginalized as theoretical or laboratory curiosities, with few working hardware prototypes being built and even fewer making it to market.

Despite their limitations, serial digital computers accommodated, shaped and eventually disrupted most markets around the world during the past 70 years. In the process they have driven the electronics design, fabrication, testing and retail industry to become one of the most valuable global markets, helping fund the rate of improvement and refinement of underlying technologies to unprecedented levels.

7.2.1.1 Deep sub-micron VLSI and neural computation

The drive toward more efficient computation, from the point of view of computational speed, power consumption, miniaturization and fabrication cost, has been directing electronics designers to increasingly smaller feature size electronic chips, operating at higher frequencies. Very large-scale integration (VLSI) in electronics design has seen a consistent trend in exponential miniaturization which has lasted for several decades, famously described as "Moore's Law". By 2018 this process had achieved minimum transistor sizes deep in the sub-micron scale and minimum feature sizes down to a few nanometers.

This persistent trend toward miniaturization and monolithic system-on-chip integration has produced cheaper, faster performing and more power-efficient mixed signal electronics, but comes at a price: as electromagnetically noisy digital components come closer to noise-sensitive analog amplifiers, cross-talk and noise propagation quickly become limiting factors preventing faster operating frequencies and further miniaturization; thermal noise and power dissipation become increasingly challenging; and quantum tunneling effects exacerbate crosstalk and can add uncertainty to fault-intolerant digital computation circuits.

The aforementioned problems have led several researchers to predict the end of the line for Moore's law. Still, the economic incentives to stay on the curve of Moore's law are strong enough that several paths forward have been proposed and currently being pursued: switching from silicon to carbon nanotube technologies, building electronic circuits in 3D rather than 2D layers, or even switching from digital deterministic to stochastic neural computation. Some of these scenarios could have a profound impact for research, development and adoption of neurorobotics technology.

7.2.1.2 Parallel distributed processing and neural computation

The option of adopting noise-tolerant stochastic neural computation elements may be of importance to the field of neurorobotics, as it is closer to the engineering solution selected by biological evolution. It could therefore turn out to be more compatible when attempting to interface analog electronics to a biological organism's nervous system. After all animal nervous systems rely on imprecise, noisy neural computation yet manage to generally outperform modern digital computers in learning, adaptation, generalization and pattern recognition tasks when, speed, fault-tolerance, volume and power consumption are considered. This idea is decades old, however the electronic components to build this type of novel hardware (especially reliable analog on-chip memory), as well as the accompanying embedded AI algorithms did not become available until recently.

Another obvious advantage of the analog neural computation paradigm is massive parallelism. By bypassing the bottleneck limitations of a serial digital CPU which is based on the von Neumann architecture, and taking advantage of fast analog hardware that can perform asynchronously, it is expected that analog hardware could approximate and eventually overcome the performance of similar biological neural circuits.

This technological option has been heavily researched since the 1980s; however, progress was delayed due mostly to technical hurdles. The most disruptive among those has been the lack of stable, reliable long-term analog storage which could be integrated on VLSI silicon chips.

7.2.1.3 Memristors, a missing piece of the technological puzzle

The lack of reliable analog memory components on silicon chips proved to be a decisive factor in the development direction for most computing systems, with profound consequences in the fields of neurorobotics, robotics and automated data acquisition systems in general. The discovery in 2008 of an electronics element which could reliably store an analog value on a silicon chip for years without power supply was therefore heralded as a game-changing discovery [3].

For biologically inspired neural systems, this could be the most substantial missing piece of the technological development puzzle. Massively parallel analog neural network computation devices on silicon had to circumvent the lack of reliable analog storage on chip, either by attempting to use analog capacitors (which leaked in a time frame of milliseconds) or by digitizing analog synaptic weights for long-term storage in digital memory. Both options however stood in the way of fast-flowing, massively parallel analog computation devices, capable of interfacing to analog sensors and actuators operating in an analog real-world environment.

Adapting current silicon chip fabrication processes to accommodate for the mass manufacturing of memristor elements has proven a challenging task, while integrating memristors onto monolithic System-on-Chip designs would be the subsequent milestone. The economic incentives are strong enough to support ongoing efforts toward both goals, as this technology could help map machine learning algorithms on silicon using neuro-inspired electronics and other forms of analog AI algorithms. Efforts to streamline the technology for mass fabrication are ongoing, including testing memristors for use in harsh environments, such as those required for industrial, field robotics and space applications. In the year 2015 the first memristor chips were offered as commercially available off-the-shelf products (Knowm Inc., USA), specifically targeting machine algorithms running of "neuromemristive processors", implemented using memristors and mixed signal electronics [4].

It is possible that in the future hybrid electronic devices will emerge, neuro-inspired analog processors mediating among digital CPUs, analog sensors and actuators. For instance, multicore system-on-chip devices could comprise both von-Neumann digital processor CPU cores and analog neural processor cores, capable of massive parallelism for machine learning tasks. Such devices could be valuable both from the point of view of fast, autonomous machine learning and for neurorobotics interfacing purposes.

7.2.1.4 From MEMS to NEMS and beyond

Nanotechnology, the chemical engineering design of substances in nanoparticle form has been touted to be a promising technology which can help tackle many a scientific challenge. Though it received less publicity, over the past two decades the miniaturization of electronics has been quietly providing an alternative route to the nanoscale world. Analog VLSI design has been adapted to shape layers of polysilicon and metal, allowing for 3D electro-mechanical systems (MEMS) to be built outside the confines of mainstream electronic components. Novel sensors, microfluidic actuators, lab-on-chip devices and DNA chips are only some of the innovations made possible using this creative repurposing of existing technology.

Deep sub-micron chip fabrication technologies have naturally helped decrease the size of MEMS, with some researchers now using the acronym NEMS (nano-electro-mechanical systems). Such miniaturized electro-mechanical devices could prove disruptive in several technological fields, including Neurorobotics.

One of the basic advantages of versatile tools such as MEMS/NEMS design is the ability for the circuits to adapt to the environment in which they are intended to operate. For instance, when chip dies shrink below a dozen micrometers per side, external wire connections become difficult to establish (at sub-micron level they become almost impossible) and therefore power supply can become a problem. The development of NEMS micropower generators can solve this problem and eventually lead to autonomous nanorobotics both within and outside the human body.

With respect to neural interfaces the ability to shape metal electrodes down to the scale of a few nanometers, while keeping them attached to accompanying signal acquisition electronics could solve several challenges in neurorobotics. Biological neurons could be probed, sampled and excited in a noninvasive fashion, nanorobotic devices could be used for neural signal acquisition and neuron excitation for robotic control, even tissue sampling and targeted drug delivery for diagnostic and therapeutic purposes with respect to the nervous system.

7.2.2 Advances in software design

7.2.2.1 Artificial intelligence

The fields of neurorobotics and AI have a long and intertwined history. AI is a research area that emphasizes the creation of intelligent machines, systems and algorithms that works and reacts like humans. Core problems of AI include certain traits such as knowledge, reasoning, learning and planning [5]. The field of AI has roots in Turing Machines initially proposed in 1936, but has evolved dramatically during the past few decades [6]. Nowadays, some of the state-of-the-art AI tools are brain-inspired such as the ANNs, which mimic the structure of the mammal brain and are able to "learn" from data to perform tasks such as Speech Recognition or image classification up to the human level and in some cases they can even outperform humans [7]. AI tools are heavily used in neurorobotics models in several applications such as the learning and memory systems and the action selection and value systems, which include the notion of learning and planning that are core problems of AI [8]. As an example, AlphaGo software trained to play the deceptively complex ancient Chinese game of Go—represents a recent breakthrough in the development of AI. On March 16, 2016, it beat Lee Sedol, one of the world's top Go players, to win the Go series by four games to one.

7.2.2.2 Machine learning

One of the core parts of AI is Machine Learning. Machine Learning is a subset of AI and it is based on the idea that systems can learn from past experience (data), identify patterns and make decisions with minimal human intervention [9]. While many machine learning algorithms have been around for a long time, the ability to automatically apply complex mathematical calculation to big data—over and over, faster and faster—is a recent development. Firstly, due to the more data available in the Internet, which allows machine learning models to train on massive datasets, and thus, generalize better in various tasks. Secondly, due to advancements in hardware, mostly in GPUs, which allowed these models to train faster that resulted in rapid prototyping and development. In addition, new methods and algorithms have emerged the last decade that opened new horizons in various fields [10]. For instance, a new field that is called deep learning has emerged due to the key advancements in ANNs. ANNs are a type of machine-learning algorithm that mimics the structure of the mammal brain and can learn from data and to model nonlinear functions in very complicated data and tasks. ANNs are considered the state of the art in machine-learning models and offer significant results in a variety of tasks such as image recognition, speech recognition and natural language understanding. Machine learning and ANNs are used to neurorobotic systems due to their ability to handle multimodal data that is present in this kind of systems. To conclude, the developments in AI and machine learning have helped Neurorobotics systems to have intelligent behavior, accuracy sensory-motor coordination, near optimal action planning and environment perception.

7.2.2.3 Robot vision

A rather promising subcategory of machine learning heavily researched and applied in neurorobotics systems is robot vision. Robot vision incorporates aspects of robotics and machine learning into techniques and algorithms, such as kinematics, scene understanding and the robot's ability to physically affect its environment. Generally, in robot vision the input visual data (usually images) is transformed to output physical actions [11]. The images can be captured using different types of cameras, such as infrared or RGB cameras, which allows robots to detect features that are not visible to the human eye. Nowadays, machine learning is the frontier technology in image recognition and scene understanding. One of the state-of-the-art machine learning models that is used in Robot Vision are the convolutional neural networks (CNNs) [12]. CNNs were inspired by neuro-biological processes in that the connectivity pattern between neurons resembles the organization of the animal visual cortex. Individual cortical neurons respond to stimuli only in a restricted region of the visual field known as the receptive field. The receptive fields of different neurons partially overlap in such a way that they cover the entire visual field [13]. CNNs can capture both low-level and high-level features in visual data due to their architecture. They have been proved to be very powerful machine learning models in image recognition, object detection and scene understanding tasks [14], which are important tasks in robotic systems. Finally, robot vision plays a significant role in neurorobotics, because it allows robotic systems to perceive their environment better and interact more effectively with it.

7.2.3 Advances in electromechanical engineering design

While much of the discussion so far has focused on key electronics and algorithmic technologies, the electromechanical aspect of neurorobotics is also expected to rapidly evolve in the next couple of decades. Modern robotic setups are controlled by increasingly faster, more efficient microcontrollers following Moore's law. They move powered by increasingly torque-capable actuators, the result of technological evolution driven by economic demand for more efficient and reliable automation in industrialized economies.

Several technological hurdles need to be overcome for this technological trend toward more efficient robotics to stay on track. The most important among those are high manufacturing and maintenance costs, the lack of reliable high-power density power accumulators (batteries) and the inability of algorithms to adapt to unpredictable environments without the support of highly specialized programming by Robotics Engineers.

The first of those hurdles is gradually being overcome through economies of scale, as more adaptable, less specialized robots are being developed, capable of handling multiple tasks, such as innovations in machine learning, improvements in digital computing power and versatility in robotics chassis.

The hurdle of adaptability is closely associated with cost. More versatile robots ought to be in higher demand, leading to increased sales, driving economies of scale which should in theory reduce manufacturing and maintenance costs.

The power autonomy issue, however, is arguably less predictable. From electric airplanes and automobiles to robotics and even medical implants, technological progress is being hampered by a lack of reliable battery technologies capable of delivering the power densities, high current-on-demand, small volume, low weight and low-cost requirements in each of these application fields. Neurorobotics is no exception, particularly in the case of exoskeletons, where the need for a large robotic setup to perform cordless for hours is proving to be a formidable task.

Lithium ion technology has been state of the art for the highest density commercially available batteries for decades. While a steady stream of minor improvements has been maintained, the technology may be close to reaching its limits with respect to the power density it can offer. Political and economic strategic concerns regarding distribution of lithium global reserves worldwide (in Chile, Argentina, Australia and China) [15] is also a factor making investors skeptical before funding large scale electromechanical design projects which could lead to manufacturing economies of scale. While looking for technological alternatives for electrical power storage, the issue of Lithium availability may be about to be partially resolved by a current shift toward mining for this metal at the bottom of the oceans [16].

7.2.4 Improvements in electronics–neuron interfaces

Electronics that can be effectively integrated with the soft, curvilinear surfaces of biological tissues offer important opportunities for diagnosing and treating diseases and for improving brain–machine interfaces (BMIs). During the last decade, some significant improvements have been emerged in the electronics–neuron interfaces. One very interesting material strategy for bio-interfaced systems was introduced by Kim *et al.*, which includes ultra-thin electronics supported by bioresorbable substrate of silk fibroin. State-of-the-art microelectrode arrays are made of sharp shanks 10 × 10 arrays of pins with base widths ~80 μm, lengths ~1.5 mm and pitch ~400 μm. Because these are arrays are made of silicon, they are rigid and inflexible. Silk is an appealing biopolymer for this application because it is optically transparent, mechanically robust and flexible in thin film form, compatible with aqueous processing, amenable to chemical and biological functionalization. The silk is biocompatible, bioresorbable and water soluble with programmable rates of dissolution. In addition, silk can be also used as a platform for transistors and various classes of photonic devices [17]. On the other hand, Minev *et al.* introduced a new class of neural implants with the topology and compliance of dura mater, the protective membrane of the brain and spinal cord. These neural interfaces, which they call e-dura, achieve chronic bio-integration within the subdural space where they conform to the statics and dynamics of neural tissue. E-dura embeds and interconnects, electrodes that sustain millions of mechanical stretch cycles and electrical stimulation pulses. These integrated modalities enable multiple neuroprosthetic applications. For instance, E-dura extracted cortical states in freely behaving animals for brain machine interface, and delivered electrochemical spinal neuromodulation that restored locomotion after paralyzing spinal cord injury (SCI) [18]. One very important type of electrode that is widely used in neural interfaces is the cuff electrode design, which is wrapped around a nerve and it can be used to record and interact with the peripheral nerves (more details about this type of electrode design will be discussed in the next subsection). One very interesting usage of cuff electrodes showed by Daniel W. Tan *et al.* to produce touch perceptions on prosthetic devices on human subjects with upper limb amputations. The implanted electrodes provided natural touch sensation to the subjects for several months [19]. Another very significant development was introduced by Rodriguez *et al.* that they developed a new tripolar spiral cuff electrode suitable for stimulation applications on peripheral nerves. By delivering single electrical pulses through the cuff electrodes graded recruitment curves of α-motor nerve fibers were obtained [20]. To conclude, electronics-neuron interfaces is a very significant aspect of neurorobotics, because it is the basis for effective interaction between the nervous system and robotic devices, and thus, it is a field that is heavily researched and in continuous improvement.

7.2.4.1 Advances in electrode design

Electrodes often serve as the primary interface between the biological tissue with which a neurorobotic device is designed to operate and the device itself. In neuroprosthetic applications for instance, electrodes interface with neural tissue either for signal recording or neural stimulation. Such neural interfaces help us improve our understanding of neurophysiology in humans and animals, while also offering the possibility for the eventual treatment of neurological disorders.

Recent advances in fabrication technology have provided the option of creating tiny, high-density electrode arrays which can be used to interface directly with neural tissues. Existing neural interface electrode designs are being further miniaturized and arranged in more compact patterns, in order to become less invasive and improve the quality of signal obtained and transmitted [21].

A type of electrode which is widely used in neural interfaces and continuously improved are cuff electrodes: they are made of elastic material which can be wrapped around a nerve and used to record electrical signals, as well as stimulate the nervous system. Cuff electrodes are designed to be implanted into the human body and are suitable for both acute and chronic implantation and are available in a large variety of sizes. Also, the electrode layout is customizable, and the arrangement options include several configurations such as clustered or multipolar. One very interesting usage of cuff electrodes showed by Daniel W. Tan *et al.* to produce touch perceptions on prosthetic devices on human subjects with upper limb amputations. The implanted electrodes provided natural touch sensation to the subjects for several months [19]. Another very significant development was introduced by Rodriguez *et al.* that they developed a new tripolar spiral cuff electrode suitable for stimulation applications on peripheral nerves. By delivering single electrical pulses through the cuff electrodes graded recruitment curves of α-motor nerve fibers were obtained [20].

Finer and more flexible electrodes, more compact electrode arrays, inorganic nanostructures and biologically inspired or biocompatible electrode coatings, more flexible electrode structures [22] and even materials that partially integrate themselves into human tissue are all research directions currently being pursued in order to optimize direct neural–electronics interfaces [23].

7.3 Neural human–robot interfaces

Interfacing with the human nervous system and its various components and subsystems is crucial for the field of neurorobotics (Figure 7.2). Neural–electronic

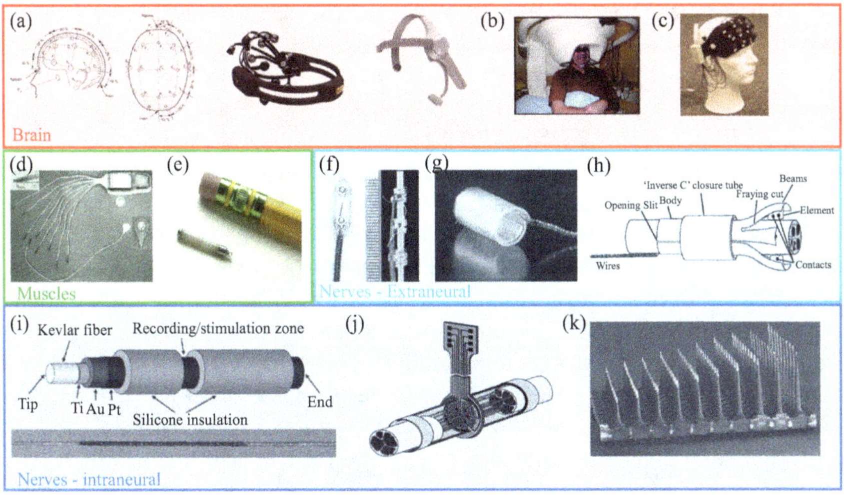

Figure 7.2 This figure illustrates a selection of the technologies described in Section 7.1. The upper part of the figure (a)–(c) shows a selection of BCIs. (a) On the left side EEG electrode placement over the scalp. On the right, low-cost portable EEG devices, with fewer electrodes (adapted from [26]). (b) MEG device (modified from [31]). (c) NIRS cap (adapted from [32]). The middle left part of the figure (d),(e) shows implantable muscle electrodes. (d) Implantable stimulator with one epimysial electrode (adapted from [28]). (e) Implantable myoelectric system (IMES, adapted from [29]). The middle right part of the figure (f)–(h) shows extraneural electrodes. (f) On the left, an epineural and on the right a helicoidal electrode (adapted from[28]). (g) Cuff electrode (adapted from [28]). (h) Interfascicular electrode (adapted from [28]). The lower part of the figure (i)–(k), shows intraneural electrodes. (i) Intrafascicular polymer-based electrode (adapted from [28]). (j) Regenerative electrode (adapted from [28]). (k) Penetrative electrode (adapted from [28])

interfaces can enable improved medical diagnosis and provide important insights into understanding various functions of the human central nervous system (motor, sensory, etc.) Additionally, they can enhance neuro rehabilitation either as stand-alone interventions (stimulation) or combined with neurorobotic technologies (motor intention decoding).

On a more indirect interface level, affective computing technology has found its way into robotics research and development, forming the relatively new fields of affective and social robotics. The ability to interpret and elicit human emotions is a higher cognitive ability and requires robots to perform complex classification and identification at a higher semantic level. The benefits can be substantial ranging from more efficient human–robot interaction to increased engagement and acceptance of robots by humans.

7.3.1 Neural–electronics interfaces

A common approach for neural interfacing for neuro robotics is with the brain. Interfacing is achieved via BCIs also referred to as BMIs. Noninvasive approaches that measure electric current, include electroencephalography (EEG) [24] and magnetoencephalography (MEG) [25]. EEG uses wet or dry electrodes, with the dry electrodes deemed most efficient, whereas for MEG the head is positioned inside a MEG machine. EEG demonstrates excellent temporal, but very poor spatial resolution and it is generally quite portable and low-cost, whereas MEG the exact opposite in those aspects. Hemodynamic effects in the brain are measured via functional magnetic resonance imaging (fMRI) with an MRI machine (non-portable, high-cost) and near infrared spectroscopy (NIRS) with the use of spectrometers (portable, low-cost) [25]. More invasive BCIs include electrocorticography, where electrode grid arrays are placed on the exposed surface of the brain [26] and intracortical neuron recordings, directly from the gray matter inside the brain [26]. Additionally, for people with Parkinson's disease, devices called neurostimulators act as pacemakers that deliver constant electrical stimulation on specific areas inside the brain, in order to attenuate the symptoms [27].

Another level of interfacing with the central or peripheral nervous system can be achieved by interfacing with the nerves and muscles [28]. Noninvasive approaches include surface electromyography (sEMG), which measures electric current on the skin surface, and mechanomyography (MMG), which measures the vibration of the muscles when they contract [17]. Additionally, real-time changes in the muscle dimension can be measured via ultrasound sensors with the use of sonomyography (SMG). The peripheral nerve activity can also be measured non-invasively with the help of magnetoneurography [28]. Recent advances in electronics, microelectromechanical systems (MEMS), and materials favor the creation of implantable muscular and neural electrodes. Epimysial electrodes are used to invasively stimulate the muscle for functional electrical stimulation (FES) applications and intramuscular to record muscular activity [28]. Additionally, implantable myoelectric systems (IMES) are currently spearheading sensor technologies for good quality measurement of muscle activation and its wireless transmission [29]. For interfacing with the nerves, there are two main types of electrodes. Extraneural electrodes are positioned around a nerve and include epineurial, helicoidal, book and interfascicular electrodes, which are mainly used for stimulation of peripheral nerves, and cuff electrodes, which are used for both stimulation and recording [28]. Intraneural electrodes are inserted inside a nerve and include intrafascicular, penetrative microelectrodes and regenerative electrodes, used for both stimulation and recording [28]. More details about the types of neural-electrodes, their design and their commercial availability can be found here [30].

The broad spectrum of neural interfaces described previously can be used in combination and create hybrid approaches in neural interfaces. Novel sensor fusion techniques enable the optimal hybridization of neural interfaces and facilitate interfacing with neurorobots [33]. According to Novak *et al.* [33], sensor fusion can be unimodal or multimodal. In case of unimodal fusion, neurorobotics utilize a single interface. This can be a neural (sEMG, BCIs, intraneural electrodes, etc.), mechanical (inertia measurement units, gait-phase detectors, limb motion estimation sensors, etc.) or parallel (eye-tracking, artificial vision, etc.) interface. Each interface can be seen as a separate sensory modality and processed with a number of tools and techniques currently available for sensor fusion [33]. Multimodal sensor fusion combines multiple interfaces and fuses them together in order to overcome their specific shortcomings. The main approaches to multimodal sensor fusion are: (1) single fusion algorithms, (2) unimodal switching, (3) multimodal switching and (4) mixing [33].

7.3.2 Affective robotics

Affective robotics involves the study, research and development of robotic devices which can identify, correctly interpret and simulate human emotions. While not physically interacting with a human nervous system through the direct flow of electrons, it is a subtle, complex and ultimately important form of human–robot interface.

There is a theoretical distinction between two aspects of emotions that are in fact closely interconnected. These are the external and the internal aspect of emotions, which deal with the social and individual emotional perspectives respectively [34,35]. Research on the external or social aspect involves the construction of robotic systems that further emotional and empathetic relations between robots and humans. This type of robots takes advantage of the human tendency toward anthropomorphism to increase their favorability on humans. On the other hand, research on the internal or individual aspect of emotions focuses on the development of robotic agents that model human emotions. This research is directed to progress both in the conceptual understanding of natural affective phenomena [36].

Affective robotics is a technology which attempts to decode and exploit one of the primary and most complex channels of human interaction: the communication of emotions. By attempting to piece together multiple behavioral queues originating from a human's voice, facial expressions and body posture, a robot faces formidable tasks such as sensor data fusion under unpredictable and often noisy circumstances; generalization, classification and prediction at a higher semantic level; as well as recognizing apparent inconsistencies among individuals, across cultures and even day to day emotional mood variations of the same individual.

A combination of methodical behavioral analysis in centralized databases available to multiple robots, and improvements in machine learning algorithms are considered promising research directions for these challenges. Data acquisition directly from the human nervous system could also prove to be valuable, provided legal and ethical issues are properly addressed. There are certainly important incentives to improve affective robotics technologies, as they could lead to improved human–robot interaction, better acceptability of robots by humans (including robotic prosthesis scenarios), as well as more effective services provided by companionship robots.

7.4 Neural rehabilitation robotics

Rehabilitation robots are becoming a valuable component of neural rehabilitation [37]. The common practice in clinical neurology is for clinicians to work with each patient individually, having one-on-one interaction [38]. However, robot-assisted neuro-rehabilitation alone or combined with conventional therapy can increase the amount of rehabilitation time [38] and help patients who are unable to regularly visit the clinic [37]. The most common approaches that are adopted include repetitive exercises, where in most of the cases the patient has a passive role, augmentation of external manipulation, FES and biofeedback [39,40]. However, the outcomes of robotic assisted neural rehabilitation, are modest at this point concerning the effectiveness of robot-assisted rehabilitation for neurological restoration of both upper and lower limbs [41,42]. The slow translation of neural rehabilitation robotics from the technical laboratories to the clinics [41], their equally slow translation to home or daily use and the unclear benefits of robotic assisted therapy compared to conventional therapy [38], are limiting factors for the broad adoption of this technology. Technical issues with joint alignment and the additional weight imposed on the existing limb are also critical challenges that limit comfort and make robotic exoskeletons complex and time-consuming to set up. The development of soft exosuits aims to make the fitting easier and provide a compact, yet lightweight design with low energy demands [43,44]. However, soft exosuits are able to provide less support compared to robotic exoskeletons with a rigid frame [43]. In this section, some representative cases of upper and lower limb neural rehabilitation robots are reported. Additionally, it is apparent from previous research that there is still a lack of knowledge on what exactly makes a therapy beneficial, and as a consequence of that how to personalize the therapy, in order to boost the effectiveness [45]. According to recent literature, the control of such robotic exoskeletons is still a major challenge related to their currently unimpressive performance [46,47]. The last part of this section will discuss the interfacing between humans and robotic exoskeletons and how those are usually controlled.

7.4.1 Robotic technologies for neural rehabilitation of the lower and upper limb

7.4.1.1 Lower limb robotic exoskeletons

Robotic exoskeletons targeting the improvement or recovering of the walking function in neurological disorders dominated most of the research on robot-assisted neurorehabilitation, starting from the early 1960s [48,49]. The Lokomat (Hocoma AG, Switzerland) is a medical wearable robot with modular design for gait rehabilitation (Figure 7.3(a)), currently coming in two variations: LokomatPro and LokomatNanos. Its main benefit is that severely affected patients, that are

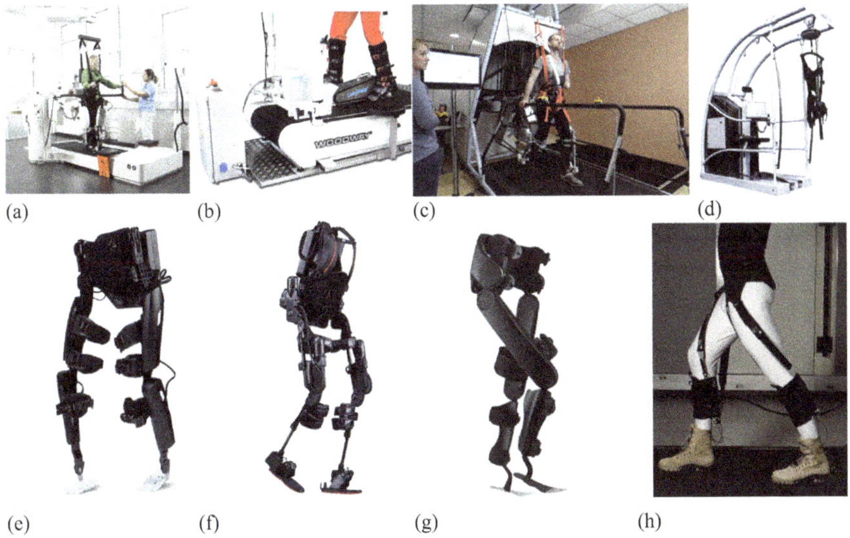

Figure 7.3 The upper part of the figure (a)—(d) illustrates nonmobile lower limb exoskeletons, which aim in rehabilitation of walking after neurological damage, in a controlled environment. From left to right: (a) Lokomat, (b) LokoHelp, (c) LOPES and (d) RehaStim. The lower part of the figure (e)–(g), shows portable wearable exoskeletons, which are used to support mobility out of the lab and promote independent walking. From left to right: (e) ReWalk, (f) Ekso GT, (g) Indego and (h) Harvard Exosuit. Photographs courtesy of their copyright owners

extremely difficult to train manually, still can receive training, while the therapist is relieved from strenuous physical work. LokoHelp (WOODWAY Inc., USA) was designed to improve the quality of therapy by disencumbering the therapists from the labor-intensive task of guiding the user's legs during locomotion training (Figure 7.3(b)). The Lower extremity Powered ExoSkeleton–LOPES (University of Twente, The Netherlands) was developed to enhance gait training in post stroke and SCI patients despite the diversity of disorders linked to the impairment of the walking function (Figure 7.3(c)). The electromechanical Gait Trainer GT II (Reha-Stim Medtec, Germany) has been used in several studies in stroke survivors, showing that training with this device may improve gait performance significantly (Figure 7.3(d)). Portable robotic exoskeletons aiming to assist the hip and knee joints can be used as an assistive device outside of the lab to enable wheelchair-users to stand and ambulate, which can benefit their mobility and social participation [50]. ReWalk (ReWalk Robotics Ltd, Israel) provides powered hip and knee motion allowing the individual to stand upright, walk, turn and climb stairs (Figure 7.3(e)). Ekso GT (Ekso Bionics, USA) is a device for gait therapy in individuals with SCI or stroke survivors and can be used as a tool to assist professional physiotherapists (Figure 7.3(f), credit: Ekso Bionics©). Indego (Parker Hannifin Corporation, USA), enables individuals with SCI or individuals with hemiplegia due to stroke, to perform ambulatory functions in rehabilitation institutions (Figure 7.3(g), Photo Courtesy of Parker Hannifin Corporation, USA). A recent review [50] concluded that such robotic exoskeletons can provide individuals with thoracic-level motor-complete SCI—the ability to walk at modest speeds. A prime example of a lower limb soft exosuit is the cable-driven Harvard soft exosuit, which is able to provide assistive walking at a reduced metabolic cost without a rigid exoskeleton frame [44].

7.4.1.2 Upper limb robotic exoskeletons

It recently became increasingly apparent that with an ever ageing population, upper limb function (arm/wrist/hand) is equally important for the independence of individuals and their ability to manipulate their environment [41,51]. Robotic devices can assist the neuro rehabilitation of upper limb motor function in neurological disorders (e.g. Stroke, SCI, Multiple Sclerosis (MS), etc.) or provide daily support to people with muscular weakness (Elderly, muscular dystrophy/atrophy, etc.). The first category includes robots that assist physiotherapists in patient neuro rehabilitation and they are either end-effector or robotic exoskeleton type [41]. Early robotic assistive devices are end-effector type, which is simpler in structure and has only one point of attachment to the user's limb. Such devices are the InMotionRobots (the InMotionARM (Figure 7.4(a)), which is the commercial version of MIT Manus, for the rehabilitation of the arm and the InMotionHAND and InMotionWRIST for the hand and wrist (BIONIK Inc., USA), as well as the AMADEO (Tyromotion GmbH., Austria) for the hand and fingers (Figure 7.4(b)). They are low-cost devices aiming in the robotic neuro rehabilitation of the upper limb in stroke survivors in the clinic. However, such devices are limited to mimic the motion of the joints of the upper limb, and are unable to measure joint torques and drive joint specific rehabilitation [41]. These issues were addressed with the use of wearable robotic exoskeletons. Such examples include the Armeo Power (Figure 7.4(c)) and Armeo Spring (Hocoma AG, Switzerland), which can assist the functional rehabilitation of multiple degrees-of-freedom (DOF) of the arm and retrain functional activities of daily living , in a laboratory or clinic. Devices that are targeting daily unsupervised upper limb support include the more portable MyoPro elbow/wrist/hand orthosis (Myomo Inc., USA), which aims to restore function of the upper limb in people with neurological or neuromuscular disorders (Figure 7.4(d)). The A-Gear arm robotic exoskeleton (Figure 7.4(e)) developed recently at the University of Twente [52] followed by the SymbiHand finger exoskeleton (Figure 7.4(f)) for the daily support of individuals with duchenne muscular dystrophy (DMD) [53]. Last but not least, the soft arm exosuit developed by Xiloyannis *et al.* for elbow rehabilitation [43] and the SaeboGlove (Saebo Inc., USA), which is a lightweight solution for assisting finger and thumb extension in neurological patients, showcase the new trend of soft exosuits for the upper limb (Figure 7.4(g)).

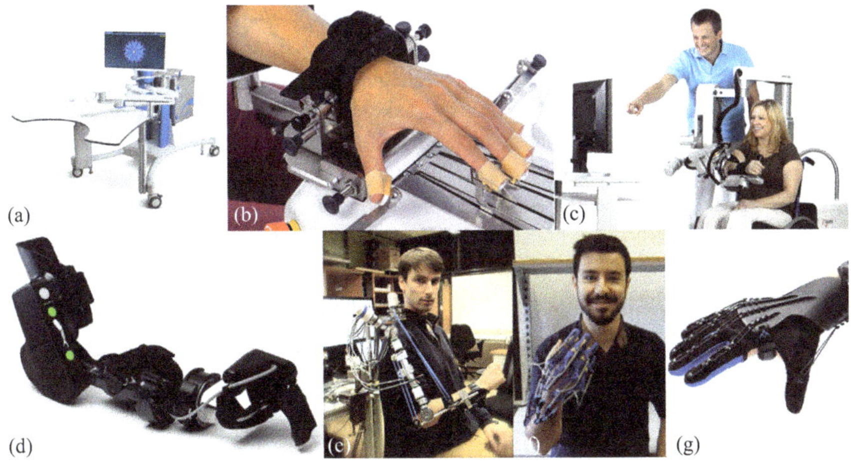

Figure 7.4 *The upper part of the figure (a)–(c) illustrates nonmobile upper limb rehabilitation robotics. (a) and (b) are end-effector type and (c) is a robotic exoskeleton. Both types aim in rehabilitation of the upper extremity function after neurological damage, in a clinic together with a physiotherapist. From left to right: (a) InMotionARM, (b) Amadeo and (c) Armeo Power. The lower part of the figure (d)–(g), shows lightweight wearable exoskeletons, which intend to support upper extremity function in daily living. From left to right: (d) MyoPro, (e) A-Gear, (f) SymbiHand and (g) SaeboGlove. Photographs courtesy of their copyright owners*

7.4.2 Motor intention decoding for robotic exoskeleton control

The optimal use of wearable robotic exoskeletons for daily use, required the robust and accurate decoding of the motor intention of the user [54]. This is achieved via a human–machine interface. The term refers to methodologies for the identification of the user's intent to move, from biological signals (i.e. sEMG) or body force and position data and its translation into robotic commands [54]. Based on the type of sensors used to interface with the human, numerous invasive and noninvasive interfaces were developed in the past and were applied to robotic exoskeletons and prosthetic limbs [46]. Many of the neural interfaces described in Section 7.1 are used in order to enable motor intention decoding and subsequently robotic exoskeleton control. As described previously, such technologies can be invasive or noninvasive. The choice is heavily depending on the specifics of each type of impairment and the intended application.

7.4.2.1 Noninvasive HMIs

The most common noninvasive HMIs are sEMG, impedance/admittance control and body-powered control [46]. sEMG-based HMIs often use model-free machine learning approaches in order to decode motor intention. However, training in a specific spatiotemporal condition using machine learning is not necessarily generalizable to different spatiotemporal conditions. A rapidly emerging alternative is the model-based approach using musculoskeletal modeling in order to predict muscle force or stiffness. It has been suggested that the combination of data-driven machine learning approaches with biomechanical models can overcome the limitations of the former [46]. Often FES is used as a stand-alone interface or combined with a wearable exoskeleton [46]. Other noninvasive HMIs include BCIs like EEG, MEG, fMRI and NIRS at the brain level and interfaces like MMG and SMG at the muscle level [25] Additionally, HMIs such as eye tracking, tongue interfaces and joysticks have been heavily used for the control of bionics, however with the clear disadvantage of sacrificing one function to support another [25].

7.4.2.2 Invasive HMIs

Invasive BCIs (i.e. intracortical grid arrays), neural interfaces (i.e. extraneural and intraneural electrodes) and implantable EMG electrodes (i.e. IMES) are the most common invasive HMIs, used in motor intention decoding for the control of robotic exoskeletons. Several of these technologies are described in Section 7.1. Noninvasive HMIs are often preferred to invasive, due to the required surgery, concerns about their ability to integrate to the human body and long-term tissue damage [55]. However, future advances in invasive HMIs, may highlight their many benefits. For EMG, IMES applications reports stronger and more reliable signals, minimal movement artifacts and reduced effect of skin properties [56]. Invasive BCIs are expected to offer improvements in signal long-term stability, biocompatibility and increased information transfer rates and lead to more intuitive interfacing with the human and better controllability over multiple DOFs [55,56]. The complete embeddability of such HMIs, depends on further technological advances and for such HMIs to be acceptable by a large number of users, their benefits must first outweigh the concerns and even then, their implementation should carefully consider societal and ethical issues of privacy, ownership, liability and self-perception [55].

7.5 Robotic prosthesis

A prosthesis is an artificial device that replaces a missing body part. The word prosthesis comes from Ancient Greek meaning addition and metaphorically speaking attachment. The missing body parts may be lost due to various causes such as accident, trauma, medical reasons, etc. The main idea is that prostheses will restore the functions of the missing part. There is an open debate on the design of prosthesis parts, though the main idea is that a person's prosthesis should have personalized character i.e. a person's prosthesis should be designed and assembled according to the person's appearance and functional needs.

There are different available options of prosthetic for selecting the most suitable depending on the desired use and outcomes and they are categorized by: electrically powered, body-powered, harness, hybrid and activity/task specific [57].

Electrically powered prostheses use motors to influence move, which are driven by a rechargeable battery. The most usual control system is through electromyography signals. Other inputs include Inertial Measurement Units , linear potentiometers or transducers, servos, rocker switches, push-button switches and harness pull switches, as well as force-sensing resistors. Body-powered prostheses utilize a harness to captivate proximate body gesture, which generates excursion of a cable. The cables end on a hand or hook and influence the move of the appurtenances. Passive prostheses do not actively move, but they rehabilitate the anthropomorphic limb-length necessary to move or consolidate an item during ipsilateral or bilateral usage and can be described as positional or static. Using a harness for a successful control depends on four concepts: a safe suspension of the prosthesis, a component cabling system to connect the TD and the harness, sufficient range of movement in the body segment, and adequate strength to create the excursion needed to activate the apparatuses. Hybrid prostheses incorporate two prosthetic options into one mechanism. Activity-specific prostheses are manufactured to enable tasks that entail an advanced level of performance [57].

Robotics can be used to improve patient's therapy outcome and customize therapies into patient's motor abilities. It also supports compliance with patient's treatment and maintain patient's records. For example, for a robotic prosthetic limb to work it must have several components such as biosensors, controllers, integrated processes, etc. Biosensors are employed in myoelectric prostheses since they detect signals for the user's nervous or muscular system [58]. The data flows via a controller that is located inside the device and processes feedback from the position or force of the part, e.g., the limb. Electrodes may detect activity on the skin, needle electrodes implanted in muscle, etc. The controller monitors also the movements of the device. An actuator mimics the actions of a muscle or parts of it such us the original muscle tissue. Other techniques are targeted muscle reinnervation (TMR) and targeted sensory reinnervation (TSR). TMR is a technique in which motor nerves are surgically rerouted such that they reinnervate a small region of a large intact muscle. Using sensors over the reinnervated muscle it may be possible to control the movement of an appropriate part of the robotic prosthesis. Recently, the ability of a robot to take signals from the human brain has been improved [59]. The driving idea is to create artificial parts that tie directly into the nervous system. TSR is an improved version of TMR where the sensory nerves are surgically rerouted to skin on the chest [58,59].

7.5.1 Neuroprosthetics

Modern robot technologies in the field of neuroprosthetics are based on user-centered, functional processes of cognitive neuroscience and perception studies. Such new devices demonstrate the capability to adapt, online and in real time, to the dynamic environment and to the user's actions. Such advanced systems allow a natural approach to interaction among people and robotics self-extensions. In many studies it has been demonstrated that there is a significant improvement in upper limb motor function after stroke using neuroprosthetics tools. The field of neuroprosthetics started in the 1970s when Vidal published an influential article on the development of bioengineering and neuroscience, in which he described how cortical responses recorded by an electroencephalogram (with very low frequencies <1 Hz up to 30 Hz) could be interfaced to a digital system [60]. His results were supported through the analysis of event-related potentials, responses evoked by sensory stimuli in certain attentional and perceptual tasks. BCI is a tool that interfaces with the EEG responses and transduces them through a digital interface. This was, at first, especially useful for individuals with motor paralysis, amyotrophic lateral sclerosis, and in some cases comma (even interfacing responses such as saccadic eye movements and reflexes). One of the main goals of prosthesis design is to improve the biomechanics of gait, by improving stability, symmetry or energy expenditure [60].

Robotic prosthesis control is a branch of control that allows the proper interaction between human parts and wearable robots. Myoelectric control is based on EMG techniques for evaluating the electrical activity produced by skeletal muscles. Pattern recognition algorithms can be applied to decode EMG signal patterns generated by muscles during certain movements [59]. Advancements in the processors used in myoelectric arms have allowed developers to make gains in fine-tuned control of the prosthetic.

7.6 Future directions

7.6.1 Expected advances in key technologies

In the past few years Neurorobotics has evolved dramatically, while an increasing number of brain-inspired technologies have emerged. A rather ambitious neurorobotic project is the human brain project; a large-scale research project focused on improving our understanding of the human nervous system. It includes a sub-project entitled Neurorobotics Platform, which enables researchers to conduct embodiment experiments on high performance computing clusters. In addition, researchers provide any simulated brain model with its own "body"—virtual or real—and explore how it controls movement, reacts to stimulus, and learns in a virtual environment [61].

The growing research interest in the field of neuro-robotics and related areas has resulted in an explosion of wearable robotic technologies for neurorehabilitation. However, several issues such as kinematic compatibility, excessive device weight and autonomous power supply still need to be resolved, as they are currently limiting robotic exoskeleton use. As mentioned earlier in this chapter, the development of soft exosuits aims to tackle these issues by providing more versatile and lightweight designs that require less power to operate [43]. In soft exosuits, the lack of a rigid frame makes sensor placement less challenging, and prevents extra strains to the body of the user [46]. However, soft exosuits generally provide a limited amount of support when compared to rigid robotic exoskeletons and they are still considered an emerging technology with a plethora of challenges to resolve before full implementation.

Neuro-robotics is already benefiting—and likely to continue benefiting—from the overall trend in miniaturization of sensors and the development of affordable portable sensor technologies focused on interfacing with the human nervous system [62]. Wearable and textile sensors can enhance the overall portability of neuro-robotics and promote home-based and self-rehabilitation. Advances in the field of the Internet of Things can lead to the creation of digital avatars representing the patient or even robotic telepresence avatars controlled by patients and caregivers, possibly leading to more efficient and highly personalized rehabilitation strategies. Such strategies will likely complement existing neurorobotics technology possibly help to create novel hybrid in neurorehabilitation.

Technological innovations and improvements of interest to neurorobotics are also being developed targeting more efficient and less invasive neural interfacing. Recently the field of optogenetics raised expectations in the direction of non-invasive stimulation of neural circuits. Optogenetic technology enables the manipulation of either specific cell types or neural circuits. This is associated with minimum tissue damage and adverse side effects. Optogenetics is an alternative approach to DBS or FES. In a nutshell, optogenetics utilizes light-sensitive proteins in the nerves, in order to optically stimulate specific neural circuits. However, the field is still in its infancy and before any broad application optogenetics become available, they would need to demonstrate their feasibility and be validated in human studies. This will likely be facilitated by advances in other related fields in science and engineering, including gene therapy, opsin engineering and optoelectronics [63].

7.6.2 Convergence of key technologies

Advances in analog neural processing and parallel distributed processing [64] could have a dramatic impact on the implementation efficiency of machine learning algorithms. In the case of neural processing, a missing piece of the puzzle—the lack of reliable on-chip analog memory—has recently been provided. In the case of parallel distributed processing, the improvement in efficiency of modern digital GPU chip architectures, research into parallel supercomputing, the provision of cloud computational power as a commodity and what has been dubbed the "big data" revolution have been converting to facilitate faster, more efficient and ultimately more capable machine learning algorithms. While ANN deep learning is not a new concept, over the past two decades it has found increasing application in industrial processes, information technologies, healthcare and robotics. The field of Neurorobotics is well positioned to take advantage of this trend in algorithmic improvement, particularly with respect to customization and adaptability of devices to their human operator, especially when in cases of direct neural–electronics interface.

Further technological improvements in electronics–neuron interfaces are also helping improve data flow size and quality and transmission between biological neurons and robotic devices. Because data quantity and quality are crucial in machine learning, it is likely to boost the efficiency of such algorithms involved in neurorobotics applications. This could prove to be particularly applicable in the case of noninvasive BCIs and deep brain stimulation scenarios, where data quality and resolution is currently poor without the use of implanted electrodes.

The recent discovery of memristors [65] and their potential use as analog on chip memory, combined with parallel distributed processing and analog neural computation technologies is likely to have substantial impact in real time computation capabilities as well as our ability to map machine learning algorithms on silicon hardware. Real time computation is fundamental in neurorobotic systems, it can possibly enable neurorobots to perform heavy computational tasks in real time. At the same time, analog neural computation is biological evolution's solution to biological organisms' computational needs: it is therefore likely that artificial analog neural computation will turn out to be more compatible to interfacing with its biological counterpart, leading to novel neurorobotic control interfaces and systems.

A clear trend in neuro robotic technologies is the combination of individual approaches and methods for the creation of hybrid systems. A prime example of that are hybrid FES-robotic technologies for the rehabilitation of the upper [66] and the lower limb [67]. This relatively new idea aims to address the main drawbacks of FES (i.e. fatigue, poor joint control, etc.) and neuro robotics (challenges for physical and cognitive interaction, energy-management and portability issues) by combining them into a hybrid system [67]. Especially for wearable robotic exoskeletons, this idea is also supported by a recent opinion paper that points out that wearable neurorehabilitation robots show very modest results and they can greatly benefit from the example of prostheses and the hybridization of various HMIs currently used independently (sEMG, admittance/impedance and body powered control) [46]. Novel sensor fusion approaches [33] also favor this standpoint and their further development will enable the optimal hybridization of already existing technologies. Recently, a study by Lobo-Prat *et al.* [68], shows one implementation of such a hybrid approach between two HMIs. In this study, an individual with DMD can control an end-effector arm support with the use of a combination of sEMG and admittance control. Additionally, the rotational movement of the robotic support endpoint, is decoupled from the active control with the use of force admittance, that measures the interaction forces/torques between the user and the robot and tries to minimize them [69].

7.6.3 Expected demand

Neuro-robotics is the area of neuroscience with robotics that deals with the study and application of technology of autonomous neural systems. The global autonomous robot market was appreciated at around USD 4,615 million in 2016 and it is estimated to reach more than USD 11,920 million by 2024. Additionally, it is valued to grow at a compound annual growth rate (CAGR) of over 14 per cent between 2017 and 2024. In terms of volume, the demand for autonomous robot is expected to reach around 115,400 units by 2024 at a CAGR of around 13% during the forecast period. With the subject of neuroscience emerging as it has, various neural treatments have arisen, from pharmaceuticals to home-based neurorehabilitation. This advance is dependent on a complicated understanding of the brain and how precisely it functions; how closely related is with the human consciousness, and what legal and ethical issues arise from the use of neuro-robotics.

7.6.4 Home-based rehabilitation

Neurological rehabilitation generally is a long-lasting procedure for patients suffering from neurological disorders. Home-based rehabilitation is an effective and promising way to provide more training time and intensity to patients without compromising comfort and help patients who are unable to visit the clinic regularly [70,71]. Various types of recently researched and developed technologies, such as serious games, smart wearable sensors, VR, augmented reality (AR) and lightweight portable robotic exoskeletons, can support a home-based rehabilitation environment. Some examples of home-based robotic rehabilitation systems are MIT-Manus for upper limb training and the ARM Guide 7 (Assisted Rehabilitation and Measurement Guide) for patients with stroke [72]. Neurological rehabilitation can benefit from the precise diagnosis of the underlying pathology and its effects on the nervous system, the rehabilitation method and the amount of time spent, and the quantitative assessment during and after physical training [70,71]. Traditional therapies are mostly coordinated by therapists, who support patients to complete the chosen training tasks. However, assistive home-rehabilitation robots will soon be in great demand due to the growing population of the patients and the lack of rehabilitation professionals. Rehabilitation robots typically refer to the devices that can work together with the voluntary motor intents of a patient, as assistants to therapists, offering intensive and safe training. The most frequently mentioned motion types offered by developed rehabilitation robots are continuous passive motion, active-assisted movements and challenge-based movement [73]. Additionally, VR/AR and serious gaming, are increasingly used in combination with rehabilitation, in order to provide motivation for the patients, with reportedly promising results [74]. Its added value though, is still unclear compared to conventional therapy as a stand-alone but shows benefits when combined with conventional therapy. Such immersive technologies combined with robotic exoskeletons can pave the road for home-based rehabilitation [75].

7.6.5 Research into consciousness

Understanding of consciousness in AI and neuroscience has to deal with several challenges affecting the attempts to synthesize consciousness in computational architectures [76]. As John Searl [77] pointed out in his “Searle’s Chinese room” [78] the complexity of consciousness is one of the most pressing problems open to biological examination and clarification. Tononi *et al.* [79] claimed that consciousness does not necessitate interaction with the outer world but rather on the assumption that a system has the correct inner architecture and forms a complex competent of distinguishing a significant number of inner states, it would be extremely conscious.

Goertzel [80] presents three aspects of consciousness which are “free will”, “reflective consciousness”, and “phenomenal self”. Tani [81] summarizes the results of three neurorobotic experiments regarding three levels of self, similar with

O'Regan's [82], namely "minimal self, social self, and self-referential self". The first experiment accounts for the presence of minimal selves in a plain robot navigation testing, the second for presence of social selves in a learning experiment which focused at the imitation between robots and human subjects, and the third for presence of self-referential selves in a more complicated environment for learning abilities. The three categories of selves distinguish from each other, but they also share a comparable form of self-organized criticality that arises in dynamic interactions. Tani *et al.* neurorobotics experiments aimed to articulate body dynamics acting on the environment as well as the reaction dynamics from the environment dynamics in a way that amounts to formal consciousness [81]. Formal consciousness represents the structural dynamics vital to consciousness [76].

7.6.6 Legal and ethical issues

The specified legal and ethical challenges posed by neurorobotics should be regarded within the framework of the wider societal impact of emerging technologies. There are different fields (neuroscience, robotics, AI, information technology, etc.) that are applicable to neurorobotics and can be used as a guidance to outline the ethical framework for this novel area, since there is currently very little work specifically for neurorobotics. The scientists and the public are usually enthused by innovative technologies, but there are also remarkable concerns as they pose dangers that are hard to foresee. The fundamental disquiets are subsumed within four interconnected topics: safety, appropriate use, capability and responsibility. Safety and responsibility have always been of importance to the engineers and enterprises who design and produce the devices. Additionally, there is a necessity for a regulatory commission that evaluates the capabilities of systems and confirms their use for certain activities, as well as ethical theorists who question which tasks should be considered appropriate for robots.

Literature is a way by which humanity expresses itself, unrestricted from restraints, and by which it can envisage forthcoming social advances. In some cases, by way of literature, significant and foresighted technological issues have been visualized and anticipated, as in Asimov's *Handbook of Robotics*, 56th Edition, 2058, Asimov introduced the three laws of robotics, which are a set of principal requirements for the development and manufacture of intelligent robots as quoted below:

1. A robot may not injure a human being or, through inaction, allow a human being to come to harm.
2. A robot must obey the orders given it by human beings, except where such orders would conflict with the First Law.
3. A robot must protect its own existence, if such protection does not conflict with the First or Second Laws.

The last technological and scientific advances have brought forward the boundaries of robotics (hence the neurorobotics as well) so that those disquiets that years ago seemed only hypothetical and a subject of literature, are becoming a real-world necessity, and even an urgent one. They are intended to ensure that robots will operate for the benefit of humanity, rather than becoming a threat to humans. Could Asimov's laws really be the ethics of neurorobotics? Surely not, but they can express some general values as a work of art and a source of inspiration. While these rules were intended for autonomous robots in general, they can form the basis for public debate toward consensus on ethically acceptable interaction between humans and their assistive robots, including those integrating neurorobotic control technologies.

The first law of neurorobotics outlines the necessity of a high benefit/risk ratio [83], indicating for instance that neurorobotics medical interventions should have a greater ratio when compared with other medical options (which should be valued before commercialization of the robotic device). Euro-rehabilitation robots may risk user's physical integrity caused either by irregular operation or even through regular operation, when their effectiveness has not been correctly estimated [84]. Thus, as harm can be considered the time spent on a nonfunctional or harmful robot instead of in a more efficient treatment. The use of a neuro-rehabilitation robot should be at least as safe and effectual as other human administered methods.

In general, the current regulatory framework for rehabilitation robots (and medical electrical devices in general) is partly in line with these laws, as it is built upon the concept of delivering maximal clinical benefit under acceptable levels of risk (as defined by the IEC 60601 family of standards). It deviates, however, by stating that the higher the clinical benefit, the more acceptable the risk level: for example, a life-saving AED (automated external defibrillator), may in unfortunate conditions kill a patient because the benefit far outweighs the risk level. It should be considered whether this approach is valid for medical robotics as well. Recently a standard is voted on, that specifically addresses rehabilitation robots (IEC 80601-2-78), but it rather addresses the specific risk nature of physical human–robot interaction, than matters of autonomy. IEC TR 60601-4-1 is a first technical report (nonbinding) addressing medical electrical equipment employing a degree of autonomy, and specifically addresses autonomy in medical robots, and aims to classify autonomy and link it to risk assessment. A general consideration is that autonomy is not directly correlated with risk. (For an understanding of degrees of autonomy, see Figure 7.5.)

The second law of neurorobotics indicates that the robot should be used as a tool in therapists' hands [83], since it is possible that therapists may see a robot as a substitute of them. As Hidler claimed, the purpose goal of rehabilitation robots is to complement therapists and not to substitute the role of therapists [85] and thus neuro-rehabilitation robots should not be considered as a standing-alone treatment device [74]. However, robot totally compliant to the human's instructions may seem but it is not obvious, as indicated from the problems in Lokomat therapy [84] and from cases of "disobedience" in Gait Trainer [86].

The third law of neurorobotics is about AI that supports human intelligence [83] regarding the presence of human control of the device. Robot's AI should not be limited to the secure control of human estimations, since there are many parameters to measure, to tune and to adjust, as well as considerations of a parameter's effects on the other ones. A neuro-robot should be capable of automatically perform all the

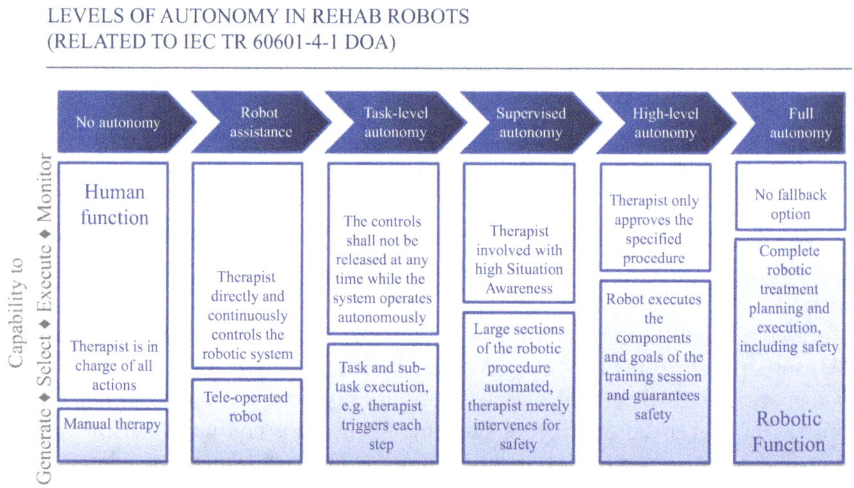

Figure 7.5 Illustration depicting the levels of autonomy in rehabilitation robotics

modifications and deliver a clear quantitative overview of them. Kinematic robotic measures can be beneficial in the assessment of proprioceptive function of hand [87], upper [88] and lower [89] limbs as well as for motor deficits [90].

These laws of neurorobotics are based on the ethical need for secure and efficient neurorobots and on the redefinition of their role as therapist assisters [83]. These laws may indicates the need to include knowledge of the MDR/FDA regulations and the ISO/IEC body of standards in engineering education in order for engineers to be more educated and integrate the legal and ethical aspect throughout the development procedure.

7.7 Conclusions

Neurorobotics is a nascent technological field with potential socioeconomic impact which is both promising and hard to predict. It is primarily fueled by scientific research and technological development in electronics, AI and machine learning, neuroscience and neurophysiology. It is further supported by economic growth in the electronics and healthcare markets, particularly in researching the human brain, exoskeletal robotic aids and robotic prosthesis.

We identified several key technological convergence and synergy points which could have substantial impact in Neurorobotics. The recent discovery of the Memristor electronic element could solve the decades-old problem of inadequate analog memory storage on electronic chips, therefore empowering analog neural electronics to become far more efficient and versatile. This type of neural computation is far more similar and compatible to a biological organism's nervous system, possibly facilitating more direct neural–electronics interfaces that are not yet prevalent to date.

On the software side, the DOF of movement involved in programming most complex robotic setups make AI algorithms a necessity. Until recently, however, we simply did not have enough computing power and data to enable more effective algorithms to tackle more ambitious problems. The increasing diversity and accessibility of sensors, computing power and what has been called the "big data" breakthrough have converged with advances in machine learning algorithms to provide novel, more efficient paths to robotic control algorithm design. We are getting close to the point where nontechnical users will become capable of training robots without expert help, possibly working in tandem with machine learning algorithms that are adapting the robot to the user's needs. This is likely to make robots more versatile and cost-effective, with profound socioeconomic impact in the near future; it is already having a dramatic impact in neural robotic interfaces, including research into robotic prostheses and exoskeletons.

Like any period of dramatic technological advances in human history, our societies will need to adapt and address novel leadership, ethical and legal issues that consequently arise. The current pace of exponential technological growth has been labelled as the path toward approaching the "singularity", the point at which AI algorithms will acquire superhuman capabilities and possibly rise to a level where humans cannot fully understand how they function. Neurorobotics technology could play an important role in such a scenario, bridging the gap between software intelligence, the physical world and biological organisms, helping facilitate human–robot teams, synergies and ultimately symbiosis.

Acknowledgments

This chapter is based upon work from COST Action CA 16116, "Wearable Robots for augmentation, assistance or substitution of human motor functions", supported by COST (European Cooperation in Science and Technology).

COST (European Cooperation in Science and Technology) is a funding agency for research and innovation networks. Our Actions help connect research initiatives across Europe and enable scientists to grow their ideas by sharing them with their peers. This boosts their research, career and innovation.

References

[1] Goodman PH, Zou Q, and Dascalu S-M. Framework and implications of virtual neurorobotics. Front Neurosci. 20087;2(1):123–8.

[2] Arbib MA, Metta G, and van der Smagt P. Neurorobotics: from vision to action. In: Siciliano B, and Khatib O, editors. Springer Handbook of Robotics [Internet]. Berlin, Heidelberg: Springer Berlin Heidelberg; 2008 [Accessed 2019 Feb 27]. p. 1453–80.

[3] Strukov DB, Snider GS, Stewart DR, and Williams RS. The missing memristor found. Nature. 2008;453(7191):80–3.

[4] Knowm.org: Memristors to Machine Intelligence [Internet]. Memristors to Machine Intelligence. [Accessed 2019 Apr 30]. Available from: https://knowm.org/about/

[5] Russell SJ, and Norvig P. Artificial intelligence : a modern approach [Internet]. Malaysia; Pearson Education Limited; 2016 [Accessed 2019 Feb 27].

[6] Li D, and Du Y. Artificial Intelligence With Uncertainty. Second edition. Boca Raton London New York: CRC Press, An Imprint of the Tylor & Francis Group; 2017. 290 p. (A Chapman & Hall book).

[7] Haykin S. Neural networks: a comprehensive foundation. 1st ed. Upper Saddle River, NJ, USA: Prentice Hall PTR; 1994.

[8] Knoll A, Röhrbein F, Kuhn A, Akl M, and Sharma K. Neurorobotics: from computational neuroscience to intelligent robots and back. Inform-Spektrum. 2017;40(2):161–4.

[9] Bishop CM. Pattern recognition and machine learning. New York: Springer; 2007. 738 p.

[10] Jordan MI, and Mitchell TM. Machine learning: trends, perspectives, and prospects. Science. 2015;349(6245):255–60.

[11] Pugh A. Robot vision. Berlin: Springer Science & Business Media; 2014. 347 p.

[12] LeCun Y, Bengio Y, and Hinton G. Deep learning. Nature. 2015;521(7553): 436–44.

[13] Fukushima K. Neocognitron: a self-organizing neural network model for a mechanism of pattern recognition unaffected by shift in position. Biol Cybern. 1980;36(4):193–202.

[14] Razavian AS, Azizpour H, Sullivan J, and Carlsson S. CNN features off-the-shelf: an astounding baseline for recognition. In: 2014 IEEE Conference on Computer Vision and Pattern Recognition Workshops [Internet]. Columbus, OH, USA: IEEE; 2014 [Accessed 2019 Apr 29]. p. 512–9.

[15] Martin G, Rentsch L, Höck M, and Bertau M. Lithium market research – global supply, future demand and price development. Energy Storage Mater. 2017;6(1): 171–9.

[16] Kim RE. Should deep seabed mining be allowed? Mar Policy. 2017;82(1): 134–7.

[17] Kim D-H, Viventi J, Amsden JJ *et al.* Dissolvable films of silk fibroin for ultrathin, conformal bio-integrated electronics. Nat Mater. 2010;9(6): 511–7.

[18] Minev IR, Musienko P, Hirsch A *et al.* Electronic dura mater for long-term multimodal neural interfaces. Science. 2015;347(6218):159–63.

[19] Tan DW, Schiefer MA, Keith MW, Anderson JR, Tyler J, and Tyler DJ. A neural interface provides long-term stable natural touch perception. Sci Transl Med. 2014;6(257):257ra138.

[20] Rodríguez FJ, Ceballos D, Schüttler M *et al.* Polyimide cuff electrodes for peripheral nerve stimulation. J Neurosci Methods. 2000;98(2):105–18.

[21] Patil AC, and Thakor NV. Implantable neurotechnologies: a review of micro- and nanoelectrodes for neural recording. Med Biol Eng Comput. 2016;54(1):23–44.

[22] Thukral A, Ershad F, Enan N, Rao Z, and Yu C. Soft ultrathin silicon electronics for soft neural interfaces: a review of recent advances of soft neural interfaces based on ultrathin silicon. IEEE Nanotechnol Mag. 2018;12(1):21–34.
[23] Fekete Z, and Pongrácz A. Multifunctional soft implants to monitor and control neural activity in the central and peripheral nervous system: a review. Sens Actuators B Chem. 2017;243(1):1214–23.
[24] Lopez-Gordo MA, Sanchez-Morillo D, and Valle FP. Dry EEG electrodes. Sensors. 2014;14(7):12847–70.
[25] Lobo-Prat J, Kooren PN, Stienen AH, Herder JL, Koopman BF, and Veltink PH. Non-invasive control interfaces for intention detection in active movement-assistive devices. J NeuroEngineering Rehabil. 2014;11(1):168.
[26] Nicolas-Alonso LF, and Gomez-Gil J. Brain computer interfaces, a review. Sensors. 2012;12(2):1211–79.
[27] Gardner J. A history of deep brain stimulation: technological innovation and the role of clinical assessment tools. Soc Stud Sci. 2013;43(5):707–28.
[28] Navarro X, Krueger TB, Lago N, Micera S, Stieglitz T, and Dario P. A critical review of interfaces with the peripheral nervous system for the control of neuroprostheses and hybrid bionic systems. J Peripher Nerv Syst. 2005;10(3):229–58.
[29] Pasquina PF, Evangelista M, Carvalho AJ *et al.* First-in-man demonstration of a fully implanted myoelectric sensors system to control an advanced electromechanical prosthetic hand. J Neurosci Methods. 2015;244(1):85–93.
[30] Russell C, Roche AD, and Chakrabarty S. Peripheral nerve bionic interface: a review of electrodes. Int J Intell Robot Appl. 2019;3(1):11–8.
[31] Buch E, Weber C, Cohen LG *et al.* Think to move: a neuromagnetic brain-computer interface (BCI) system for chronic stroke. Stroke. 2008;39(3):910–7.
[32] Kopton IM, and Kenning P. Near-infrared spectroscopy (NIRS) as a new tool for neuroeconomic research. Front Hum Neurosci [Internet]. 2014 [Accessed 2019 Jun 7];8.
[33] Novak D, and Riener R. A survey of sensor fusion methods in wearable robotics. Robot Auton Syst. 2014;73(1):155–70.
[34] Arbib MA, and Fellous J-M. Emotions: from brain to robot. Trends Cogn Sci. 2004;8(12):554–61.
[35] Ziemke T, and Lowe R. On the role of emotion in embodied cognitive architectures: from organisms to robots. Cogn Comput. 2009;1(1):104–17.
[36] Pfeifer R, and Scheier C. Understanding Intelligence. Cambridge, MA: MIT Press; 2001. 724 p.
[37] Chang WH, and Kim Y-H. Robot-assisted therapy in stroke rehabilitation. J Stroke. 2013;15(3):174–81.
[38] Laut J, Porfiri M, and Raghavan P. The present and future of robotic technology in rehabilitation. Curr Phys Med Rehabil Rep. 2016;4(4):312–19.
[39] Krebs HI, Hogan N, Aisen ML, and Volpe BT. Robot-aided neurorehabilitation. IEEE Trans Rehabil Eng Publ IEEE Eng Med Biol Soc. 1998;6(1):75–87.
[40] Díaz I, Gil JJ, and Sánchez E. Lower-limb robotic rehabilitation: literature review and challenges. J Robot. 2011;2011:1–11.

[41] Maciejasz PP, Eschweiler J, Gerlach-Hahn K, Jansen-Troy A, and Leonhardt S. A survey on robotic devices for upper limb rehabilitation. J Neuroengineering Rehabil. 2014;11(1):3.

[42] Gorgey AS. Robotic exoskeletons: The current pros and cons. World J Orthop. 2018;9(9):112–9.

[43] Xiloyannis M, Chiaradia D, Frisoli A, and Masia L. Physiological and kinematic effects of a soft exosuit on arm movements. J NeuroEngineering Rehabil. 2019;16(1):1.

[44] Quinlivan BT, Lee S, Malcolm P *et al.* Assistance magnitude versus metabolic cost reductions for a tethered multiarticular soft exosuit. Sci Robot. 2017;2(2):eaah4416.

[45] Marchal-Crespo L, and Reinkensmeyer DJ. Review of control strategies for robotic movement training after neurologic injury. J NeuroEngineering Rehabil. 2009;6(1):20.

[46] Nizamis K, Stienen AHA, Kamper DG *et al.* Transferrable expertise from bionic arms to robotic exoskeletons: perspectives for stroke and duchenne muscular dystrophy. IEEE Trans Med Robot Bionics. 2019;1(2): 88–96.

[47] Young AJ, and Ferris DP. State of the art and future directions for lower limb robotic exoskeletons. IEEE Trans Neural Syst Rehabil Eng. 2017;25(2): 171–82.

[48] Tefertiller C, Pharo B, Evans N, and Winchester P. Efficacy of rehabilitation robotics for walking training in neurological disorders: a review. J Rehabil Res Dev. 2011;48(4):387.

[49] Zhang X, Yue Z, and Wang J. Robotics in lower-limb rehabilitation after stroke. Behav Neurol [Internet]. [Accessed 2019 Apr 15]; 2017:1–13. Available from: https://www.ncbi.nlm.nih.gov/pmc/articles/PMC5480018/

[50] Louie DR, Eng JJ, and Lam T, Spinal cord injury research evidence (SCIRE) research team. Gait speed using powered robotic exoskeletons after spinal cord injury: a systematic review and correlational study. J NeuroEngineering Rehabil. 2015;12(1):82.

[51] Bos RA, Haarman CJW, Stortelder T *et al.* A structured overview of trends and technologies used in dynamic hand orthoses. J NeuroEngineering Rehabil. 2016;13(1):62.

[52] Kooren PN, Dunning AG, Janssen MMHP *et al.* Design and pilot validation of A-gear: a novel wearable dynamic arm support. J NeuroEngineering Rehabil. 2015;12(1):1–12.

[53] Bos RA, Nizamis K, Plettenburg DH, and Herder JL. Design of an electro-hydraulic hand orthosis for people with duchenne muscular dystrophy using commercially available components. In: 2018 7th IEEE International Conference on Biomedical Robotics and Biomechatronics (Biorob). 2018. p. 305–11.

[54] Nizamis K. Hand Neuro-Motor Characterization and Motor Intention Decoding in Duchenne Muscular Dystrophy [PhD Thesis – Research UT, graduation UT]. [Enschede]: University of Twente; 2019.

[55] Waldert S. Invasive vs. non-invasive neuronal signals for brain-machine interfaces: will one prevail? Front Neurosci [Internet]. 2016 Jun 27 [Accessed 2019 Jun 16];10.

[56] Pasquina PF, Evangelista M, Carvalho AJ *et al.* First-in-man demonstration of a fully implanted myoelectric sensors system to control an advanced electromechanical prosthetic hand. J Neurosci Methods. 2015;244(1):85–93.

[57] Trent L, Intintoli M, Prigge P *et al.* A narrative review: current upper limb prosthetic options and design. Disabil Rehabil Assist Technol. 2019;1–10.

[58] Lenzi, Tommaso H L, and Sensinger, J. Speed-adaptation mechanism: robotic prostheses can actively regulate joint torque. IEEE Robot Autom Mag. 2014;21(4):94–107.

[59] Artur Saudabayev AV. Sensors for Robotic Hands: A Survey of State-of-the-Art.

[60] Cangelosi A, and Invitto S. Human-robot interaction and neuroprosthetics: a review of new technologies. IEEE Consum Electron Mag. 2017;6(3):24–33.

[61] Krichmar JL. Neurorobotics—a thriving community and a promising pathway toward intelligent cognitive robots. Front Neurorobotics [Internet]. 2018 Jul 16 [Accessed 2019 Jan 31];12. Available from: https://www.ncbi.nlm.nih.gov/pmc/articles/PMC6054919/

[62] Haghi M, Thurow K, and Stoll R. Wearable devices in medical internet of things: scientific research and commercially available devices. Healthc Inform Res. 2017;23(1):4–15.

[63] Mahmoudi P, Veladi H, and Pakdel FG. Optogenetics, tools and applications in neurobiology. J Med Signals Sens. 2017;7(2):71–9.

[64] Rumelhart DE, Hinton GE, and McClelland JL. Parallel distributed processing: explorations in the microstructure of cognition, Vol. 1. In: Rumelhart DE, and McClelland JL, PDP Research Group C, editors. Cambridge, MA, USA: MIT Press; 1986 [Accessed 2019 Feb 27]. p. 45–76.

[65] Baumann D, Ascoli A, Tetzlaff R, Chua LO, and Hild M. Memristor-enhanced humanoid robot control system – Part II: circuit theoretic model and performance analysis. Int J Circuit Theory Appl. 2018;46(1):184–220.

[66] Resquín F, Cuesta Gómez A, Gonzalez-Vargas J *et al.* Hybrid robotic systems for upper limb rehabilitation after stroke: a review. Med Eng Phys. 2016;38(11):1279–88.

[67] Anaya F, Thangavel P, and Yu H. Hybrid FES–robotic gait rehabilitation technologies: a review on mechanical design, actuation, and control strategies. Int J Intell Robot Appl. 2018;2(1):1–28.

[68] Lobo-Prat J, Nizamis K, Janssen MMHP *et al.* Comparison between sEMG and force as control interfaces to support planar arm movements in adults with Duchenne: a feasibility study. J NeuroEngineering Rehabil. 2017;14 (1):1–17.

[69] Nizamis K, Lobo-Prat J, Keemink AQL, Carloni R, Stienen AHA, and Koopman BFJM. Switching proportional EMG control of a 3D endpoint arm support for people with duchenne muscular dystrophy. In: IEEE International Conference on Rehabilitation Robotics. 2015. p. 235–40.

[70] Reed KB, Handžić I, and McAmis S. Home-based rehabilitation: enabling frequent and effective training. In: Artemiadis P, editor. Neuro-Robotics: From Brain Machine Interfaces to Rehabilitation Robotics [Internet]. Dordrecht: Springer Netherlands; 2014 [Accessed 2019 Jan 31]. p. 379–403. (Trends in Augmentation of Human Performance).

[71] Hsieh Y-W, Chang K-C, Hung J-W, Wu C-Y, Fu M-H, and Chen C-C. Effects of home-based versus clinic-based rehabilitation combining mirror therapy and task-specific training for patients with stroke: a randomized crossover trial. Arch Phys Med Rehabil. 2018;99(12):2399–407.

[72] Pignolo L. Robotics in neuro-rehabilitation. J Rehabil Med. 2009;41(12): 955–960.

[73] Hu X, Wang Y, Zhao T, and Gunduz A. Neural coding for effective rehabilitation. BioMed Res Int. 2014;2014:1–17.

[74] Laver KE, Lange B, George S, Deutsch JE, Saposnik G, and Crotty M. Virtual reality for stroke rehabilitation. Cochrane Database Syst Rev. 2017;11(11):CD008349

[75] Amirabdollahian F, Ates S, Basteris A *et al.* Design, development and deployment of a hand/wrist exoskeleton for home-based rehabilitation after stroke – SCRIPT project. Robotica. 2014;32(08):1331–46.

[76] Oizumi M, Albantakis L, and Tononi G. From the phenomenology to the mechanisms of consciousness: integrated information theory 3.0. PLoS Comput Biol. 2014;10(5):e1003588.

[77] Searle JR. The problem of consciousness. Conscious Cogn. 1993;2(4):310–9.

[78] Preston J, and Bishop M J M. Views into the Chinese Room: New Essays on Searle and Artificial Intelligence [Internet]. 2002 [Accessed 2019 Feb 21]. Available from: http://research.gold.ac.uk/15123/

[79] Tononi G, Edelman GM, and Sporns O. Complexity and coherency: integrating information in the brain. Trends Cogn Sci. 1998;2(12):474–84.

[80] Novamente BG. Hyperset models of self, will and reflective consciousness. In 2011.

[81] Tani J. The dynamical systems accounts for phenomenology of immanent time: an interpretation by revisiting a robotics synthetic study [Internet]. 2004 [Accessed 2019 Feb 21]. Available from: https://www.ingentaconnect.com/contentone/imp/jcs/2004/00000011/00000009/art00001

[82] O'Regan JK. How to build a robot that is conscious and feels. Minds Mach. 2012;22(2):117–36.

[83] Iosa M, Morone G, Cherubini A, and Paolucci S. The three laws of neurorobotics: a review on what neurorehabilitation robots should do for patients and clinicians. J Med Biol Eng. 2016;36(1):1–11.

[84] Datteri E. Predicting the long-term effects of human-robot interaction: a reflection on responsibility in medical robotics. Sci Eng Ethics. 2013;19(1): 139–60.

[85] Hidler J, Nichols D, Pelliccio M, and Brady K. Advances in the understanding and treatment of stroke impairment using robotic devices. Top Stroke Rehabil. 2005;12(2):22–35.

[86] Iosa M, Morone G, Bragoni M *et al.* Driving electromechanically assisted Gait Trainer for people with stroke. J Rehabil Res Dev. 2011;48(2):135–46.

[87] Ingemanson ML, Rowe JB, Chan V, Wolbrecht ET, Cramer SC, and Reinkensmeyer DJ. Use of a robotic device to measure age-related decline in finger proprioception. Exp Brain Res. 2015;234(1):83–93.

[88] Cappello L, Elangovan N, Contu S, Khosravani S, Konczak J, and Masia L. Robot-aided assessment of wrist proprioception. Front Hum Neurosci. 2015;9:198.

[89] Domingo A, and Lam T. Reliability and validity of using the Lokomat to assess lower limb joint position sense in people with incomplete spinal cord injury. J Neuroengineering Rehabil. 2014;11(1):167.

[90] Otaka E, Otaka Y, Kasuga S *et al.* Clinical usefulness and validity of robotic measures of reaching movement in hemiparetic stroke patients. J Neuroengineering Rehabil. 2015;12(1):66.

Chapter 8

Mobile apps for neuroscience

Albert-Jan Plate[1] *and Pieter Kubben*[1,2]

This book chapter will provide the reader with an overview of recent developments in mobile device technologies and development opportunities, focusing on iOS and Android. It will offer an overview of scientific evidence regarding mobile clinical decision support systems and elaborate on upcoming areas for innovation such as machine learning (ML) and augmented reality (AR).

8.1 Introduction

For several years already, smartphone applications are an upcoming wave of new technologies within the neuroscience sector. The popularization of smartphones among the general population and health-care providers has resulted in an increasing amount of available medical smartphone applications.

First, we will check out what platforms mobile apps are available on. We will explore the Internet of Things (IoT) and their possible role within the medical sector. How are those IoT-devices connected to the Internet? Hardware add-ons for smartphones add specialized features to smartphones.

After we introduced how apps can be distributed, we will discuss how apps can be used in the neuroscience sector. We will do this using real-life examples which highlight some possible use cases. There are still a lot more possibilities apps could be used. We will cover the following domains: research, diagnostics, pre-surgical planning, predictive modelling, training, communication and patient education.

Obviously there are risks and limitations involved. What regulatory issues apply regarding security? How did the EU and the FDA react to the sudden rise of this new medium? How can quality be ensured? When is an app considered a medical device?

Then we will shortly cover what are the benefits of using mobile apps within the medical sector. Although this part is mostly based upon the use-cases of applications discussed earlier, we will also talk about how apps are saving money and other benefits in general.

[1]School for Mental Health and Neuroscience, Maastricht University, Maastricht, The Netherlands
[2]Department of Neurosurgery, Maastricht University Medical Center, Maastricht, The Netherlands

We finish the chapter with a comparison of different types of smartphone application and how to build them.

8.2 Platforms for apps

Portable devices have become more common over the years. According to the data of statista.com, there are two major operating systems (OS) for mobile devices: iOS and Android. Globally, Android is installed at 88% of all mobile devices, whereas iOS is on 11% [1]. This difference can be explained by the cheaper smartphones which are a massive success in upcoming countries like India. But on the other hand, the statistics about the mobile OS market share in the United States show us that iOS is more popular among the local smartphone users. More specifically, according to data from statcounter.com 54.47% of Americans used iOS and 45.27% used Android in May 2019 [2]. The percentage therefore seems to depend very much on the target group. We were unable to find specific numbers for the healthcare system about the most used OS.

Both significant platforms have a wearable variant which runs on smartwatches. According to the data of statista.com, the smartwatch market overall takes up around 34.7% of the wearable market. WatchOS (16.2% market share) is the wearable counterpart of iOS, whereas Android Wear (4.3% market share) and WearOS (6.4% market share) originate from Android. Watch OS is the second generation of Android Wear. Other OS (7.8% market share) include but are not limited to Tizen, Ubuntu Touch, Asteroid OS and others. Dedicated fitness trackers are the remaining 65% of wearables.

Due to Android being open source, it also has been modified to target the IoT. This platform, which is called Android Things, enables developers to develop their portables/wearables. Of course, there are other platforms besides Android Things which can be used to develop connected devices. Due to the reason they share data to the cloud, which on its turn can be accessed through mobile apps, we will mention them shortly.

Additionally, as not all tools are necessary for the general consumer, there is an option to expand the functionalities of a smartphone. Extending the functions can be achieved by using add-ons. They can be electrical or nonelectrical and serve a particular medical function whereas the computational part is the smartphone.

8.2.1 Smartphones and tablets

On the website from Google's new Digital Wellbeing service, the following quote can be found: "Great technology should improve life, not distract from it." In 2017, the average consumer spent around 3.3 h looking at its phone [3]. The statistics do not include the 2.1 h and the 0.6 h we spent on our desktop/Laptop and our other connected devices (Gaming consoles and others). In total 5.9 h, we spent looking at a screen.

When in 2008 the first smartphones which could install third-party applications were launched a new market began to rise. The app market has become a multibillion dollar business within a few years.

Nowadays, the average smartphone user spends around 40 dollars [4] while downloading on average ±40 free/premium apps from the app stores every year. The app stores are more booming than ever. In total there are around 2.1 million apps available in the Google Play store on Android and 2 million apps available in iOS' App Store.

Due to the commonness of more powerful smartphones in the society, there has been a rise of use cases for smartphones as medical devices. Nowadays, a smartphone is more than just an environment for apps. Recently most known smartphone brands have started to include local, specialized chips for security reasons but also quicker processing of big data and local ML. These enhancements result in smartphones that are in most cases just as powerful as an ordinary computer.

Medical and health apps are also on an inevitable rise with huge potentials. Although the medical sector is slow in adopting new technologies, the mobile Health (mHealth) market is very dynamic. Due to the easy accessibility of development tools for mobile applications, it is not hard to include the latest developments, such as including local ML, in personal specialized applications. In total there were over 325,000 mHealth applications available in the app stores in 2017. These apps are published by over 84,000 app development companies specialized in the health-care sector [5].

The proliferation of unregulated mHealth applications also brings untested claims and false advertisement. This cluttering of uncertified apps also result in a competitive app market where professional apps are suppressed. Altogether, the mHealth sector is changing the health care with new workflows and new business models. Which calls for better regulations.

During this chapter when we mention smartphones, we associate tablets under this same category. Due to the small differences between them.

8.2.2 Smartwatches and fitness trackers

Fitness trackers are wearables which focus primarily on health. Compared to fitness trackers, smartwatches offer more diverse functionality. The OS of smartwatches is a stripped version of their smartphone counterpart. This "lite"-version enables them to run an optimized version of the app as a smartwatch application. Essentially the user is wearing a specialized smartphone with almost the same features, but optimized for a small screen. It is possible to make phone calls, send text messages and receive notifications from a phone. More recent smartwatches even have their own sim-card tray so they can be used independently from a smartphone. The new specialized app stores result in the availability of third-party applications.

After the release of the Apple Watch in 2015, there has been a rise in global consent about smartwatches. In 2018 an estimate of 181 million new smartwatches were shipped [6]. Every year dozens of new smartwatches are introduced with new features and higher accuracy. This last is often a point of critique on smartwatches as most people still see them more like a gimmick rather than a medical observer.

The latest smartwatches include basic features like a microphone, an accelerometer, a barometric pressure sensor, a GPS and a gyroscope [7]. However, some of these smartwatches also include advanced sensors like an ambient light sensor, a contact sensor, a temperature sensor, an optical sensor and now even a "cleared" ECG [8].

- Accelerometer: measures the body's movement,
- Barometric pressure sensor: measures the elevation according to the pressure,
- Gyroscope: measures the body's rotation,
- Ambient light sensor: measures the intensity of the light in the surroundings,
- Contact sensor: detects if the device is on the skin,
- Temperature sensor: measures the temperature of the users' body,
- Optical sensor: measures the heart rate,
- ECG: measures the electric impulses of the heart.

8.2.3 IoT and wearables

8.2.3.1 What is IoT?

The IoT is a connected network of machines and objects which contain data-collecting technologies. This data will be transmitted to the cloud using wireless connections like Wi-Fi or satellite connections. The objects in the IoT can communicate with each other, the so-called machine-to-machine talk (M2M). The generated data will be stored and can be used for analysis to determine the status of living and nonliving objects or persons.

The IoT is comparable to the Intranet of things connected to an external network. The difference is in short that the data generated by the devices are stored online and are being monitored and made available to third-party apps while the Intranet is only for inside use [9].

To make sure there is inter-connectivity between the devices developed by multiple vendors, there are a few proposed open standards. Although some vendors still choose to limit their inter-connectivity. Due to these limitations, the measured data cannot be shared with a central algorithm which on its turn can send a specific command to the other connected devices.

For example, Android Things is a modified version of Android optimized to run on connected microcontroller units. It runs on devices with RAM as low as 32–64 MB. These low-cost devices can be used as part of the IoT network. Its communication of data goes through Wi-Fi or Bluetooth LE to a central database.

Another example is Arduino or Raspberry-pi which are already much longer available. These MCUs have open-source sensors available which enable developers and engineers to build custom IoT-devices. These can send the measured data to the cloud for analysis. Examples of open source platforms are [10]:

- **MyOpen**: open-source hardware for myoelectric control of a prosthesis
- **OpenBCI**: an EEG signal capture platform and intended for human brain–computer interface technologies

- **NightscoutxDrip**: platform which allows real-time access to a continuous glucose monitor connected to a mobile phone

One significant advantage of these open-source projects is the allowance of participation from anywhere in the world without the need of being recruited by an R & D company. As the costs of health-care development are on an all-time high, this might open up new opportunities.

8.2.3.2 What are connected wearables?

The difference with the above paragraph is in a gray area. This ambiguity is because connected wearables can connect to a smartphone but also to the Internet. Thus the same effect can be achieved. To be clear, although a smartwatch is also applicable in this section, we decided to create a particular paragraph for this type of device.

According to Gartner [11], the worldwide consumer acceptance of wearables is rising due to three primary reasons. First, there are more active clinical trials for mHealth solutions. Secondly, insurance companies are slowly starting to cover, partially or wholly, the expenses of wearable health-care solutions. Finally, there is a new pilot program of the FDA which starts to license wearables with pre-certificates so they can become available much quicker.

The next wave of wearables will not only be able to record health data but also process this data. This on-device processing will enable them to notify the user when they detect abnormal behavior. This next wave will most likely also be able to communicate with the users' electronic health record (EHR). The new inter-connectivity will result in a connected health and wellness ecosystem which will be beneficial to the patient. On their turn, this creates an immediate availability of vital data to the appropriate parties. To enhance this principle, the European Union is working on a single market. This will be discussed further in Section 8.4.3.2.

8.2.3.3 Use cases of wearables in the medical sector

Examples of use cases in the medical sector of connected devices of the IoT and connected wearables are very wide-spread. Below are a few examples, but this is only the top of the iceberg of what is and will be possible with them.

Hearables are microcomputers which can fit in an ear. Using wireless technology they can enhance the users' listening experience, but more importantly, they also offer different medical measurements. They offer a more accurate measurement point of vital signs.

A hearable is a wireless in-ear computational earpiece. Principally the patient has a microcomputer that fits in their ear canal and utilizes wireless technology to supplement and enhance the listening experience.

- Hearables
- Ingestible sensors
- Moodables (mood enhancing devices

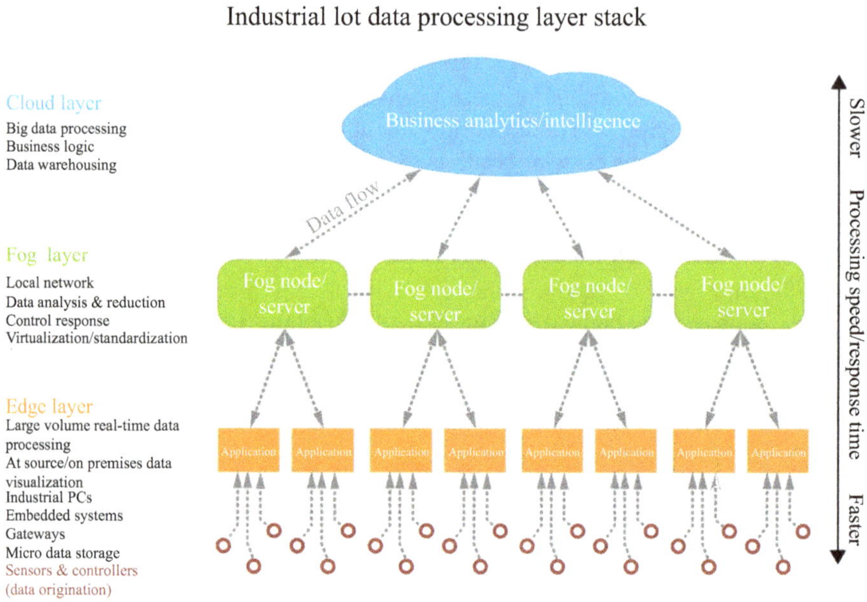

Figure 8.1 Conceptual map of IoT infrastructure

8.2.4 Cloud vs. edge layer

The difference between the cloud layer and the edge layer is already in its name, see Figure 8.1*. When the data is being processed in the cloud, it is cloud computing, which is done in major data warehouses and data centers. On the contrary, if it happens on the outer part of the Internet (e.g. the devices connected to the Internet which are used to generate the data, such as smartphones, wearables and embedded systems in sensors), we are talking about the edge layer. These systems pass on their data to the fog layer which on their turn save the data or decide to pass them on to the cloud layer.

More concretely, sensors can have an implemented system, the edge layer, that can transfer the generated data to the fog layer where it gets stored for data analysis. Other use-cases for the edge layer are but are not limited to: real-time data processing, data visualization, data caching and buffering, data filtering and optimization, and M2M communications.

One of the advantages of edge-computing is that sensitive data will remain within the walls of the organization and will not be transferred to the web. As processing power is on site, it is reliable and always accessible for the device without the requirement of being connected to the outer world. It is also straightforward to scale the edge/fog layer.

*Figure 1 from: https://www.winsystems.com/wp-content/uploads/2017/12/cloud-fog-edge_infographic.jpg

The cloud-infrastructure offers, on the contrary, the ability to access its data from any location as it is stored on a higher level of the infrastructure with more devices connected to it. It has a higher density of computing, storage and networking resources.

8.2.5 Hardware add-ons for smartphones

Smartphones are becoming more advanced every year. However, it is impossible to include the right hardware for every use case in a device that small. That is why there are hardware add-ons available for smartphones. Some add-ons are also capable of communicating with the smartphone over Bluetooth and do not need a direct connection using a cable to the smartphone. Following are some examples in which some specific hardware features are ported to a smartphone.

MobiSante has developed a portable ultrasound called the MobiUS SP1 Ultrasound Imaging Device [12]. The device is capable of assisting doctors during injections, line placements or diagnosis. The ultrasound can be displayed on the device using an application. Images can be stored and shared with a specialist.

Although this kind of hardware has gotten a bit redundant due to its integration in a smartwatch, there is a smartphone case with an ECG monitor built-in [13]. Although there will only be measurements using a single lead, the case will still be able to detect abnormal heart rhythms. The version used in hospitals, 12 leads, is much more advanced and will be able to provide more information. Most symptoms of palpitations do not occur regularly, the patient and the doctor need to be lucky the patient is on an ECG when the palpitations occur. This smartphone add-on allows the patient to monitor himself regularly at home. The app is also able to share the ECG with the doctor and log all measurements in a calendar.

Early versions of the Whitings blood pressure monitor [14] needed to be connected to the smartphone using a cord; nowadays all communication goes through a Bluetooth connection. The inflatable cuff can measure systolic and diastolic blood pressure and pulse rate.

Another example is a camera lens developed by D-EYE which can be mounted upon a smartphone [15]. By doing so, the smartphone is transformed into a retinal imaging system or more specific, a digital direct ophthalmoscope. With the assistance of the correct application, the smartphone camera is now much more capable. In Section 8.3.2.4, more about the use of this lens.

To port pulse oximetry to nonclinical settings, the UBC has developed the phone oximeter [16]. Patients can easily use the device at home and non-specialized health-care workers. The results can be displayed to a specialist using the Internet who can advise the patient or health-care workers about what steps to take next. The device has been exported to the third world, to help patient monitoring during anesthesia [17]. Now, the doctors can use this low-cost device to monitor hypoxemia, low blood oxygen levels, during an operation. Eventually, the UBC hopes to reduce the risk of dying of hypoxemia in a third world country. Currently, the risk is around 100–1,000 times higher compared to developed countries.

Although the results cannot be compared to a high-tech EEG, researchers have been able to develop a portable EEG which transfers data to a smartphone [18]. On this device, an application will produce a real-time 3D reconstruction. According to the researchers, the disadvantages of a lower resolution will be put aside by the advantages of having an EEG.

8.3 Use cases of mobile apps

There has been an incremental number of apps available for medical purposes. In this paragraph, we try to cover some use cases of mobile applications within the neuroscience sector.

8.3.1 Research

In 2015, Apple introduced ResearchKit [19]. This framework for developers can be used by researchers to develop research-centered applications easily. Ever since this launch, an increasing number of apps dedicated to research has been published in the different app stores. Many commercial and academic institutions develop these applications.

If an assessment requires more specific data, a smartwatch which has more sensors than a regular smartphone can be used. The data of the smartwatch can be synced to the mobile phone and then imported into the specific research-application.

"Sea Hero Quest" is a game for your smartphone that is designed to generate an essential crowd-sourced data set regarding human spatial navigation. This data set enables researchers to develop new diagnostic tests to detect dementia. Anonymous research data is sent and stored to a secure server as the gamer navigates through the game. The University College London will use this data for analysis regarding dementia research. Other researchers can request permission to the data set via Deutsche Telekom. Early results, dating from 2016 though, are auspicious according to the researchers. The game has provided the researchers with over 9,500 years of data from over 3 million participants. The total of session length is over 63 years. With a regular study, this would have been nearly impossible. The researchers found differences in how special navigation changes over the lifespan and there are significant differences in how genders and geographical origin influence navigation.

"DreamLab" is an application that uses your smartphone's processing power to perform cancer research. According to the app's description, this works as follows: "*Researchers need supercomputers to scan DNA in immense details and crunch complex data. When you plug your phone in and power DreamLab, it downloads tiny parts of a huge research project from the cloud to calculate, and sends the results back to the research team. The more people who power DreamLab, the faster we can complete critical research.*"

8.3.1.1 Advantages

The following advantages and limitations are based upon an article published in Academic Medicine [20].

First, performing research over the Internet removes all geographic barriers to research participation. As demonstrated earlier, apps are nowadays very widespread, and thus they are applicable as a new tool for communication. The smartphone is a more personal device and by using this personal device as a communication tool, might get the researchers more engagement in the study. There is no need to invite participants over to the research clinic. The removal of this geographic barrier results in a much broader reach of the research. Possible participants who live in more remote areas are still able to participate. With a more widespread group of participants the results can be conducted from a more diverse test-group.

Second, as an always-on and always connected device, the smartphone allows the researchers to have daily interactions with the participants. They are sending them regular notifications to fill in the daily form or tips regarding their performance in their assessments. The real-time connection also allows the participants to reach out to the researchers more easily, even integrating a real-time chat line, could lower the barrier for asking questions.

Third, as we will mention later on in the chapter, the smartphone includes several sensors which can be used to monitor the participant. These measurements can happen passively in the background or actively at the foreground with interactions of the participant. There are more data-points for the research to be based on due to the fact participants can perform (some of) their assessments individually. This sense of individuality is of course only real if there is no requirement to use advanced medical equipment. Otherwise, the researcher can ask for regular follow-ups between two clinical sessions.

Fourth, assessments are conducted objectively. Some studies require to follow-up on common fields like symptoms, condition, performance or weight of the participant. An app can help the researchers with quantifying the progression and the variability of disease. This can result in a more precise overview of the efficiency of new treatments.

Fifth, this new model of research is much more engaging for participants compared to a clinical trial where there is barely any interactivity between researchers. The real-time feedback of the assessments can give insights to the participants. This feedback can eventually lead in them to be more engaged, as they want to follow along with their progress. Indirectly this results in a more informed participant.

The above advantages imply that the participant has got clear instructions and is familiar with using a smartphone.

8.3.1.2 Apple's ResearchKit

Apple has a few initiatives regarding health. During the World Wide Developer Conference in 2014, Apple announced a new application called "Health." This app is a centralized service to collect all the users' health data. Using HealthKit, apps can share their data with each other. Before this can happen, the app requires, of course, the explicit permission of the possible participant in the study. In the process of setting this up, the app even requires the user to draw a signature.

One year later, in April 2015, Apple's new ResearchKit [19] introduced tools for researchers to perform digital research using iPhone-apps. Users can opt-in or out whenever they want. The users decide which data will be shared with the researchers. The participant is entirely in charge. It is hard for researchers to get access to a diverse world-wide population, and as the iPhone is a wide-spread phone, many people will be able to participate in these studies. This additional openness results in more qualitative data for a more significant study. A lot of people are already familiar with keeping up a digital record of medicine use, their weight, blood pressure, hearth beat, cholesterol level and much more.

ResearchKit also contains tools that allow the researchers to ask the participant to do specific exercises. With the built-in sensors, the assignment can be recorded. One application that got available at launch is an app that can estimate the participants' chance on Parkinson using a spoken exercise.

ResearchKit got a major update to version 2.0 in 2018. Current tools available:

- Institutional research board consent form [21]
- Survey tools
- Interfaces for sensors
- Highest file protection

There are other variants of ResearchKit available, including an Android port, called ResearchStack.

8.3.2 Diagnoses

Apps can be used to get a quicker overview or determination of the symptoms of a disease. Using an app can thus result in an easier diagnosis for the patient. Over the years there has been an ongoing shift toward more technology in the diagnosis. Nowadays, many diagnosis are not based upon symptoms but rather on data provided with technology. The smartphone is pretty new in this sector, as the first app stores only opened in 2008, and thus there is still a lot of room for unexplored territory [22].

8.3.2.1 Tremor detection

Before diagnosing a patient with Parkinson's disease, the patient needs to suffer from at least two of the following four symptoms: rest tremor (vibration), slow movement, balance problems and rigidity (stiffness).

There are several apps developed to classify tremors by their intensity [23]. An orthostatic tremor (trembling in the lower extremities) is in most cases invisible for the naked eye. To be sure there is a tremor, the doctor would require to perform an electromyogram for a conclusive diagnosis. A smartphone contains a gyroscope which detects any movement on any of the three axes. The accelerometer logs all the movements of the leg to an excel-sheet if the phone is attached to the leg of the patient (e.g. in the sock of the patient) [24]. For example, the app TREMOR12 is an open-source iOS app for tremor quantification [25].

Research [26] has proven the miniature accelerometers in phones nowadays are precise enough for medical use. The calculated frequency can determine if there is a tremor.

Another example for classifying a tremor on a smartphone is to measure the hand dexterity with a test on a smartphone [27]. The patients were asked to do tapping exercises on buttons with varying sizes. They were also asked to draw spirals and circles.

The results from the above tests can also be used to track treatment results after some time. Due to the simplicity, there is no need to have the patient return to the clinic for regular exercises. The patient can do more self-examinations over time between two appointments. With more data available, the patient's medication can be adjusted more accurately by the doctor.

8.3.2.2 Detection of a stroke

In Spain, Valencia's Polytechnic University (UPV), there is ongoing research regarding an app that can pre-diagnose possible patients with a brain stroke. The user needs to fulfill three simple tasks. If the user fails 2 or all of the three tests, the possible patient will be redirected to the emergency services.

First, the patient smiles at the phone with the rear-camera open. Using computer vision in combination with an advanced trained model, the coordinates of the left- or right-hand side of the mouth can be retrieved. While comparing these coordinates using a model created by machine learning the app can determine if there is paralysis. Not being able to smile at one side is often an indication for a stroke.

New technologies introduced the possibility to perform real-time tracking of facial expressions and features to determine if there is a stroke. All tracking happens on the device, so there is no need to be connected to the Internet or worry about the privacy of the patient's data. If a newer version of the ML model (the neural network that gets compiled after ML did its job) is available, it will automatically update over the air.

As for the second assignment, the app will ask the user to repeat a simple sentence which will then be compared to a pre-defined model. If the sentence were not spoken coherently, the app would detect this as one of the symptoms of a stroke.

The final test is where the user has to lift both hands to their maximum. By comparing the two results, the app can determine if the user has paralysis on one of the two sides.

If the user turns out to be a patient, the app will redirect the user to the according to emergency services. The app is currently still under development.

8.3.2.3 Motor performance testing

Childhood dystonia is defined as "a movement disorder in which involuntary sustained or intermittent muscle contractions cause twisting and repetitive movements, abnormal postures, or both" [28]. Children who suffer from Dystonia often rely on augmented and assistive communication (AAC) devices for communication with the outside world (e.g. touch screens). It is essential to test these children upon their sensory-motor abilities to check if they will be able to push on virtual buttons.

Current scales for determining the severity of dystonia are often subject to interpretation. That is why looking for a more objective and quantitative test could improve the health care for children with a movement disorder such as dystonia.

Using an application where a user has to drag an item across the screen from one specific point to another, the app can test the motor performance [29]. Fitts' law is a predictive, mathematical model which is used in human–computer interaction [30]. The law is used to predict the time a human needs to move a finger or a mouse across the screen. The relation between accuracy and time to move the finger is also included in the law. That is why it can be used to test the upper extremity motor performance of children with Dystonia.

The University of Southern California and the Children's Hospital Los Angeles performed some research between two groups of children to determine if Fitts' law can be used to differentiate two groups of test subjects. One group included children with primary or secondary dystonia, and the other existed out of healthy children.

According to the researchers, "*Children with dystonia showed a significantly lower index of performance than healthy children and adults, resulting in lower information transfer rate. In other words, at the same index of difficulty children with dystonia perform the task more slowly.*" (See Figure 8.2.) More research should eventually enable the researchers to come up with a scale to quantify motor-functions for dystonic children.

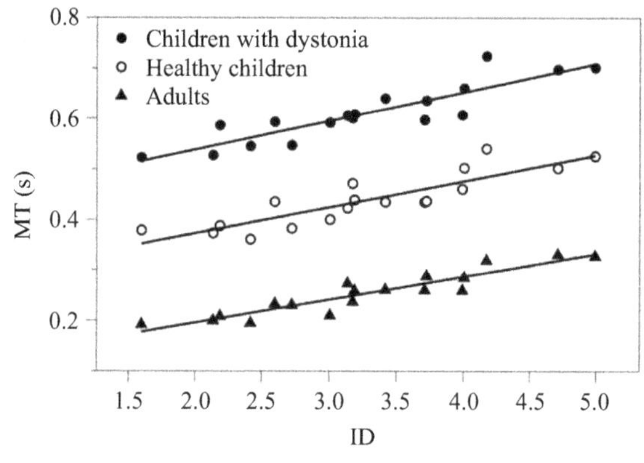

Figure 8.2 Mean movement time (MT) for all subjects as a function of the index of difficulty (ID). The straight lines show the best fits by the least squares method for each group [28]

8.3.2.4 Mobile clinical decision support systems

Due to the high portability, always at hand and low cost of smartphones, they are a rising tool for mobile clinical decision support systems (CDSS) [31]. Although CDSS's are already longer around, the recent implementations in smartphones are beneficial for the systems as they will get much more refined with this new pulse.

Today's smartphones are powerful enough to fulfill the heavy required computational algorithms.

CDSS are the link between medical observations and medical knowledge as they influence clinicians in making the right health decision. They can play a role from diagnoses to patient follow-up. There are two major types of CDSS: rule-based systems and models created using ML although ML can also design a rule-based system [32].

A CDSS can vary a lot according to type and complexity. There are passive, semi-passive and active systems. Whereas the passive systems require input from a human being for assistance, the semi-passive is automatically activated. This observational tool only shows data on request. Finally, there is an active system which gets automatically activated and presents data without waiting for a request. Sometimes, it also can decide without any human interaction.

General use cases are but are not limited to: order facilitators, point-of-care alerts/reminders, medication dosing support, relevant information display, workflow support systems and expert systems.

Benefits from using a mobile CDSS are according to [31]:

- Cheap accessibility of comfortable wide-spread hardware,
- Wide-range of available applications (free and paid),
- Applications are built for ease of use,
- Time-saving which results in better performance,
- Worldwide information is available at hand when needed.

Of course, besides having loads of advantages, there are also some points of critique according to [31]:

- The software is in most cases developed by nonmedical engineers,
- Sometimes requires explicit numerical data,
- Some applications are not medically tested, which results in unknown accuracy and reliability, so discretion is recommended,
- Lack of ability to see algorithm, which can be required to check if specific conditions are taken into account,
- Internet connection may be required, but this might not always be available,
- The patient could feel suspicion toward the doctor as he is using software for the diagnoses.

The CDSS applications are nearly endless for the future. Possible specific applications could be the following. This is an overview of possible applications found in [31].

In a lot of medical domains CDSS are used, these include but are not limited to internal medicine, surgery, psychiatry, neurology, pediatrics. The CDSS can be used to interpret diagnostic or laboratory tests. They can perform risk assessment and help to select the appropriate diagnostic or surgical procedure. It is also possible to use them for classification of diseases like cancer or congenital heart defects. In the pharmaceutical field, they should also help to determine which pharma logical interactions might happen between medications.

Computer-aided diagnosis (CAD)

CAD is, what the name already suggests, a system where the doctor can input a list of symptoms of the patient. The application will then compare these symptoms with a model. This model is in most cases generated by ML. Each possible diagnosis will be presented to the doctor with a score. The score will provide the doctor with an overview of the likelihood of the suggested diagnose of disease or medical condition [33].

As more and more specialized neural chips are getting integrated directly in the smartphone, there is no need for specialized external training devices for neural networks. These new developments enable app developers to train new models and compare models on-device without the need to send sensitive data to a server.

Another use case of a CAD-system is to input an image instead of medical symptoms. Now the application will analyze the image and compare it with images available in its data-set. Again, the application will output the most likely diagnose.

Viz.ai, for example, has developed a mobile application and a cloud-platform which can process a CT-scan. The application can determine if the patient has had a stroke or not. Using artificial intelligence and deep-learning the product is capable of reducing the average picture to notification time from 66 min down to 6 min. If an LVO, large vessel occlusion, has been detected, the algorithm will automatically notify the specialist. Currently, the specialists are still required to review the image in a professional workstation. The app should only be used as an assistant, and should not replace the full patient evaluation. A human review is still required.

Another major problem in the neuroscience sector in the third world is that it is hard for patients to see a qualified physician, let alone a specialist. People who live in the countryside are far away from the cities where most physicians and doctors live. In India, there are many people with untreated epilepsy due to this reason. There are currently two solutions trying to close the gap between seeing a health-care worker and a physician or a neurologist.

The first is an initiative to educate local doctors and make them play a more significant role in the treatment of epilepsy. Of course, specialists will remain in touch with them to support them with advice. The operational details are not documented of this system of care, but they are out of reach of this chapter.

Secondly, there is an initiative to provide health-care workers, who are much more numerous in rural areas, with apps [34]. These smartphone applications have a survey of 11 questions about the episode which will be filled in together with the patient. The results of these questions will then be sorted through an algorithm which calculates one general number. This result is the likelihood of the episode to be epileptic. According to these results, the health-care worker could refer the patient to a specialist (Table 8.1).

Smartphones with their applications are nowadays that powerful that they are capable of performing heavy computational calculations within no time. This strength enables them to run complex image analyses such as face recognition and

Table 8.1 Agreement between app and clinician diagnoses in 132 patients using an app score of >90 as epilepsy and one <30 as not epilepsy [34]

Diagnosis by phone app	Diagnosis by neurologist			Total
	No epilepsy	**Uncertain**	**Epilepsy**	
No epilepsy	13	4	1	18
Uncertain	0	5	12	17
Epilepsy	0	0	97	97
Total	13	9	110	132

Weighted *kappa* = 75.3% (95% CI: 60.9%, 89.7%).

detection or object identification algorithms. An example of this is a mobile app which can detect an optic nerve head , which is crucial for diagnosing conditions such as diabetic retinopathy [35]. The only requirement is a unique hardware add-on, a specialized lens which can be mounted on a smartphone. The ONH is shown as a yellowish region in a retinal fundus image. In the future, this real-time view could then be analyzed for containing the symptoms for, for example, diabetic retinopathy.

Interactive decision-trees

An interactive decision-tree is a system where the clinical worker is asked a few questions which follow up on each other. A different question may follow each question according to previous answers. Eventually, the system will prompt the user with a suggested action or diagnose. There are two major types of decision-trees, the classification tree and the regression tree. Within the classification tree, a set of logical if-then questions will be used to determine whether which classification fits the best. A regression tree, on the other hand, is used when the target variable is numerical instead of classification.

The concept of an interactive decision-tree can be implemented in a smartphone application to make the tree more engaging and less overwhelming when there are a lot of possible outcomes.

Disadvantages of using an interactive decision-tree are the same as the disadvantages of using a regular decision-tree. When there is not a clear, abrupt difference between the possible outcomes, it is difficult to set up a tree. Another factor to keep in mind is that there needs to be some correlation between possible outcomes, as it would otherwise be pretty challenging to find a question which differentiates both outcomes. The final point of critique is the fact that depending upon what starting point has been chosen; the outcomes will differ too.

8.3.3 Pre-surgical planning

During neurosurgical operations, precise planning is of great importance. Experienced neurosurgeons are often capable of correlating the patient's anatomy to the 2D scans and thus localizing the intracranial lesion without neuronavigational systems. Due to the fact, this is highly correlated with experience and neuronavigational tools are very expensive and difficult to learn. Another reason why it is difficult to find the right spot for incision is due to the lack of available landmarks. Therefore researchers have tried to develop a smartphone application which is capable of fusing the virtual images with the real surgical field. That is where AR comes in the picture.

AR is capable of displaying virtual images onto the real world. The overlaid sensory information can be either constructive or deconstructive. With the launch of advanced tools, it is easier than ever to develop applications with AR capabilities.

Within the application that is the current subject, it is the main focus to help with pre-surgical planning for the scalp flap and the localization of the lesion. The accuracy of the applications overlay was with a mean $\pm$ SD of 4.4 $\pm$ 1.1 mm and a range of 2.3–6.3 mm [36]. More of the use of this application and research can be seen in Figure 8.3.

8.3.4 Predicting

Mathematical models can be compared to patient data in real-time to predict the need for intervention. Platforms which are capable of doing this are currently on the rise, but a few obstacles still have to be avoided. As personal data will be transferred, it is essential to ensure data is safely stored and transmitted between the users and the databases reliably. Integration with electronic record systems is also critical for not losing/always having at hand the critical data. These platforms are also capable of providing clinical staff with an interface to show patient data where deterioration of symptoms and the risk for relapse can be spotted. Following are some use cases of these mobile platforms.

8.3.4.1 Self-monitoring, follow-up and treatment intervention

Self-monitoring allows for more personalized treatment. Traditional treatments are based on the "average" patient, but as patients change during their treatment, their treatment needs to change with them. This need for adaptability is mainly essential for mental illnesses like schizophrenia and psychosis. The experience sampling method consists of contextual mental state data and allows us to monitor the mental state of the patient [37]. As this was mainly a psychology tool, nowadays combined with a mobile smartphone application, this is becoming a main-stream instrument for personalizing the healthcare, self-insights and supporting self-monitoring. By a wide range of randomly timed assessments ($\pm$ 10 times/day) throughout 10 days up to 1 year and maybe even more. Due to the fact of more regular follow-ups, instead of sporadic monitoring during consultations, more personalized care can be offered. A consultation can thus be more organized toward the patient instead of collecting data.

Limitations that get overcome when using smartphone applications for the management of severe mental illnesses are the following [38]. Patients feel disengagement when there is no constant reminder for their care planning. Late detection

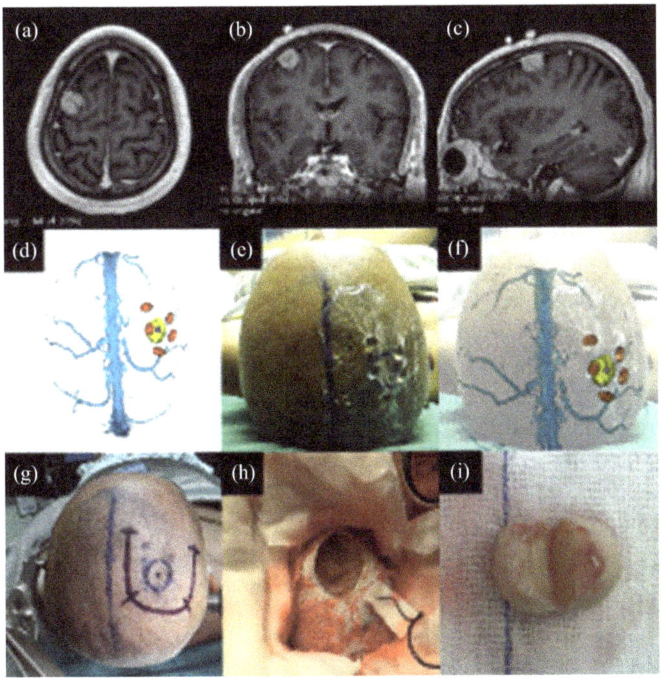

Figure 8.3 *Axial, coronal and sagittal MRI image showing a superficial supratentorial meningioma (Figure 8.3(a), (b) and (c)), the superior view which highlights (the lesion: yellow, marker: red, center of the lesion: dark blue, sagittal sinus with cortical vein: light blue) (Figure 8.3(d)), superior view of the patients' head with marker (Figure 8.3(e)), screen capture of the SINA during registration (Figure 8.3(f)), U-shaped skin incision was marked according to location of lesion (Figure 8.3(g)), intraoperative photography verified the location of lesion (Figure 8.3(h)), lesion was removed (Figure 8.3 (i). © 2017. Reprinted, with permission, from [36]*

of relapse symptoms can provide more stress on the home-situation, and unscheduled hospital admissions can be a considerable cost for the care system.

As mentioned earlier, modern treatment methods adapt to the patient over time. This requires intervention at the right time, to re-shape the treatment and eventually to shift the course.

Another well-known and widespread disease is diabetes, where the patient needs to track their insulin levels regularly [39]. When their insulin-levels drop below standards, they need to inject insulin. Applications can play different roles in the self-management of diabetes: (1) logging of test results, (2) alerts for tests, (3) calculators for insulin doses. Of course, there are more options available.

EpiWatch is an Apple Watch app developed for a research study which collects data about oncoming epileptic seizures [40]. If the accelerometer and heart rate sensors are triggered, an alert is automatically sent to a designated family member or caregiver. The app also logs all medication intakes. If this is put together with the intensity and timing of the seizure, the doctor may change the treatment plan.

8.3.4.2 Monitoring a stroke

Researchers of the University of Tennessee Chattanooga aimed to develop a smart system called mStroke [41]. The software is specialized in monitoring the risk of a stroke. They developed several wearables which are connected to a smartphone. According to the researchers, follow-up data is required for correlating successful and unsuccessful recovery and decide whether intervention is needed. Providing health-care workers with movement data of the specific patient besides the directly observed movement will enable them to prescribe specific treatment-interventions. This will over the long term enable the patient to recover much more efficiently.

The goal of the project is to improve care on three phases for an acute stroke patient:

- Enhancing acute management: 2–4 days after a stroke when physicians are working actively toward achieving a stable condition of the patient. Also, the start of physical rehabilitation where the smart system can provide the health-care worker with new activity and movement data. The health-care worker will be able to understand the magnitude of the movement dysfunction and provide the most optimal acute stroke rehabilitation setting.
- Enhancing acute rehabilitation: after being discharged, the smart system will continue with monitoring of the patient's movements. Physicians will be able to access data that previously was not available to them.
- Extending care: to support long-term follow-up, physicians will be able to keep track of patients even though appointments are less frequent.

8.3.5 Training

As a medical professional, "scholar" is one of the CanMEDS specialties a doctor is supposed to meet. Due to the high development rate of new technologies, an engaged doctor/health-care worker should always be following recent discoveries or advances in their specialty. Applications can play a significant role in this. Specialized news applications can suggest specific articles according to your interests. It is also possible to read journals through a mobile phone. The search function can come in very handy, as this is much more efficient to select what articles you want to read in a specific journal. The same applies to references.

There are also specialized apps available which can be an extension of a classic textbook or might even replace it [42]. Apps offer interactivity with images which results in a better visual understanding of the anatomy. They can contain schematic figures, pictures of real cases, scales related to the disease, training videos and interactive drug pre-subscription models. Another benefit from this is the ability to add interactive decision-trees within the learning space so the student can learn what each characteristic have for influence.

Major conferences nowadays tend to have an application dedicated to guidance during a conference [43]. These applications also enable the participants to look back any speeches they were not able to attend due to a double booking. This also increases the learning possibilities of the conference.

Another example is the availability to run simulators on a mobile device. Neuronify is an educational simulator which allows the user to create their neural circuits, which can be accomplished by combining different types of neurons [44]. The simulation is then able to predict the volt over two specific points. The application is much more advanced than what can currently be described in this chapter.

8.3.6 Communication

Communication is vital for an efficiently organized patient-care system—a secure, smooth communication between all parties that play a role for the patient. Mobile applications can play a significant role in this digital interaction.

Communication between colleagues as physicians is not only crucial for patient transfers but also for when another physician takes over a shift [45]. It is also a way of asking for advice right on the spot. Efficiently communicating is also essential for preventing misunderstandings. There is no need to explain trust from both sides could get damaged. When a patient is under the care of multiple physicians, all of them tend to stick to their specialty and sometimes there is a lack of communication [46].

Keeping a professional relationship with patients is essential and can be achieved in using specialized applications. In most cases contact via email is sufficient, but sometimes more intense contact is wanted. For example, if there is a need to monitor a patient.

Communication between patients can enhance their acceptance of the disease and can promote sharing tips and tricks to live with chronic diseases. The feeling of not being alone is key to the acceptance of a disease. More about this in Section 8.3.7.

Mobile applications can also play a role in administrative tasks for physicians and nurses as well.

8.3.7 Patient education

A secured, controlled environment is required to make sure the correct, easily understandable, verified data is within reach of the patient whenever they need the data. This need for data can be when they are at home doing some personal research but also when in conversation with another person who has got some unverified quality data. In both these cases, the doctor wants to make sure the patients get hold of the right information. A mobile application is an environment-friendly alternative to providing paper folders. There is also a more significant chance the app will stick around within reach of the user. Applications feel more personal compared to websites, but above all, they provide offline access. Applications can also offer interactive elements such as 3D-models.

To reach back to our example of diagnosing epilepsy in Nepal, a mobile application is also able to close the gap before seeing a doctor. For less developed countries it is often a significant step before they go toward a doctor [34]. With educating patients about the disease and the consequence of no treatment combined with a more accessible health-care worker, a lot more patients might seek help.

Child-friendly communication about their disease is vital to make sure they understand what is going on and why everything is changing. The child's disease might also be explained by providing educational videos.

It is essential to keep the patient engaged with their treatment. This can be achieved to make sure that the patient knows better what is going on. That is why some patients want to access their digital health reports. This also offers them the ability to ask more in-depth questions they have stuck in their mind.

As referred to earlier it is important to make sure patients can communicate with each other to talk about how to live with their disease. This can happen through patient support networks where tricks and possible tools for pain relief can be exchanged.

8.4 Risks and limitations

8.4.1 Risks

8.4.1.1 General

Patients whose confidence has been damaged and refuse their personal information to be stored online, may not be able to get the right care they deserve. Thus breach of patient confidentiality for one patient might result in less efficient care for another patient as bad news spreads faster than good.

Every mobile device might be a possible victim of viruses and malware. Regular security checks are thus recommended [47].

The use of applications which are developed by a specific medical device manufacturer or a pharmaceutical company can result in a biased recommendation which will lead to the use of the company's products. They might promote their own products as the possible treatment for the diagnosis of the software. At the moment of writing there are no regulations about this [48].

As a physician might have to use their smartphone instead of a medical device of the clinical institution, this might result in the distraction of the physician. Although this can easily be avoided by providing professional smartphones. Another possible side effect is that the patient takes the doctor less seriously as it seems that he has to look things up on his smartphone.

Although a clinical institution is very likely to have the appropriate policies in place to prevent these risks, it is still good to stand still to these risks. Possible workshops might prevent the simple mistakes mentioned above.

8.4.1.2 Research

If a researcher only relies on an anonymous survey taken using an app, there is a possibility the identity of the participant is as they pretend they are [20]. The total population can also be biased as there is less control of who is participating in the study. Due to the limits of technology in a device as small as a smartphone, there is not room for full-size medical equipment. Which on its turn, limits the number of tests and clinical trials that can be conducted using an application. In the future, it should be possible to perform blinded tests just as other Internet-based technologies are capable of now.

Participants who do not own a smartphone or a smartphone which is not supported by the application, are excluded from the research and thus resulting in a limited population.

77% of the average smartphone users abandon an application after using it within 72 h according to Chen [49]. There has been no substantial research how this impacts this type of apps. However, when personal face-to-face communication is absent, the retention rate might be higher as there is no human face connected to the research.

8.4.2 Privacy and security

8.4.2.1 General

Now and then a news story pops up about some scandal where a significant company leaked personal data [50]. These leaks raise the question of how this affects digital health. Being treated by a doctor is a matter of trust that they will be able to cure the patient with the correct treatment that suits their wishes. The doctor can select the correct treatment using the provided confidential information collected through the anamneses. In a modern age, this data often gets stored in a digital database, which requires the patient, again, to trust the doctor with our most intimate details.

It is important for health-care workers to understand these issues regarding privacy and digital health. Although every doctor is aware of the principles of confidentiality, there are new standards regarding confidentiality digitally. Trust is very vulnerable, and once broken it takes years to rebuild. If trust is lost in the health sector and the patient will not share all information due to this reason, there will be irreversible consequences.

The health-care system itself should be very aware of the big data capitalism for whom the multi-billion dollar health industry is an exciting market opportunity.

According to a study conducted by the European Union, 90% of the surveyed would be interested in accessing their health data. 80% would allow their data to be shared digitally, only if privacy and security are assured.

There is a need for security guidelines to ensure privacy in mHealth applications. That is why the European Union has written a code of conduct on privacy for mobile health applications [51]. The code provides developers with guidelines about how their apps must be designed for an optimal experience for their users. Contained in the code are the following significant points:

- **User's consent**: Before the application starts collecting data, the user must have agreed with the consent of the application. If the user denies the consent, at start-up or later, all user data needs to be deleted [21].
- **Purpose limitation and data minimization**: The collected data may only be used according to the purpose of the application. Only collect data which is necessary for the app.
- **Privacy by design and by default**: The app-developer needs to consider privacy consequences every time the user is prompted with a form. The least privacy sensitive has to be selected by default.
- **Data subjects rights and information requirements**: The app user should be able to access their data, and request changes to it. The developer should also be clear about how the data is processed.
- **Data retention**: Data should not be stored longer than necessary.
- **Security measures**: The developer should take actions to ensure integrity, confidentiality and access of the data and protect the data against accidental or unlawful destruction, alterations, loss, disclosure, access or other forms of processing. This by using an appropriate type of encryption.
- **Disclosing data to third parties for processing operations**: If the developer relies on data-processing using a third party, the user should be aware of this and always be notified.
- **Data transfers**: Data transfers from inside to outside the EU/EEA requires legal permission.
- **Personal data breach**: The code of conduct also includes guidelines for when a data breach occurs.
- **Data gathered from children**: The developer needs the consent of the parents before it is allowed to collect data from a child. What limits on ages are valid, can be checked at the national legalization.

Other variants of this code of conduct are the Health Insurance Portability and Accountability Act and U.S. Federal Information Processing Standards.

It is recommended for developers to encrypt their data on two different levels: Device-based and on the transmitting level.

The European Union is also working toward a single market to increase cybersecurity [52]. Significant milestones of this are, but not limited to enabling European inhabitants to get access to their digital health data across EU, enabling researchers to access resources across Europe (data, expertise, computing processing and storage capacities), personalized medicine through a shared European data infrastructure.

8.4.2.2 GDPR and HIPAA

The EU's GDPR (General Data Protection Regulation) and the HIPAA (Health Insurance Portability and Accountability Act) of the United States both define legal frameworks for data processing and health applications [53].

HIPAA is specifically meant for the health-care sector in contrary to the GDPR. It applies to all data collected which is related to the health-care sector, health insurance and all other relevant matters. The definition for PHI (Protected health information): "*Information that relates to the past, present, or future physical or mental health or condition of an individual, which identifies the individual and that is transmitted or maintained by electronic media or in any other form or medium.*"

GDPR covers any data related to the residents of the European Union. One of the categories the GDPR focuses on is the health-care sector. Health data is defined as: "All data that is related to the health status of a subject and is collected for the purpose of deducting the health status of someone."

8.4.3 Quality control

At first medical applications were designed to help educate patients about their health. Later apps began to help users track their health with measures such as weight, quality of sleep and the amount of cigarettes smoked. Due to this sudden but rapid wild grow of new medical smartphone applications, the FDA and the European Union needed some time to investigate the best way how to handle these new "medical tools." As time passes developers started to use the extra computing powers modern smartphones got and started to develop tools for self-diagnosis and treatment. This resulted in apps with false claims and wrong self-diagnosing tests. As of 2011, the US Food and Drug Administration (FDA) published draft guidelines for quality control. In 2013 the FDA published Mobile Medical Application Guidance which should clarify the rules regarding these apps.

It is especially important to ensure quality control for applications that have the intention to transform a mobile smartphone into a medical device. Another kind of applications that are controlled via the FDA are apps that claim to have therapeutic benefit. It is essential to have active control over these apps instead of pro-active control. For example, a Therapeutic Goods Administration service only investigates apps when they get reported via its users.

8.4.3.1 FDA

The FDA considers mHealth applications as medical devices [54]. Though not all apps need to be verified by the FDA. According to their website, they will only test applications that meet the regulatory definition of "device" and that:

- are intended to be used as an accessory to a regulated medical device, or
- transform a mobile platform into a regulated medical device.

The fact that not all apps are tested is due to the exploding app market, which makes it impossible to regulate everything. As a result, only apps that fall under the status "Medical device" will be tested.

There is a significant variation of mHealth applications, and most of them carry minimal risk. However, those that carry a more substantial risk for the patient requires FDA review. Applications that:

- Help users to self-manage their disease or condition without recommending specific treatment suggestions.
- Help patients with simple tools to track and organize their health data.
- Provide patient education related to health conditions or treatments.
- Help patients show or communicate possible medical conditions to their health-care providers.
- Help health-care providers with simple automated tasks.
- Allows patients or providers to access personal health records or EHR systems.

A more complete and up to date list of apps that do not require FDA clearance or approval can be found on their website.

The difference between "clearance" and "approval" is also critical to determine the stability of the application before you decide to start to use them. The most advanced is FDA approval, which is only used for Class III products or technologies. Devices/technologies in this class might have higher risks but at the same time a higher benefit (e.g. Implantable pacemakers). Approval is very beneficial for companies, but much testing is also required before achieving this designation.

For example, the Apple Watch Series 4 is classified as a Class II device [8]. These devices are not able to receive "approval," but they get a clearance. Small note regarding the Apple Watch Series 4, the EKG got classified as "de novo," which tells the FDA there is no comparable standard available at the market. So all testing needs to be started from the ground up.

There is a third possible certificate: the "pre-cert" [55]. This certificate is based upon the status of the company who wants to publish the application instead of the quality of the app. This allows the company to publish and market the application while it is still under review. More information can be found in the press release of the FDA.

8.4.3.2 The European Union

The European Union considers some specific software, just like the FDA, a medical device. More specific, it is considered a medical device if it has at least one functionality which allows the physician to prescribe or calculate a dosage with patient-specific information.

Starting from May 26th, 2020 the new medical device regulations is in force under which most eHealth companies are most likely to fall. From this point on all applications which fulfill the following definition of the MDR are considered a medical device.

> "*any instrument, apparatus, appliance, software, implant, reagent, material or other article intended by the manufacturer to be used, alone or in combination, for human beings for one or more of the following specific medical purposes:*
>
> - *diagnosis, prevention, monitoring, prediction, prognosis, treatment or alleviation of disease,*
> - *diagnosis, monitoring, treatment, alleviation of, or compensation for, an injury or disability,*
> - *investigation, replacement or modification of the anatomy or of a physiological or*
> - *pathological process or state,*
> - *providing information by means of in vitro examination of specimens derived from the*
> - *human body, including organ, blood and tissue donations*"

All service providers an app relies on, are required to comply with the GDPR, the MDR and other stringent standards which also apply to the application (such as ISO13458).

8.5 Benefits

Following is a brief enumeration of the possible benefits of using mobile applications within the neuroscience sector. Some of the examples are a summary of one of the use cases mentioned earlier.

8.5.1 Data collecting and analysis

Health-care analysis is used to interpret big data and transform it into refined and relevant information that is necessary for the doctor. The analysis of big data has the potential to predict the spread of epidemics or less specifically: improve the general quality of life.

With the use of mobile apps and connected wearables, more data can be collected. This might be through research or an opt-in feature for patients. We have talked about data collection and its risks earlier and how this might have a negative influence if not done correctly.

Another example of data-collection which benefits the research of human spatial navigation can be found in Section 8.3.1.

The results of the performed analysis can be used to assist in the process of coming up with the right diagnosis or treatment. Clinical decision support software analyzes the EHR and relevant medical data on the spot, providing doctors with advice as they make prescriptive decisions.

8.5.2 Simultaneous reporting and monitoring

Self-monitoring allows for more personalized treatment. Traditional treatments are based on the "average" patient, but as patients change during their treatment, their treatment needs to change with them. This need for adaptability is mainly essential for mental illnesses like schizophrenia and psychosis.

The sooner treatment can be (re)-started, the better the expectations. Thus speed is crucial in emergencies, and the sooner the medical staff gets notified, the sooner they can work on a solution. Apps or smartphones, in general, can play a significant role in notifying health-care practitioners.

Doctors can follow-up their patients while they are not required to return to the hospital as monitoring their developments are still being monitored.

8.5.3 End-to-end connectivity

In remote areas, it is not always possible to see a qualified physician. In most cases, a general health-care worker is the closest they will be able to get to a qualified medical specialist. With the use of telemedicine or specialized apps (Interactive decision-trees or CDSS) the required expertise can be. Of course, a consult between the health-care worker and the specialist is required, as although the information is available, the experience is still mandatory to make a good call.

As discussed in Section 8.3.6, mobile apps can enhance communication between doctors, patients and between both of them. They can be used for communication between doctors as for asking advice, patient transfers and others. Efficient communication between physicians can save lives.

8.5.4 Reducing costs and time

Mobile applications are capable of replacing specific medical equipment. Especially on missions to the third world, it is a functional addition to have a multidisciplinary device which reduces the purchase of specific equipment.

When data is being processed using off-site supercomputers to make advanced algorithmically calculations, there is no need for the research center or hospital to invest in such advanced computers.

Researchers will be able to perform tests without the need of having test subjects to come by the facility every time. They will be able to perform their tests individually which results in a savings of time spent by researchers which can be devoted to more critical parts of the research.

8.6 Developing apps

8.6.1 Native vs. Hybrid

Apps can technically be classified into two different types: native and hybrid applications [56]. The difference is made upon the way the app is coded. Depending on the expectations of the developer and the expectations of the users, the developers must carefully make a decision.

In this chapter, pure web-apps are not considered smartphone applications, as we consider them websites. A web-app is an app that is accessed through the Internet browser of a smartphone. They differ from websites as there is interaction possible with the user. They are not downloadable and always require an Internet connection.

8.6.1.1 Native apps

A native application is an app that has been written for one platform specifically. This has been done using official development languages and software development kits (SDK). It is not possible to export the app for multiple OS. The app needs to be developed for each platform individually.

For iOS, the available programming languages are Objective C, and the modern programming language developed by Apple called swift. Swift is an open-source programming language which gets regularly updated. Both languages can be combined if the developer wishes to do so.

Android apps are developed in Java or Kotlin. Kotlin is a new programming language which is simpler to read when compared to Java. Just like Apple, Google has made sure that Kotlin can run side-to-side with Java. So developers can easily mix code from two different languages together.

Applications are, in general, developed using an Integrated Development Environment (IDE). This is a specialized program for building new programs and in this case an app. For iOS this is XCode and Google has also developed their IDE. This is based on IntelliJ IDEA.

	iOS	**Android**
Supported programming languages	– Objective C – Swift	– Java – Kotlin
Extra development requirements	– MacOS – SDK	– JDK – SDK
IDE	XCode	Android Studio

Advantages of developing a native application are in general that they are written in the most efficient way for the OS. They are superior regarding speed and stability as this is the way apps are tested while the OS was being built. As the app is not running through an emulator or a WebView, this results in the app being able to access all sensors and APIs. An API is an application programming interface, which is used to make the application communicate with other parts of the OS. For example, on Android, an "Intent" is used to switch between activities. By calling a function of this API, the OS knows it should start looking for the according java class and open the required layout-file.

Apps from appstores are more often considered save compared to web-apps. This is a result of the owners of the appstores being very strict about what kind of apps are permitted to be sold through their store.

Maintaining an application that relies on native features requires more work, as they are subject to changes to the OS. Every update the developer publishes also has to be reviewed by the owner of the appstore.

8.6.1.2 Hybrid apps

Hybrid apps, as the name already suggests, are a mixture of native apps and web-apps. It essentially is a website in a WebView which got modified so it can access native features of the smartphone. They are written in HTML (structure), JavaScript (interactivity) and CSS (styling). A hybrid app is much easier to develop since applications are developed from 1 codebase using common web development techniques.

There are a few frameworks available which enable a developer to write hybrid apps: React Native and Ionic are two major players, but of course, there are a lot more frameworks available. React Native is developed by Facebook together with other members of their community and is used by significant apps like Facebook, Instagram and Uber.

All frameworks rely on a native wrapper which converts the web-app into an installable application for the specific platform. The big advantage of developing a hybrid app instead of a native application is that saves much time. This is because the developer only has to build one single code-base for the app to run on multiple platforms.

The down-side of this is that hybrid apps are not responsive and most of the time feel out of place on the device regarding poor design. As every activity is just a web-page, they load slower than a native component as the WebView has to be loaded first.

8.6.2 Native apps from a single source code

Another approach to app development is combining the best of both worlds. Take the simplicity of working with one code-base and combine this with the speed of native components. This is what, for example, Flutter does.

Within Flutter the developer writes code in Dart, a programming language developed by Google, and compile this down to native code. One advantage of this is that the design of the apps looks native too, which results in higher user experience.

Like setting up Bluetooth connections, some device-dependent features requires the developer to include this per platform individually.

Downsides of using Flutter at the moment of writing is the limited amount of libraries which are available. A library is a software package a developer can import to re-use the code of another developer, and this speeds up the development process.

Another downside of developing a Flutter or Hybrid app is that they are way bigger when compared to a native application.

References

[1] Mobile operating systems' market share worldwide from January 2012 to December 2018. (2019, March 16). Retrieved from Statista: https://www.statista.com/statistics/272698/global-market-share-held-by-mobile-operating-systems-since-2009/

[2] Mobile Operating System Market Share United States Of America. (2019, March 16). Retrieved from http://gs.statcounter.com/os-market-share/mobile/united-states-of-america
[3] Meeker, M. (2018, May 30). *Internet Trends 2018*. Retrieved from https://www.slideshare.net/kleinerperkins/internet-trends-report-2018-99574140
[4] Lovejoy, B. (2017, February 22). The Average US iPhone User Spent $40 on Apps Last Year, Downloaded 33 Apps. Retrieved from 9To5Mac: http://9to5mac.com/2017/02/22/us-iphone-users-average-number-of-apps-average-spend-on-apps/. Accessed on March 16, 2019.
[5] The Rise of mHealth Apps: A Market Snapshot. (2018, March 26). Retrieved from Liquid State: https://liquid-state.com/mhealth-apps-market-snapshot/
[6] Smartwatch unit sales worldwide 2014-2018. (2019, March 16). Retrieved from statista.com: https://www.statista.com/statistics/538237/global-smart watch-unit-sales/
[7] Reeder, B. (2016). Health at hand: A systematic review of smart watch uses for health and wellness. *Journal of Biomedical Informatics*, 269–276. Retrieved from https://www.sciencedirect.com/science/article/pii/S1532046416301137. Accessed on March 16, 2019.
[8] Chen, A. (2018, September 13). *What the Apple Watch's FDA clearance actually means*. Retrieved from The Verge: https://www.theverge.com/2018/9/13/17855006/apple-watch-series-4-ekg-fda-approved-vs-cleared-meaning-safe
[9] Botelho, B. (2019, March 16). *IoT Explained: What is the Internet of Things?* Retrieved from internetofthingsagenda.techtarget.com: https://internetofthingsagenda.techtarget.com/feature/Explained-What-is-the-Internet-of-Things
[10] Niezen, G. (2019, March 16). *Open-Source Hardware for Medical Devices*. Retrieved from innovations.bmj.com: https://innovations.bmj.com/content/2/2/78
[11] Gartner®. (2018, March 8). Wearables Hold the Key to Connected Health Monitoring. Retrieved from: gartner.com: https://www.gartner.com/smarterwithgartner/wearables-hold-the-key-to-connected-health-monitoring
[12] Smartphone Ultrasound. (2019, March 16). Retrieved from mobisante.com: http://www.mobisante.com/products/product-overview/
[13] Bogle, R. (2019, March 16). *The AliveCor iPhone ECG Monitor*. Retrieved from richardbogle.com: https://www.richardbogle.com/alivecor-ecg-monitor.html
[14] Wireless Blood Pressure Monitor. (2019, March 16). Retrieved from withings.com: https://www.withings.com/nl/en/blood-pressure-monitor
[15] D-EYE Retina. (2019, March 16). Retrieved from https://www.d-eyecare.com/product
[16] The Phone Oximeter. (2019, March 16). Retrieved from phoneoximeter.org: http://www.phoneoximeter.org/the-phone-oximeter/
[17] Monitoring anesthesia in the developing world. (2019, March 16). Retrieved from Electrical & Computer Engineering in Medicine: http://www.phoneoximeter.org/projects/anesthesia/

[18] Stopczynski, A. (2014, February 5). *The Smartphone Brain Scanner: A Portable Real-Time Neuroimaging System.* Retrieved from https://www.ncbi.nlm.nih.gov/pmc/articles/PMC3914802/

[19] Introducing ResearchKit. (2019, March 16). Retrieved from http://researchkit.org/

[20] Dorsey, E. R. (2017, February). *The Use of Smartphones for Health Research.* Retrieved from https://journals.lww.com/academicmedicine/fulltext/2017/02000/The_Use_of_Smartphones_for_Health_Research.15.aspx. Accessed on March 16, 2019.

[21] Moore, S. (2017, August 30). *Consent Processes for Mobile App Mediated Research: Systematic Review.* Retrieved from https://www.ncbi.nlm.nih.gov/pmc/articles/PMC5597795/

[22] Jutel, A. (2015, March 5). *Digitizing Diagnosis: A Review of Mobile Applications in the diagnostic process.* Retrieved from https://www.degruyter.com/view/j/dx.2015.2.issue-2/dx-2014-0068/dx-2014-0068.xml

[23] Mónica Huerta, A. M. (2014, October). *Smartphone Application for Quantitative Measurement of Parkinson Tremors.* Retrieved from https://www.researchgate.net/publication/267779811_Smartphone_Application_for_Quantitative_Measurement_of_Parkinson_Tremors. Accessed on March 16, 2019.

[24] Balachandar, A. (2017, July 18). *Characterizing Orthostatic Tremor Using a Smartphone Application.* Retrieved from https://www.ncbi.nlm.nih.gov/pmc/articles/PMC5623759/

[25] Kubben, P. L., Kuijf, M. L., Ackermans, L. P. C. M., Leentjens, A. F. G., and Temel, Y. (2016). *TREMOR12: An Open-Source Mobile App for Tremor Quantification.* Retrieved from https://www.ncbi.nlm.nih.gov/pubmed/27395052. Accessed on March 16, 2019.

[26] Culhane, K. M., O'Connor, M., Lyons, D., and Lyons, G. M., (2005, November). Accelerometers in rehabilitation medicine for older adults. *Age and Ageing*, 34(6):556–60.

[27] Lee, W., Evans, A., and Williams, D. R. (2016, April 2). *Validation of a Smartphone Application Measuring Motor Function in Parkinson's Disease.* Retrieved from https://www.ncbi.nlm.nih.gov/pubmed/27061062

[28] Sanger, T. D., Chen, D., Fehlings, D. L., *et al.* Definition and classification of hyperkinetic movements in childhood. *Movement Disorder* 2010;25:1538–1549. http://dx.doi.org/10.1002/mds.23088

[29] Bertucco, M. and Sanger, T. D. (2014, November). *Speed-Accuracy Testing on the Apple iPad provides a Quantitative Test of Upper Extremity Motor Performance in Children with Dystonia.* Retrieved from https://www.ncbi.nlm.nih.gov/pubmed/23965400. Accessed on March 16, 2019.

[30] Fitts's law. (2019, March 16). Retrieved from Wikipedia: https://en.wikipedia.org/wiki/Fitts%27s_law

[31] Masic, I. (2016, July). *Mobile Clinical Decision Support Systems in Our Hands – Great Potential but also a Concern.* Retrieved from https://www.researchgate.net/publication/304676572_Mobile_Clinical_Decision_Support_

Systems_in_Our_Hands_-_Great_Potential_but_also_a_Concern. Accessed on March 16, 2019.

[32] Martínez-Pérez, B. (2014, January 4). *Mobile Clinical Decision Support Systems and Applications: A Literature and Commercial Review.* Retrieved from https://www.researchgate.net/publication/259627633_Mobile_Clinical_Decision_Support_Systems_and_Applications_A_Literature_and_Commercial_Review

[33] Kiszka, K., Haduch, J., and Pasowicz, M. (2019, March 16). *Application of Computer Aided Diagnosis (CAD) in Clinical Imaging.* Retrieved from https://www.ncbi.nlm.nih.gov/pubmed/20043591

[34] Patterson, V. (2015, May 12). *Validation of a Phone App for Epilepsy Diagnosis in India and Nepal.* Retrieved from https://core.ac.uk/download/pdf/82266324.pdf

[35] Yaroub Elloumi, M. A. (2018, May 22). *A Mobile Computer Aided System for Optic Nerve Head Detection.* Retrieved from https://hal-upec-upem.archives-ouvertes.fr/hal-01796762/file/A%20Mobile%20Computer%20Aided%20System%20for%20Optic%20Nerve%20Head%20Detection.pdf

[36] Chen, J.-G. (2017, June 24). *Presurgical Planning for Supratentorial Lesions with Free Slicer Software and Sina App*. Retrieved from https://www.sciencedirect.com/science/article/pii/S1878875017310434

[37] Os, J. v. (2017, May). *The Experience Sampling Method as an mHealth Tool to Support Self-monitoring, Self-insight, and Personalized Health Care in Clinical Practice*. Retrieved from https://www.researchgate.net/publication/317189462_The_experience_sampling_method_as_an_mHealth_tool_to_support_self-monitoring_self-insight_and_personalized_health_care_in_clinical_practice. Accessed on March 16, 2019.

[38] Whelan, P. (2015). *Mobile Early Detection and Connected Intervention to Coproduce Better Care in Severe Mental Illness*. Retrieved from http://ebooks.iospress.nl/publication/40181

[39] El-Gayar, O., Timsina, P., Nawar, N., and Eid, W. (2013, January 1). *Mobile Applications for Diabetes Self-manage ment: Status and Potential*. Retrieved from https://www.ncbi.nlm.nih.gov/pubmed/23439183

[40] EpiWatch: Frequently Asked Questions. (2019, March 16). Retrieved from https://www.hopkinsmedicine.org/epiwatch/faqs.html#.XE9F5c9KhmA

[41] Fell, N. L. (2017, April 30). *mStroke: Mobile Technology for Post-Stroke Recurrence Prevention and Recovery*. Retrieved from grantome.com: http://grantome.com/grant/NIH/R15-EB015700-01A1

[42] Chaudry, N. S., F. U. Ahmad, M. R. Bullock, and M. Y. Wang (2014) *Traumatic Brain Injury at Your Fingertips! Neurosurgery*, 74(2):N19–N22. hhttps://doi.org/10.1227/10.1227/01.neu.0000442980.84206.dd.

[43] CM, W. (2016). *Measuring Participants' Attitudes Toward Mobile Device Conference Applications in Continuing Medical Education: Validation of an Instrument*. Retrieved from https://www.ncbi.nlm.nih.gov/pubmed/26954248

[44] Dragly, S.-A. (2017, March 9). *Neuronify: An Educational Simulator for Neural Circuits*. Retrieved from http://www.eneuro.org/content/4/2/eneuro.0022-17.2017

[45] Shannon, D. (2012, February). *Effective Physician-to-Physician Communication: An Essential Ingredient for Care Coordination.* Retrieved from http://www.mdwriter.com/uploads/1/8/0/3/18033585/md_to_md_communication_pej.pdf. Accessed on March 16, 2019.

[46] Nyborg, M. (2013). *SmartNursing – A mobile application to improve communication in home care.* Retrieved from http://orbit.dtu.dk/files/59420810/Telemed2013SmartNursing.pdf

[47] Choudhri, A. F. (2015, October 14). *Security Issues for Mobile Medical Imaging: A Primer*. Retrieved from https://pubs.rsna.org/doi/full/10.1148/rg.2015140039

[48] Zaki, M. (2014). *Smartphone Use in Neurosurgery? APP-solutely!* Retrieved from https://www.ncbi.nlm.nih.gov/pmc/articles/PMC4123261/

[49] Chen, A. (2018). *New Data Shows Losing 80% of Mobile Users is Normal, and Why the Best Apps Do Better*. Retrieved from https://andrewchen.co/new-data-shows-why-losing-80-of-your-mobile-users-is-normal-and-that-the-best-apps-do-much-better/

[50] Trent Yarwood, J. W. (2018, July 3). *Privacy in digital health: Matters of trust in a scandal-plagued era*. Retrieved from https://www.healthcareit.com.au/opinion/privacy-digital-health-matters-trust-scandal-plagued-era

[51] Privacy Code of Conduct on mobile health apps. (2018, December 10). Retrieved from https://ec.europa.eu/digital-single-market/en/privacy-code-conduct-mobile-health-apps

[52] Yasemin Şener, G. A.-B. (2019, March 16). *Electronic/Mobile Health.* Retrieved from https://www.healthparliament.eu/wp-content/uploads/2017/09/Electronic_mobile-health.pdf

[53] Building GDPR & HIPAA compliant eHealth applications. (2019, March). Retrieved from https://docsend.com/view/32isvcj

[54] Mobile Medical Applications. (2019, March 16). Retrieved from U.S. Food & Drug Administration: https://www.fda.gov/medicaldevices/digitalhealth/mobilemedicalapplications/default.htm

[55] FDA selects participants for new digital health software precertification pilot program. (2017, September 26). Retrieved from U.S. Food & Drug Administration: https://www.fda.gov/NewsEvents/Newsroom/PressAnnouncements/ucm577480.htm

[56] Patro, N. (2018, March 25). *Native App vs Hybrid App*. Retrieved from codeburst.io: https://codeburst.io/native-app-or-hybrid-app-ca08e460df9

Chapter 9

Ideas for a school of the future

Sidarta Ribeiro[1], Valter Fernandes[2], Natália Bezerra Mota[1,3], Guilherme Brockington[4], Sabine Pompeia[5], Roberta Ekuni[6], Felipe Pegado[7], Ana Raquel Torres[1], Patrick Coquerel[8], Angela Naschold[9], Andrea Deslandes[2], Mauro Copelli[4], and Janaina Weissheimer[1,10]

Schooling has changed surprisingly little since its origins 5 millennia ago, despite efforts toward massification and inclusion. Research on a wide arch of fields indicates that schools must undergo major transformation in order to benefit from the current science of learning. This chapter outlines some of the seemingly utopic main kernels of this necessary transformation. The school of the future will fully embrace the physiological aspects of learning, and strive to optimize sleep, nutrition, and exercise for children and adults of all ages, so as to improve intrapersonal and interpersonal relationships. Learning will have computer games as scaffold for the crystallization the link between phonemes and graphemes during literacy acquisition, and for the generation of individual learning curves for the personalized tracking of performance. The retention of academic contents will be greatly improved by frequent cumulative tests and other opportunities for retrieval practice, and the use of novelty as an extrinsic adjuvant of learning will be emphasized. Learning from either teachers or peers will mostly occur within "flipped classrooms", and their interactions shall be guided in real time by simultaneous recordings of brain activity. The efficacious dissemination of high-quality

[1]Brain Institute, Federal University of Rio Grande do Norte, Natal, Brazil
[2]Graduate Program in Psychiatry and Mental Health, Psychiatry Institute, Federal University of Rio de Janeiro, Rio de Janeiro, Brazil
[3]Physics Department, Federal University of Pernambuco, Recife, Brazil
[4]Center of Natural and Human Sciences, Federal University of ABC, Santo André, Brazil
[5]Psychobiology Department, Federal University of São Paulo, São Paulo, Brazil
[6]Department of Biological Sciences, State University of Northern Paraná, Jacarezinho, Brazil
[7]Center for Developmental Psychiatry, Department of Neurosciences, KU Leuven, Leuven, Belgium
[8]Department of Physical Education, Federal University of Rio Grande do Norte, Norte, Brazil
[9]Department of Education, CERES, Federal University of Rio Grande do Norte, Norte, Brazil
[10]Department of Modern Foreign Language and Literature, Federal University of Rio Grande do Norte, Natal, Brazil

education is our best hope to decrease educational gradients and promote world equity and peace.

9.1 Introduction

The school is a 5,000-year-old institution that has changed little since its origin. Consider, for instance, all the emphasis placed by Sumerian *edubas* on copying, keeping quiet and "behaving properly" [1,2]. Scarcity of resources is surely another hallmark of schooling across millennia. While poverty is not the sole determinant of deficient education, it is certainly responsible for a large portion of the burden [3–5]. According to the United Nations Human Settlements Program, the population that lives in shantytowns or "favelas" already surpasses 1 billion people [6], a figure estimated to double by 2030 [7]. And still, despite the many conspicuous shortcomings of formal education across the globe [8–10]—most critically in communities with low socioeconomic status—people of all walks of life continue to believe that education represents our best hopes for sustained development and social justice.

Accomplishing this daunting task has begun to seem possible in the past few decades, thanks to a large body of experimental research that has opened major holes on the blanket of ignorance on human learning that covers humanity since the dawn of our time. The planet is overwhelmed with problems that are ultimately rooted on the far from desirable education provided to the masses—and in many ways—to the elite as well. It is high time we developed the schooling of the future, and this chapter outlines some main kernels of what must soon blossom on a large scale.

9.1.1 Mens sana in corpore sano

When the physiological conditions of learning and teaching drop below minimum thresholds, nobody can learn properly, let alone remember details after many years. Low-SES schools regularly experience budget cuts, classroom overfill and a low-paid workforce. If students or teachers have not eaten breakfast, if they had to skip a night of sleep, or had to deal with domestic violence, if they are never offered a chance to stretch, work out and play, formal education is sabotaged from the onset. For similar reasons, if teachers are not paid enough to make ends meet, stress and deprivation will spoil the very effort of schooling. The underfunding of schools and teachers goes against the core point of education: it must always strive to be as good as possible. This sounds like common sense but across the globe education is marked by physiological deprivations. Here we will focus on three aspects: nutrition, sleep, and exercise.

9.1.2 Mangia que te fa bene

The notion that schools need to provide healthy meals is quite intuitive, but not universally accepted by public or private schools under the relentless pressure to shrink their budgets. Yet, the link between nutrition and health should by itself suffice to argue for the importance of healthy meals in schools. Malnutrition and food scarcity are associated with diabetes [11,12], obesity [13], hypertension, heart disease and hyperlipidemia [14,15], mental illness [16,17], particularly depression [15,18]. Not surprisingly, food insufficiency is linked to negative psychosocial and academic outcomes [19–21].

Consider for instance the key importance of glucose availability for learning. Because the brain all but lacks glycogen, it uses much of the glucose consumed by the body. High levels of circulating glucose are essential for mental functions, and their increase after glucose intake benefits a variety of cognitive tasks, in comparison to saccharin controls [22–24]. The results suggest that glucose benefits learning because of its caloric value rather than its sweet taste, a notion in agreement with the fact that the nutritional and hedonic values of sugar are encoded in mice by separate neural circuits—with a preference for energy-seeking in comparison with taste quality [25].

Caloric intake by itself is no guarantee of adequate nutritional status for the purposes of school learning. There is now a body of research in animal models as well as with humans supporting the notion that Western-style diets comprising high-fat and high-sugar contents lead to obesity and a variety of cognitive deficits [26–31]. Diets rich in polyunsaturated fatty acids correlate with enhanced cognitive skills, but the optimal ratios of Omega-3 and Omega-6 fatty acids remain to be determined [32,33].

In England, the Department for Education and the Education Endowment Foundation funded a long-run classroom randomized controlled trial with approximately 8,600 students to estimate the impact of free healthy breakfasts delivered to primary schools. The program called "Magic Breakfast" ran between September 2014 and July 2015 and focused on schools with a high proportion of disadvantaged students. Independent evaluation by the Institute for Fiscal Studies and the National Children's Bureau observed that Year 2 children in schools with a breakfast club made two additional months' progress in reading, writing and mathematics compared to a similar control group of schools without support for breakfast. There was also improvement in the students' concentration and behavior. Overall, the results indicate that breakfast clubs improve outcomes for all children, even those that do not participate, by creating better classroom environments [34].

As persistently stressed by public figures such as Michelle Obama, the inadequate quality of school meals, rich in sugars and fat, may explain why a developed country like the United States fares comparatively worse in worldwide academic comparisons with other similarly wealthy countries. In the future, school meals will be chosen based on the empirical classroom research on the effects of caloric intake and meal composition, micronutrients and hydration, portion sizes, eating frequency, timing and the reward value of food. There is much to gain in schools from the optimization of all these aspects of nutrition.

9.2 Sleep before and after learning

There is a strong need to create spaces and positive attitudes able to guarantee access to adequate sleep in the school environment, because students and teachers alike tend to be sleep deprived. The night's invasion by electrical and electronic devices had a profound impact on both the quality and the quantity of human sleep [35–38]. Insufficient sleep affects approximately one third of the adult population in the United States [39], and sleep deficits are risk factors for health disorders related to socioeconomic disadvantage [40], including malnutrition [41], obesity [42], diabetes and hypertension [43]. Sleep debt is much more noticeable among the poor, with total sleep per 24 h dropping to as little as 3.8 h in some blue-collar jobs [44,45]. In Finland, unemployment has been linked to insomnia, use of hypnotics, and other markers of low sleep quality [46]. Housing overcrowding, hopelessness and exposure to verbal and physical violence all contribute to hinder sleep in low SES communities [47–52], especially in females [53–55]. An assessment of more than 20,000 college students from 26 low-income or middle-income countries found that 10.4% of the students suffer from major sleep problems, in association with diseases such as depression and obesity [56–58]. Health and socioeconomic gradients are often mediated by sleep [59], particularly when adolescents are engaged in the work force [60,61]. Among the US preschoolers, inadequate bedtime routines has been linked to poverty, overcrowded households and low maternal education [62].

Sleep benefits learning in a variety of ways, with underlying psychobiological mechanisms that have been comprehensively characterized, including specific brain oscillations and calcium-dependent molecular cascades associated with synaptic plasticity [63–66]. Poor sleep is correlated with academic deficits across a wide range of intellectual quotients (IQ) [67], but it was not until recently that researchers began to experiment with sleep in the classroom. The pioneering studies show that post-learning naps are associated with better and longer retention of declarative memories [68–71], which comprise most of the memories acquired through formal education. Altogether, the evidence strongly supports the notion that adequate sleep must be provided both before and after a learning event, and should be enthusiastically embraced by schools [72]. The optimization of sleep schedules and the determination of the role of self-regulation is an urgent problem for psychobiological research. In the schools of the future, opportunities for sleep will be provided when needed.

9.3 Move on!

The increase of screen time on TV, smartphone or other electronic devices is enhancing physical inactivity and the associated physiological deprivations [73]. Overcrowded homes and the scarcity of sports facilities in low SES communities produce negative consequences for health, due to the lack of physical space for stretching or exercising, but weak sport habits in high SES communities may have similar effects. The lack of adequate exercise afflicts individuals from all socio-economic levels, leading to increased body mass indices and severe health costs [74,75]. The body requires regular exercise to sustain the physiological home-ostasis that scaffolds various dimensions of cognition in children and adolescents [76–81]. Thus, schools must offer a variety of physical activities, able to promote aerobic capacity, motor skills, decision-making and executive control in a friendly and collective environment [82].

The prefrontal cortex (PFC) is the key for executive functions and has a prolonged development, so that full maturation only occurs within the third decade of life [83]. For this reason, the PFC is particularly prone to be influenced by the environment [84]. The protracted development of the PFC over time strongly constrains children's learning skills, particularly because of limitations in attention span, and the management of short-term memory [85]. Physical exercise typically occurs within enriched environments capable of associating novel physiological stimuli to the motor and cognitive tasks typical of sports practice. The formation of new memories through interactions with the world, as well as the fine motor control required for its navigation, both depend on specialized brain regions—respectively, the frontal cortex and the hippocampus—that are activated by exercise [86]. Moreover, the parieto-frontal circuit is also activated, an important network related to attention, decision making and intelligence. Indeed, exercise protects and enhances brain function through several neurobiological mechanisms.

Studies in animal models revealed that voluntary exercise promotes acute and chronic effects on the brain, contributing to improve neuroplasticity, mental health and learning processes. During exercise, myokines (e.g., BDNF, irisin, IL-15) are secreted from skeletal muscle cells and exert endocrine effects on several areas, such as central nervous system. Immediately after a single bout of exercise, there is an increased synthesis and release of trophic factors (e.g., BDNF, IGF-I, VEGF), cytokines (IL-6, IL-10) and neurotransmitters (e.g., norepinephrine, dopamine, serotonin). After several weeks of training (chronic effect), exercise is associated to increased neurogenesis, angiogenesis and synaptic plasticity in brain regions such as the hippocampus [87–91].

As expected from these studies, the assessment of school students has established a tight correlation between cognition and cardiorespiratory fitness [92–95]. Aerobic exercises are linked to enhanced executive functions, greater activation of the PFC, and improved school performance [96]. Hand-eye motor coordination and visual selective attention seem to have a strong influence on academic achievement and cognitive function. Fernandes and collaborators (2016) found a moderate relationship between a hand-eye motor coordination test and cognitive measures, such as academic achievement, processing speed, selective attention and perceptual reasoning—the faster and more accurate the coordination, the better were the cognitive test results. In fact, the stronger correlations between cognitive and motor skills are with those that demand greater cognitive engagement, such as fine motor skills, hand eye coordination, bilateral body coordination and timed performance motor tasks [97]. The positive effects of physical activity on cognitive function have been shown even after a single bout of exercise [98]. Neuroimaging studies have shown an enhancement of neural activity after a single bout of exercise, showing that parieto-frontal networks are modulated during exercise [99]. These effects are also observed in behavioral responses (mainly improvements in inhibitory control, attention and working memory). However, the great majority of studies investigated the acute effects of physical activity on cognition immediately after the end of exercise [100]. Acute physiological responses to exercise should be used as a tool to enhance attention, bringing the exercise to the classroom, enabling concretization of theoretical concepts and facilitating learning. From kindergarten to university, studies that use exercise integrated to other subjects in classroom have shown positive effect on behavior, attention and learning [101–103].

While the majority of the research has focused on the acute effects of exercise, several studies point to its long-term psychological and cognitive effects [104–108]. Some studies have claimed that aerobic exercise produces major cognitive benefits [109], but other studies point to minimal or nonsignificant effects [110]. To verify whether this discrepancy could be related to differences in the intensity of the training, a randomized controlled study was designed to measure nonverbal and verbal ability, abstract reasoning, numerical and spatial abilities, and school grades in different groups of students exposed to moderate or high-intensity physical training: participants in the high-intensity group showed significantly better cognitive and academic performance than participants in the moderate exercise group [111]. Forty minutes of moderate to vigorous physical activity were more efficient than 20 min, in enhancing executive function and mathematics achievement in a study with overweight children [112]. There is a pressing need to better understand factors such as exercise intensity, duration, frequency, type, motivation, cognitive demand, and in which moment exercise should provide the best influence on children cognition.

A recent publication suggests that areas of the PFC (involved with executive functions) and the cerebellum (involved with motor coordination and balance) show reduced activation as exercise intensity increases and, conversely, that the increased activation of these areas appears to be dependent on the reduction of effort intensity [113].

While the majority of the research on the cognitive role of exercise focuses on aerobic workout, other aspects of exercise may be equally important. Coordinative exercises such as sports games, rhythmic activities, gymnastics and fitness activities are more effective in enhancing concentration than physical exercises focused only on cardiovascular health [114]. Exercises that require hand-eye and foot-eye coordination, as expected in sports like soccer or volleyball, have potential to reduce learning delays in children with special needs. Coordination exercises are more effective in boosting attention than normal physical exercise of identical duration [115].

Another important aspect to consider is the relationship between motor coordination and cognitive skills usually required by the schooling process of typical and atypical students [116,117]. Math and reading achievement are strongly predicted by fine motor skills [97,118]. In children and adolescents, motor coordination predicts academic performance [119]. Some of the positive results of exercise on academic achievement [120] likely reflect a positive effect of exercise on attention [96].

Exercises rooted on open-skill tasks, in which the organism responds with continuous adaptation to unstable environments, yield better executive functioning in overweight children than restricted aerobic exercises [121]. In a randomized control trial, improvements in executive functions at school were found after 6 weeks of team games and not in aerobic exercise intervention [122]. Evidence of the importance of cognitive engagement in enhancing executive function inspired the design of enriched physical education (PE) programs—especially tailored to impact cognition. An enriched PE intervention showed best results in inhibition than a regular PE program and better inhibition outcomes in enriched PE were mediated by improvements on ball skills [123]. The critical difference between them is that enriched PE emphasizes deliberate play, cognitively challenging variability of practice, motor coordination and cognitive processing.

The multiplicity of stimulus that can be related to physical activities makes it especially difficult to specify their characteristics and cognitive outcomes. Until now, there is strong evidence of the effect of physical activity on mathematics performance, but researchers still seek for more conclusive evidences on executive functions and overall academic achievement [120,124]. Nonetheless, recent studies have shown that PE programs aimed to increasing daily physical activity, that integrate complex motor skill training and cognitive engagement have greater impact on executive functions [125,126].

Yoga-based exercises, for instance, have proved to be beneficial to the management of stress and anxiety in adults, as well as in children at school. Recently there is a growing evidence that different yoga-based interventions such as *asanas* (postures), *pranayama* (breathing techniques), *dharana* (concentration) and *dhyana* (meditation) have a positive effect in managing and reducing stress and anxiety [127]. As this practice has different focuses, it can be smartly incorporated to school programs not only to manage stress, but to boost attention and learning.

Unfortunately, evidence-based practices are not often seen in education programs, as reflected in the tendency to reduce physically active behavior and increase insufficiently active and sedentary behavior among US adolescents and adults [128]. The implementation of high-quality PE programs in schools is hindered by constrained schedules, limited physical space, deficient equipment and shortage of teachers [129]. Schools must offer adequate physical exercise—at least 60 min per day, 5 days per week—in order to effectively support formal education [129]. Considering that there are beneficiaries of schooling across all age groups and health conditions, evidence-based international recommendations for physical activity have been set [130], as shown in Table 9.1.

Exercise prescription should focus on improvement of motor skills and physical capacities, as well as social, emotional and cognitive enhancement. Although the physiological and psychological health benefits associated to an active lifestyle are well established, overall children are not sufficiently active. Barriers, motivators and facilitators of physical activity should be identified. The pleasure and motivation during exercise is a key factor to adherence and factors as self-efficacy,

Table 9.1 Recommendations for physical activity for different age cohorts

Age cohort	Age	Recommendation
Preschool children	Ages 3 through 5 years old	Should be physically active throughout the day to enhance growth and development
Children and adolescents	Ages 6 through 17 years old	Should do 60 min or more of moderate-to-vigorous physical activity daily
Adults	Ages 18 through 59 years old	Should do at least 150 min to 300 min a week of moderate-intensity, or 75 min to 150 min a week of vigorous-intensity aerobic physical activity, or an equivalent combination of moderate- and vigorous-intensity aerobic activity
Older adults	60 years old or more	Should engage in multicomponent physical activity that includes balance training as well as aerobic and muscle-strengthening activities
Pregnant and postpartum women	Within the fertile period	Should do at least 150 min of moderate-intensity aerobic activity per week
Adults with chronic conditions or disabilities	Any age	Should follow the key guidelines for adults and do both aerobic and muscle-strengthening activities
General and complementary observation	Recommendations emphasize that moving more and sitting less will benefit nearly everyone. Individuals performing the least physical activity benefit most by even modest increases in moderate-to-vigorous physical activity. Additional benefits occur with more physical activity. Both aerobic and muscle-strengthening physical activity are beneficial	

family social support, and available opportunities influence the participation on sports and exercises. Exercise executed immediately before attending classes is prone to improve cognition acutely due to a rise in alertness and motivation, and a drop of reaction times. The interactions of exercise with nutrition and sleep must be carefully evaluated, and it remains important to seek the integration of physical exercise with other school subjects.

Recommendations to enhance the effect of exercise on executive functions

- Promote pleasure in physical exercise
- Promote the improvement of cardiorespiratory capacity
- Develop motor coordination and visual space attention
- Diversify sensory stimuli
- Provide physical and cognitive challenges, possible and thought provoking
- Enable the gradual, adequate and constant increase of difficulty
- Offer opportunity for social interaction and group work
- Transmit moral values and respect to rules

Despite the lack of sufficient teacher-guided exercise in school, children typically have access to short periods of free playing (recreation), when they spontaneously engage in high-intensity exercise (e.g., running games), with social aspects, rules, pleasure and a dynamic alternation of games. The recreation time fulfills many of the requirements cited above. Future studies shall determine whether post-recreation time is a privileged period for learning in school, and characterize the optimal schedule of recreations (duration, frequency).

9.4 Game-based education and assessment of individual learning

Technological changes in society have led us to rethink common practices, including teaching. Game-based education has been a recurring proposal to promote individuals' cognitive engagement and motivation in the classroom [131]. The term *Gamification* is defined as the use of game design elements in a nongame context (in our case, the classroom), with the intention of boosting one's engagement and attention [132].

The mechanics of games is composed of several elements such as: points, levels, challenges and missions, reinforcement and feedback, surprise, fantasy, clear goals, among others. Competition and collaboration are also inseparable aspects of the gamification scenario [133]. Although the feeling of defeat (inherent in the game) may have a negative effect on the satisfaction of learners, competition is nevertheless a key element that motivates learners to engage in playful activities [134]. Collaboration, in turn, may be used to lessen the burden of competition in game-based interventions.

Digital game designers have relied on the "theory of flow" for the construction of more immersive games [135]. The ultimate goal is to help the user "disconnect" from the outside world and begin to concentrate almost exclusively in the game. Applying this premise to the educational setting may be promising according to a number of researchers, for it is possible to use gamification as an alternative to provoke a feeling of satisfaction and well-being in learners, boosting their attention and concentration levels, and motivating them to carry out activities in and outside the classroom. Because the gaming environment brings along an immense amount of information and stimuli, a challenge for teachers is to draw attention and focus learners on what is relevant. In this context, they need to plan carefully how, when, where and why to gamify their practice.

Feedback is a feature pertaining to game design that has proven highly effective in education. A critical requirement for retrieval learning to occur is specific, individualized feedback, a condition that cannot be adequately supplied by most schools. Large class sizes and excessive student-teacher ratios are typical of most countries [9]. In this context, there is a major need for computational strategies able to provide frequent behavioral and cognitive assessments in a productive manner. Intensive and extensive data gathering will allow the schools of the future to forecast learning curves, and detect handicaps in time for early intervention.

One limitation in that regard is the fact that most behavioral and cognitive assessments apply norms that fail to represent people from developing countries, but rather describe features common in western educated industrialized rich and democratic (WEIRD) countries [136,137]. The construction of population-specific norms in non-WEIRD countries remains a major challenge because of the budget constraints that typically hound education and science in developing countries [138]. But help is on the way, in the form of new analytical tools for Big Data and various novel technologies able to adequately represent complexity [139–144], such as adaptive collaborative learning support (ACLS) [145,146].

Adaptation to the student's particular features pushes the envelope of academic performance in a personalized manner, since the frequent and constructive feedback gradually guides the student through a unique individual learning path. The data can then be compared to the track record of the same individual, or to data from past and present students. For instance, in a non-semantic graph-theoretical characterization of literacy acquisition among low-SES children, we have identified multiple structural aspects of discourse that correlate with intelligence quotient (IQ), theory of mind and math and reading scores [147].

Just as important as it is for typical students to count with a computational scaffold of learning, it is equally vital to detect learning disabilities as soon as possible, so as to allow early intervention. In the case of attention-deficit/hyperactive disorder (ADHD), characterized by cognitive handicaps related to deficits in the temporal processing of stimuli, symptom severity has been adequately measured by a computational game [142]. Among the several disorders that can negatively impact learning in school students, such as dyslexia, disorthography, dyscalculia, and dyspraxia, schizophrenia stands out as a disease with relatively late onset but with very damaging cognitive consequences. In the past decade we developed a computational strategy to provide an early yet accurate diagnosis of schizophrenia based on the non-semantic graph-theoretical analysis of oral reports [148–151], a strategy of special value for low-SES countries where access to psychiatric services is extremely limited [147]. Another successful approach, able to forecast psychotic breaks more than 24 months in advance in at-risk youth, involves measuring semantic incoherence and the abundance of determiners [152].

Well-designed, low-cost educational software can support adaptive tutoring, collaborative learning, assessment and intervention within smart learning environments, even in low-SES countries [139,141,144]. This strategy is poised to generate large context-specific datasets over time, a *sine qua non* condition to avert the error of interpreting cultural traits as deviances from the norm. In the school of the future, schooling will occur in a more flexible environment in which individual performance will be monitored while collaboration will be boosted by a plethora of computational games and platforms. Activities will be tailored to the needs and desires of each student, learning will be more fun, and no student will be left behind.

Gamification in education is not an exclusive process of so-called digital platforms. The teaching and learning process based on games extends to classic and/or innovative recreational and leisure activities, such as games with a cognitive predominance (e.g., chess, backgammon, etc.), with a physical-motor predominance (e.g., individual and collective sports), or with an affective predominance (e.g., storytelling or symbolic games), which allow for a blend between cooperation and competition. Outdoor games shall be combined with green exercise (i.e., body practices and adventure in natural environments or areas of environmental preservation or recovery), co-education and social inclusion strategies, where people of different genres and social class play collaboratively [153,154].

9.5 To read, perchance to learn

Although the access to formal education has greatly increased over the last decades, the lack of proper schooling in developing countries has led to a large number of nonliterate people. Schooling can generate three different stages of literacy development. At the first stage, known as rudimentary literacy, individuals can read and understand text titles and short sentences; despite being able to count, they can barely understand large numbers and make simple calculations. At the second stage, basic literacy, individuals are able to read shorts texts, but can only extract sparse information from them and cannot draw conclusions from their content; they can understand large numbers, and make simple calculations, but have difficulties to perform a longer series of calculations, or mathematical operations that are more complex. At the third stage, full literacy, individuals have acquired all the reading and writing skills, and are able to perform basic and complex mathematical operations.

In Brazil, a demographical census is performed by the Brazilian Institute of Geography and Statistics (IBGE) every decade. The last edition showed that one out of five people (20.3%) are at the first stage of literacy development. Regarding the several regional differences in Brazil, the geographic census evidenced that the literacy failure problem is bigger in the Northeastern region, where that rate is 30.8%.

In this hardly promising scenario, there have been disputes on which path should be followed to improve literacy development. Some researchers have stressed a focus on teaching grapheme and phoneme associations, an approach based on phonics methods. On the other hand, grounded on the global methods, others have advocated for the use of real-life texts to encourage children to develop knowledge of the writing system [155]. Although the pendulum has swung back and forth between holistic, meaning-centered approaches and phonics approaches without much hope of resolving disagreements, today's educators should agree that phonemic awareness must receive much attention. Substantial scientific evidence has accumulated purporting to shed light on the effectiveness of phonemic awareness instruction and hopefully providing some closure to the so-called literacy wars.

Decades before the methods discussion emerged in Brazil, the situation was not different in other places around the world. In the United States, for example, the National Reading Panel (NRP) was summoned in 2000 to examine the scientific evidence relevant to the impact of phonemic awareness instruction on reading and spelling development. The conclusion reached by the panel after conducting a meta-analysis was that students who were taught systematic phonics outperformed students who were taught a variety of nonsystematic or non-phonics programs, including basal programs, whole language approaches, and whole word programs [156].

In France, the last conference on reading consensus, involving experts, teachers, parents and other professionals in the reading area, pointed out aspects that go far beyond the dichotomy between the methods. In the introduction of the report, Jean Émile Gombert, president of the conference, signaled their alignment with the main findings in the cognitive and neurobiological science of reading, evidencing that writing and reading should be explicitly taught, including repetition, as our brain is not naturally programmed for reading [157]. For this reason, becoming a good reader requires being exposed to deliberate, repetitive teaching interventions, until the skill becomes automatized. Such statement may seem trivial, but certain teaching practices would show the opposite.

The identification of the visual word form area (VWFA) [158–162] in the left occipital-temporal region of the brain was central for understanding the early reading processes occurring in the brain. Even though reading is a global process depending on the conjoint, synchronized activity of several brain areas, it starts with the individual analysis of the components of letters or orthographical signs in primary visual areas [158], and proceeds to more specialized visual areas, like the VWFA, and regions related to language processing. Reading can activate both the lexical and phonological routes:

(a) The visual areas of the occipital lobe of both right and left hemispheres perform the first analysis of the image to extract elementary forms from it, such as the straight and curve lines of the letters;
(b) After approximately 50 ms, the result of that analysis is sent to the VWFA in the left ventral occipital-temporal region;
(c) Soon after that, still in the left hemisphere, in peri-Sylvian fissure regions, meaning and sound are accessed through the activation of two main circuits: a phonological circuit (e.g., including the *planum temporal*) that connects to the sounds of language and a lexical circuit (e.g., including Broca's area) that connects to the meaning of language.

The mirror invariance of visual processing in the VWFA has been implicated with the mirror letter errors that are common during literacy acquisition, such as the confusion between letters "b" and "d", or between letters "p" and "q", which are horizontally mirrored. This intrinsic property of the visual cortex is present at all ages. Illiterate adults perform worse than literate adults in distinguishing between nonlinguistic lateral mirror images [163,164].

Breaking mirror letter invariance requires teaching how to read to include deliberate and repetitive practice [165–168], until these processes gradually become automatized. In fluent adult readers, this unlearning of the mirror invariance becomes visible in the VWFA, which displays a mirror priming for objects, animals and fruits, and a lack of mirror priming for words [169] and letters [166].

Recent evidence indicates that multisensory training is very effective at breaking mirror letter invariance during literacy acquisition, and that post-training sleep makes the learning endure [170]. In the school of the future, learning how to read and write will be based on sound scientific evidence, following the major trends in the science of reading, making every effort to assist children in becoming fully literate at the right time.

9.6 Improving retention of academic content by practicing retrieval

Educating students usually involves getting information into their heads. To this end, teachers give classes by exposing content, students take notes and revise them, read and reread texts. This makes sense. After all, knowing things involves having stored them in the first place. This seems to work well when students are asked to retrieve information in the same context as it was initially learned. This happens in most exams in which students are merely required to repeat back the content they were taught since their last exam. It so happens, however, that this is not enough to ensure that they will be able to use knowledge in other contexts. Indeed, there is a strong body of scientific evidence that students who focus on cramming information into their heads quickly forget what they thought they already "knew". Furthermore, they seem to have a lot of difficulty in using information in different scenarios. There is no reason for this to be so, except for our tendency to follow traditional ways of teaching. Several possible changes would not be expensive, time consuming, neither overburden teachers and students with extra work. Simple modifications in the way people are taught and/or in their choice of study techniques can significantly improve abilities to use knowledge more effectively.

These changes involve doing something that seems counterintuitive: instead of studying by cramming information *into* our heads, what needs to be done is the opposite, studying by taking information *out*. In other words, students must be stimulated to practice reaching into their minds to try to come up with answers or solutions to questions and problems. This is seldom done in the classroom even though countless studies have shown that, to use information that we were previously exposed to in unexpected ways and in real life settings, we must practice remembering information under varying conditions. In the literature on science-based learning this is known as retrieval practice [171].

Retrieval practice is a fancy name for "trying to remember information". This can be done in multiple ways, such as by discussion with peers about certain topics, answering or even making up questions about the taught material, doing practice problems, writing summaries about what was learned in our own words, answering quizzes and tests, thinking about what was taught and, yes, even by doing exams. These activities have been shown to boost learning considerably more than being passively exposed to information, both in the laboratory and in real classrooms environments [172]. Benefits are usually lower if answers are too easy, so some level of difficulty is desirable when encouraging students to practice retrieval. For instance, people tend to learn better when they have to come up with a written answer to a problem than by having to select one among a limited set of multiple-choice answers.

Another counterintuitive thing about retrieval practice is that it seems to work even if students get things wrong, especially when feedback is provided regarding what they were expected to answer. Feedback is, in fact, essential for students to be able to confirm that they understood concepts and, eventually, to correct mistakes they may have made. By practicing retrieval followed by feedback the learning benefits are maximized, irrespective of the correct/incorrect nature of the answer [173]. Teachers may feel that feedback is too time consuming, but for it to promote better learning it does not have to be immediate and can be given collectively in class. When studying alone, students are also advised to always check whether they got things right.

The next issue to consider when applying retrieval practice is how often (frequency) and when to do so (timing) after content was taught in class. Each time a piece of information is retrieved, memory traces become stronger. By practicing retrieval in different contexts, such as answering quizzes, doing projects about the applicability of what was just taught, drawing diagrams about the relations among pieces of information, etc. concepts become better integrated with: (a) other bits of information that are already known; and (b) new information that may be specific to that context, but can also be applicable in other situations. By doing this repeatedly, more precise delimitation of the meaning of each content is achieved, which reduces the likelihood that concepts will be confused with similar information, another crucial aspect of learning. Scientists still do not have a definitive answer on how many times a specific content should be retrieved to ensure lasting learning. If it is not feasible to carry out several retrieval practices, students must have this opportunity *at least once* for every content explored in the classroom. This is so because the most learning benefits are reaped by the first retrieval practice attempt [174]. If repetitions of retrieval practices are possible, then ideally they should be spaced out. This is known as *spaced* or *distributed* retrieval practice [175,176]. Distributing these opportunities over time leads to better learning than repeating retrieval practice in short periods, known as "massing". However, as mentioned regarding the ideal number of retrieval practice attempts, it is still unknown which is the best time interval between them. This likely varies according to the familiarity with and complexity of the content. Hence, the take home message in this case is: repeat retrieval practice as much possible, in varying ways, and do so spread out over time. However, it has been shown that it does not seem to matter if this spacing out occurs at a frequent interval (e.g., every other day or week) or at irregular intervals.

If spaced retrieval practice has so many benefits, the natural expected consequence should be that most teachers and students use this strategy to promote learning. This is far from the truth. Worldwide, most students use time-consuming and inefficient ways of studying [177]. There are many possible reasons for this that may be exclusive but are probably cumulative. The first is that teachers focus on presenting students with academic content and do not realize that they should also teach them *how* to study to improve their chances of remembering and using information. To do so effectively they would have to be aware of the counter-intuitive findings from the scientific literature regarding the importance of promoting retrieval practice opportunities in academic environments. As this is rarely the case, each student has to reinvent the wheel when it comes to deciding which study techniques works best. They end up favoring passive ways of studying, such as rereading notes and chapters, or going over practice exercises that they have already solved because of various factors: (1) this is the way they are taught; (2) these ways of studying involve less cognitive effort, in contrast to thinking about things, which is essential when practicing retrieval; (3) they usually get good results in classic exams, which only enquire about the content taught last, and do not realize that information is quickly forgotten if they do so; and (4) rereading notes and going over previously solved problems, for instance, instills a feeling of familiarity with the content which tricks us into believing we can retrieve that information at will.

This last factor deserves a more detailed explanation because we believe it is the main reason why most students study in inefficient ways. It is also a good way of showing how retrieval practice works. Consider this: when we reread a passage, our brain detects that it is familiar (was read before) and generates a "feeling of knowing". Feeling of knowing is not the same thing as being able to remember or recall information. It is another type of cognitive ability, which identifies things as previously encountered, but does not help much when it comes to recalling details about them [178]. In other words, we have cognitive abilities related to recognizing things as familiar and other abilities, which involve distinct brain systems and mechanisms, used to recall things or bring them back to mind. What we do not have is a good sense of when each of these abilities is at play and often get them muddled up. Because of this confusion, students tend to favor passively exposing themselves to content by rereading, for example, as this leads to a sense of familiarity with the material that gives them an illusion of competence, which grows as they reread the text again and again. They believe they can "remember by recognition". This cannot be done, however. A good example of this is trying to remember details about an image that we are bombarded with, such as a logo of an important company. Although we can quickly identify this logo upon seeing it (recognition), we are usually incapable of drawing it (remember). This is not surprising. One thing the brain does is to recall things when necessary. This is what we are supposed to be doing when learning academic content and which is greatly improved by retrieval practice. Another cognitive ability is identifying the commonness of things. The latter is explored in advertisements, mostly because we tend to prefer things that are more familiar [179]. After all, to decide to buy a product one does not have to go around remembering all the details of its logo; it is enough that it be identified faster/better than other products. This is rather difficult to understand without a practical example. Try differentiating recognizing from remembering by drawing the Apple logo. Do it now and check if you got it right (as "feedback") before you continue reading.

We can bet you were disappointed with yourself. A study has shown [180] that a very small number of people, including owners of Apple products, can successfully draw the shape of the apple itself, know on what side and how big the bite mark on the apple is, what is the inclination of the leaf (left or right?), etc. Even if you got it all wrong in your first attempt, you will notice that if you draw it again in the future, chances are that the drawing will be almost perfect. This is a simple way of convincing people that being familiar with something is rather useless when we have to remember details about it. It also shows that a single retrieval practice attempt works wonders and that making mistakes does not hinder the benefits of using this way of learning, as long as feedback is available.

If educational practices are to be improved, both students and teachers should be made aware of the importance of regular retrieval practice (via tests, projects, diagrams and spaced in time), especially because this is easy to implement, is not costly and does not involve extra study or teaching time [181], just a slight change in the way that teaching and/or studying takes place [for more information, see [182,183]]. Very useful guides on how to study based on scientific findings are provided Henry L. Roediger III (http://psychnet.wustl.edu/memory/publications/), Jeffrey Karpicke (https://www.purdue.edu/hhs/psy/directory/faculty/Karpicke_Jeffrey.html), and Pooja K. Agarwal (www.retrievalpractice.org).

9.7 Repeat yet surprise

While repetition is important for learning, variations and surprise also play a key role, as seen above in the context of physical exercise and indeed across fields. A classroom-based study with 1,676 children from 7 to 9 years old has shown that the memory of literary or graphical activities is enhanced by novel experience that occurs in temporal proximity with these activities [184]. The study points to the educational use of novelty in the school setting as an "extrinsic adjuvant of other information acquired some time before or after it" [184].

The influential "flipped classroom" strategy (i.e., flipping learning and homework places) combines retrieval practice and surprise by delivering contents outside of the classroom, and using the classroom to actively engage on conceptual discussions with mentor and peers [185,186]. This method, which is likely to prevail in the school of the future, is based on the need to use class time to build meaning, rather than transmit information.

9.8 Brains in synchrony: a bridge between neuroscience and education

The idea that neuroscience can influence what teachers do in the classroom is not yet widespread. Some researchers are very skeptical about the possibility that results obtained in laboratories through technologies like functional magnetic resonance imaging (fMRI) or electroencephalography (EEG) will have any impact on the daily life of schools [187,188]. One of the critiques regarding the potential impact of brain research that concerns educators has to do with the artificial nature of the experiments used in neuroscience, especially the simplicity of tests designed to investigate the cognitive processes as compared to the multifarious complexity of classroom learning.

Aware of the disparity, various researchers see as their goal to design experiments that simulate situations students actually deal with. Noninvasive methods that employ psychophysiological measurements like changes in cardiac frequency or increases in endocrine gland production, images obtained by fMRI, EEG and positron emission tomography (PET) are being used to study the brain, challenging outdated concepts about various cognitive functions with important implications for education [189–191]. In this arena, the crescent use of near-infrared spectroscopy (NIRS), a technique that has evolved as a neural imaging instrument capable of revealing how a variety of cognitive tasks activate different parts of the brain, is gaining attention. Similar to an MRI, this technique measures the concentration and saturation of oxygenated and deoxygenated hemoglobin in the cortical tissue, leading to estimations about neuronal activations. However, NIRS uses specific wavelengths of light to make such measurements, which are correlated to the BOLD signal. Moreover, NIRS has a better temporal resolution than fMRI although it is not as precise as the EEG.

The greatest advantages of using NIRS are its portability and low cost compared to the MRI. This is a lightweight, silent piece of equipment that can be used in most distinct tasks, whether motor or cognitive. Unlike the fMRI, NIRS does not require that the subject is immobile. This permits data to be collected while he or she is ambulant, a more comfortable way of conducting experiments and one that can gauge human mobility [192,193]. This may be another reason for the growing interest in the use of functional near-infrared spectroscopy (fNIRS) to study different cerebral processes in children [194]. Other attributes of fNIRS call attention to studies of what is known as "hyper-scanning", which involve simultaneous data acquisition in multiple persons [195]. This technique is beginning to be used to understand cerebral activation during the social paradigm [196] and interaction among students [197].

fNIRS has already been employed to investigate the cerebral regions involved in the numeric processing of pre-verbal infants [198]. It also produced structural changes in specifically linguistic circuitry in the left hemispheres of bilingual children learning to read [199]. Perhaps the most recurrent use of the technique is in studies of the part the PFC plays in remembering how to work and to execute functions, especially in children in sufficiently natural situations [200], such as storytelling [201].

A recent study presented a way of utilizing the theoretical and methodological knowledge of cognitive neuroscience to understand the mechanisms underlying the educational interaction between teachers and students [202]. The researchers presented conceptual proof of the use of fNIRS in three case studies that sought to simulate school circumstances, leading to the perception that the technique is completely capable of furnishing a deeper understanding of the process of teaching and learning during teacher-student, student-student and student-knowledge interactions.

In the first experiment, the researchers monitored the interaction between a 4-year-old child and a teacher. The interactive task was meant to teach the operation of adding two whole numbers (less than or equal to six). The researchers created a board game that considered the perspective of education as a social process occurring in the zone of proximal development [203]. The teacher explained to the child how to add two whole numbers with the use of some small wooden sticks (each one of the sticks represented a number), by counting the total number of sticks. The board game had a path formed by squares with designs. Two small, differently colored plastic pieces stood for the child and the teacher. The child threw two dice (each one numbered from 1 to 6) and the pair of numbers obtained by each throw had to be added to determine how many squares on the path each piece would "move". So the teacher and the pupil were competing in a mathematical "race". The child and the teacher took turns throwing the dice. The child always added the numbers with the teacher's help only if necessary.

This interaction was monitored, using NIRS's simultaneous acquisition function. The results showed that the signal from the right frontal cortex of the child correlated positively with the right anterior temporo-parietal junction (TPJ) of the teacher. The researchers also found a correlation between the teacher's and the child's right frontal cortices. In other words, they discovered that the neural activity of both child and teacher were synchronized during their interaction. We know that the pre-frontal cortex is involved in high-order cognition processes, while the TPJ is involved in social processes like empathy and imagination [204,205].

In practice, every teacher knows the importance of connecting with the student. This experiment is perhaps the first to show empirical evidence of an empathic mechanism that arises in the process of instruction and learning, thus corroborating the heretofore empirically untested theory created by Lev Vygotsky in the 1920s.

In their second experiment, the researchers monitored the hemodynamic signals of various graduate students during a class. Four graduate students (two of each gender) participated in an "Introductory Epigenetics" class. The content of this lecture was about the principal mechanisms (methylation, histones, etc.) involved in basic epigenetic concepts. The results showed that a statistically significant synchronization of oxyhemoglobin among the students was limited to the first 8 min of class. This cerebral connection faded quickly, diminishing to almost zero as the class ended. While we found some connectedness in the students' cerebral activity, our discoveries indicate that this effect was strongest only for the first few minutes. Their attention drifted as the lecture continued. All teachers and most students recognize the paramount importance of focus in the learning process. However, we know very little about strategies having to do with attentiveness in academic settings involving multiple distractions. Attention is a crucial element in learning and educators frequently raise the question of how to sustain it, particularly with groups of students. This study measured the synchronicity of cerebral activity among four students regarding their attention span during classes. Sustained attention is an important component of scholastic success [206,207] and the mPFC-basal forebrain system plays a crucial role in this process [208]. The results showed that mPFC synchronization between the four subjects was strong only for the first eight minutes, diminishing significantly thereafter. What causes this lack of synchronization remains unknown and requires subsequent investigations to better understand the phenomenon.

In their third and final study, the researchers undertook the simultaneous monitoring of the frontal signals of the NIRs wearing eye-tracking glasses. The movement of a student's gaze furnishes at least two important measurements for research in education: the first step is to be able to monitor where in the classroom the student is looking. The second step is the temporal variation of the diameter of the pupil, which is associated with the student's emotional response and changes in his or her cognitive effort [209]. In this study, the eye movements of a 10-year-old boy were simultaneously recorded by NIRs during an "Introduction to Astronomy" class where the content was unfamiliar. Basic components of the solar system, like planets and asteroids, were introduced. The teacher just used the white board to draw figures.

By using this combination of techniques it was possible to measure the student's attention during the class from a peripheral perspective. The results revealed that the student maintained visual contact with the teacher when she was in dialogue with him, and that their contact broke only when he was trying to form a response to her questions. Simultaneously, the researchers found no augmentation in brain activity accompanied by pupil dilation, a phenomenon that occurs when there is an attention overload [210].

9.9 Conclusions

In this chapter we addressed some of the main transformations that schools must undergo in order to benefit from the current findings in the science of learning and teaching. First of all, physiological aspects cannot be overlooked by the school of the future. Healthy nutrition, adequate sleep and systematic exercise benefit cognition and learning in a variety of ways and must, therefore, be available to children and adults of all ages.

Another tendency considered in this chapter is game-based learning and teaching. In the school of the future, learners will have computer games as catalysts for the automation of decoding skills during literacy acquisition, for instance. Individual differences will be catered for through the generation of learning curves to track personalized performance and, if necessary, to inform remediation so that no one is left behind.

Regarding the retention of scholastic and academic content, knowledge building will be greatly improved by frequent cumulative tests and other opportunities for retrieval practice, such as the use of novelty as an extrinsic adjuvant of learning. In addition, there will be a balance between presenting content to students and teaching them strategies on how to remember and use this content. Lastly, learning will happen through collaboration between teachers and peers in "flipped classrooms", and their interactions will be guided in real time by simultaneous recordings of brain activity.

Despite all these advances, there is still much to be investigated and a long way to go in order to translate findings from the science of learning into the classroom, especially in unprivileged societies, where educational science itself still strives to survive. In any case, we are convinced that the ultimate path to high-quality education is a clear focus on sound evidence-based interventions and the belief that education is the best hope for social justice.

References

[1] S. N. Kramer, Schooldays – A Sumerian composition relating to the education of a scribe. *Journal of the American Oriental Society* **69**, 199–215 (1949).

[2] Å. Sjöberg, in *Sumerological Studies in Honor of Thorkild Jacobsen on His Seventieth Birthday*, June 7, 1974, S. J. Lieberman, Ed. (University of Chicago Press, Chicago, 1975), vol. **20**, pp. 159–179.

[3] S. J. Lipina, Critical considerations about the use of poverty measures in the study of cognitive development. *Int J Psychol* **52**, 241–250 (2017).

[4] S. Lipina, S. Segretin, J. Hermida, *et al.*, Linking childhood poverty and cognition: environmental mediators of non-verbal executive control in an Argentine sample. *Dev Sci* **16**, 697–707 (2013).

[5] S. J. Lipina, and M. I. Posner, The impact of poverty on the development of the brain networks. *Front Hum Neurosci*, 1–12 (2012).

[6] UN-HABITAT, *The Challenge of Slums: Global Report on Human Settlements*. (Earthscan, London, 2003).

[7] UN-HABITAT, *State of the World's Cities 2006/7*. (Earthscan, London, 2007).

[8] UNESCO, World data on education (UNESCO, Paris, 2011), vol. 7th Edition 2010/2011.

[9] OECD. (OECD, Paris, 2014).

[10] OECD, in *PISA in Focus*. (OECD Publishing, Paris, 2016).

[11] M. Ding, N. L. Wilson, K. B. Garza, and C. A. Zizza, Undiagnosed prediabetes among food insecure adults. *Am J Health Behav* **38**, 225–233 (2014).

[12] H. K. Seligman, A. B. Bindman, E. Vittinghoff, A. M. Kanaya, and M. B. Kushel, Food insecurity is associated with diabetes mellitus: results from the National Health Examination and Nutrition Examination Survey (NHANES) 1999–2002. *J Gen Intern Med* **22**, 1018–1023 (2007).

[13] F. A. Tayie, and C. A. Zizza, Height differences and the associations between food insecurity, percentage body fat and BMI among men and women. *Public Health Nutr* **12**, 1855–1861 (2009).

[14] F. A. Tayie, and C. A. Zizza, Food insecurity and dyslipidemia among adults in the United States. *Prev Med* **48**, 480–485 (2009).

[15] H. K. Seligman, B. A. Laraia, and M. B. Kushel, Food insecurity is associated with chronic disease among low-income NHANES participants. *J Nutr* **140**, 304–310 (2010).

[16] P. Casey, S. Goolsby, C. Berkowitz, *et al.*, Maternal depression, changing public assistance, food security, and child health status. *Pediatrics* **113**, 298–304 (2004).

[17] B. A. Laraia, A. M. Siega-Riz, C. Gundersen, and N. Dole, Psychosocial factors and socioeconomic indicators are associated with household food insecurity among pregnant women. *J Nutr* **136**, 177–182 (2006).

[18] R. C. Whitaker, S. M. Phillips, and S. M. Orzol, Food insecurity and the risks of depression and anxiety in mothers and behavior problems in their preschool-aged children. *Pediatrics* **118**, e859–868 (2006).

[19] K. Alaimo, C. M. Olson, and E. A. Frongillo, Jr., Low family income and food insufficiency in relation to overweight in US children: is there a paradox? *Arch Pediatr Adolesc Med* **155**, 1161–1167 (2001).

[20] K. Alaimo, C. M. Olson, and E. A. Frongillo, Jr., Food insufficiency and American school-aged children's cognitive, academic, and psychosocial development. *Pediatrics* **108**, 44–53 (2001).

[21] K. Alaimo, C. M. Olson, and E. A. Frongillo, Family food insufficiency, but not low family income, is positively associated with dysthymia and suicide symptoms in adolescents. *J Nutr* **132**, 719–725 (2002).

[22] P. E. Gold, and D. L. Korol, Making memories matter. *Front Integr Neurosci* **6**, 116 (2012).

[23] D. L. Korol, and P. E. Gold, Glucose, memory, and aging. *Am J Clin Nutr* **67**, 764S–771S (1998).

[24] C. A. Manning, W. S. Stone, D. L. Korol, and P. E. Gold, Glucose enhancement of 24-h memory retrieval in healthy elderly humans. *Behav Brain Res* **93**, 71–76 (1998).

[25] L. A. Tellez, W. Han, X. Zhang, *et al.*, Separate circuitries encode the hedonic and nutritional values of sugar. *Nat Neurosci* **19**, 465–470 (2016).

[26] G. Winocur, and C. E. Greenwood, Studies of the effects of high fat diets on cognitive function in a rat model. *Neurobiol Aging* **26**, 46–49 (2005).

[27] H. M. Francis, and R. J. Stevenson, Higher reported saturated fat and refined sugar intake is associated with reduced hippocampal-dependent memory and sensitivity to interoceptive signals. *Behav Neurosci* **125**, 943–955 (2011).

[28] I. Valladolid-Acebes, B. Merino, A. Principato, *et al.*, High-fat diets induce changes in hippocampal glutamate metabolism and neurotransmission. *Am J Physiol Endocrinol Metab* **302**, E396–402 (2012).

[29] H. Francis, and R. Stevenson, The longer-term impacts of western diet on human cognition and the brain. *Appetite* **63**, 119–128 (2013).

[30] M. R. Yeomans, Adverse effects of consuming high fat-sugar diets on cognition: implications for understanding obesity. *Proc Nutr Soc* **76**, 455–465 (2017).

[31] D. M. D. Tran, and R. F. Westbrook, A high-fat high-sugar diet-induced impairment in place-recognition memory is reversible and training-dependent. *Appetite* **110**, 61–71 (2017).

[32] H. Demmelmair, J. Øyen, T. Pickert, *et al.*, The effect of Atlantic salmon consumption on the cognitive performance of preschool children – A randomized controlled trial. *Clin Nutr* (2018).

[33] K. E. Assmann, M. Adjibade, S. Hercberg, P. Galan, and E. Kesse-Guyot, Unsaturated fatty acid intakes during midlife are positively associated with later cognitive function in older adults with modulating effects of antioxidant supplementation. *J Nutr* **148**, 1938–1945 (2018).
[34] C. Crawford, A. Edwards, C. Farquharson, *et al.*, "Magic Breakfast: Evaluation Report and Executive Summary," (Education Endowment Foundation, UK, 2016).
[35] J. E. Broman, L. G. Lundh, and J. Hetta, Insufficient sleep in the general population. *Neurophysiol Clin* **26**, 30–39 (1996).
[36] C. A. T. Peixoto, A. G. T. da Silva, M. A. Carskadon, and F. M. Louzada, Adolescents living in homes without electric lighting have earlier sleep times. *Behavioral Sleep Medicine* **7**, 73–80 (2009).
[37] C. R. C. Moreno, S. Vasconcelos, E. C. Marqueze, *et al.*, Sleep patterns in Amazon rubber tappers with and without electric light at home. *Sci Rep* **5**, 14074 (2015).
[38] P. F. Lima, A. L. Medeiros, and J. F. Araujo, Sleep-wake pattern of medical students: early versus late class starting time. *Braz J Med Biol Res* **35**, 1373–1377 (2002).
[39] CDC, Vital and health statistics. Health behaviors of Adults: United States, 2008–2010 (Centers for Disease Control and Prevention, 2013), vol. 257 (10).
[40] E. Van Cauter, and K. Spiegel, Sleep as a mediator of the relationship between socioeconomic status and health: a hypothesis. *Ann N Y Acad Sci* **896**, 254–261 (1999).
[41] D. W. Beebe, S. Simon, S. Summer, *et al.*, Dietary intake following experimentally restricted sleep in adolescents. *Sleep* **36**, 827–834 (2013).
[42] D. C. Jarrin, J. J. McGrath, and C. L. Drake, Beyond sleep duration: distinct sleep dimensions are associated with obesity in children and adolescents. *Int J Obes (Lond)* **37**, 552–558 (2013).
[43] K. Spiegel, R. Leproult, and E. Van Cauter, Impact of sleep debt on metabolic and endocrine function. *Lancet* **354**, 1435–1439 (1999).
[44] D. L. Bliwise, Historical change in the report of daytime fatigue. *Sleep* **19**, 462–464 (1996).
[45] M. M. Mitler, J. C. Miller, J. J. Lipsitz, J. K. Walsh, and C. D. Wylie, The sleep of long-haul truck drivers. *N Engl J Med* **337**, 755–761 (1997).
[46] M. T. Hyyppa, E. Kronholm, and E. Alanen, Quality of sleep during economic recession in Finland: a longitudinal cohort study. *Soc Sci Med* **45**, 731–738 (1997).
[47] T. D. Hill, A. M. Burdette, and L. Hale, Neighborhood disorder, sleep quality, and psychological distress: testing a model of structural amplification. *Health Place* **15**, 1006–1013 (2009).
[48] X. Liu, L. Liu, and R. Wang, Bed sharing, sleep habits, and sleep problems among Chinese school-aged children. *Sleep* **26**, 839–844 (2003).
[49] C. D. Solari, and R. D. Mare, Housing crowding effects on children's well-being. *Soc Sci Res* **41**, 464–476 (2012).

[50] E. C. Chambers, M. S. Pichardo, and E. Rosenbaum, Sleep and the housing and neighborhood environment of urban Latino adults living in low-income housing: the AHOME study. *Behav Sleep Med* **14**, 169–184 (2016).

[51] S. Arber, M. Bote, and R. Meadows, Gender and socio-economic patterning of self-reported sleep problems in Britain. *Soc Sci Med* **68**, 281–289 (2009).

[52] G. Simonelli, Y. Leanza, A. Boilard, *et al.*, Sleep and quality of life in urban poverty: the effect of a slum housing upgrading program. *Sleep* **36**, 1669–1676 (2013).

[53] M. G. Umlauf, A. C. Bolland, K. A. Bolland, S. Tomek, and J. M. Bolland, The effects of age, gender, hopelessness, and exposure to violence on sleep disorder symptoms and daytime sleepiness among adolescents in impoverished neighborhoods. *J Youth Adolesc* **44**, 518–542 (2015).

[54] M. L. Okun, M. Tolge, and M. Hall, Low socioeconomic status negatively affects sleep in pregnant women. *J Obstet Gynecol Neonatal Nurs* **43**, 160–167 (2014).

[55] K. A. Matthews, M. Hall, and R. E. Dahl, Sleep in healthy black and white adolescents. *Pediatrics* **133**, e1189–1196 (2014).

[56] K. Peltzer, and S. Pengpid, Depressive symptoms and social demographic, stress and health risk behaviour among university students in 26 low-, middle- and high-income countries. *Int J Psychiatry Clin Pract* **19**, 259–265 (2015).

[57] K. Peltzer, and S. Pengpid, Nocturnal sleep problems among university students from 26 countries. *Sleep Breath* **19**, 499–508 (2015).

[58] K. Peltzer, and S. Pengpid, Sleep duration and health correlates among university students in 26 countries. *Psychol Health Med* **21**, 208–220 (2016).

[59] D. C. Jarrin, J. J. McGrath, and E. C. Quon, Objective and subjective socioeconomic gradients exist for sleep in children and adolescents. *Health Psychol* **33**, 301–305 (2014).

[60] L. R. Teixeira, F. M. Fischer, R. Nagai, and S. L. Turte, Teen at work: the burden of a double shift on daily activities. *Chronobiol Int* **21**, 845–858 (2004).

[61] L. R. Teixeira, F. M. Fischer, M. M. de Andrade, F. M. Louzada, and R. Nagai, Sleep patterns of day-working, evening high-schooled adolescents of Sao Paulo, Brazil. *Chronobiol Int* **21**, 239–252 (2004).

[62] L. Hale, L. M. Berger, M. K. LeBourgeois, and J. Brooks-Gunn, Social and demographic predictors of preschoolers' bedtime routines. *J Dev Behav Pediatr* **30**, 394–402 (2009).

[63] S. Diekelmann, and J. Born, The memory function of sleep. *Nat Rev Neurosci* **11**, 114–126 (2010).

[64] B. A. Mander, S. Santhanam, J. M. Saletin, and M. P. Walker, Wake deterioration and sleep restoration of human learning. *Curr Biol* **21**, R183–184 (2011).

[65] R. Stickgold, and M. P. Walker, Memory consolidation and reconsolidation: what is the role of sleep? *Trends Neurosci* **28**, 408–415 (2005).

[66] S. Ribeiro, Sleep and plasticity. *Pflugers Arch* **463**, 111–120 (2012).

[67] S. A. Erath, K. M. Tu, J. A. Buckhalt, and M. El-Sheikh, Associations between children's intelligence and academic achievement: the role of sleep. *J Sleep Res* **24**, 510–513 (2015).

[68] N. Lemos, J. Weissheimer, and S. Ribeiro, Naps in school can enhance the duration of declarative memories learned by adolescents. *Front Syst Neurosci* **8**, 103 (2014).

[69] L. Kurdziel, K. Duclos, and R. M. C. Spencer, Sleep spindles in midday naps enhance learning in preschool children. *PNAS* **110**, 17267–17272 (2013).

[70] T. Cabral, N. B. Mota, L. Fraga, M. Copelli, M. A. McDaniel, and S. Ribeiro, Post-class naps boost declarative learning in a naturalistic school setting. *NPJ Sci Learn* **3**, 14 (2018).

[71] J. N. Cousins, K. F. Wong, B. L. Raghunath, C. Look, and M. W. L. Chee, The long-term memory benefits of a daytime nap compared with cramming. *Sleep* **42**, (2019).

[72] S. Ribeiro, and R. Stickgold, Sleep and school education. *Trends in Neuroscience and Education* **3**, (2014).

[73] N. Hankonen, M. T. Heino, E. Kujala, *et al.*, What explains the socio-economic status gap in activity? Educational differences in determinants of physical activity and screentime. *BMC Public Health* **17**, 144 (2017).

[74] S. W. Ng, and B. M. Popkin, Time use and physical activity: a shift away from movement across the globe. *Obes Rev* **13**, 659–680 (2012).

[75] C. Fiuza-Luces, N. Garatachea, N. A. Berger, and A. Lucia, Exercise is the real polypill. *Physiology (Bethesda)* **28**, 330–358 (2013).

[76] S. Vaynman, and F. Gomez-Pinilla, Revenge of the "sit": how lifestyle impacts neuronal and cognitive health through molecular systems that interface energy metabolism with neuronal plasticity. *J Neurosci Res* **84**, 699–715 (2006).

[77] A. Deslandes, C. Ferreira, H. Veiga, *et al.*, Exercise and mental health: many reasons to move. *Neuropsychobiology* **59**, 191–198 (2009).

[78] L. Chaddock, M. B. Pontifex, C. H. Hillman, and A. F. Kramer, A review of the relation of aerobic fitness and physical activity to brain structure and function in children. *J Int Neuropsychol Soc* **17**, 975–985 (2011).

[79] A. Diamond, and K. Lee, Interventions shown to aid executive function development in children 4 to 12 years old. *Science* **333**, 959–964 (2011).

[80] E. A. Haapala, A.-M. Poikkeus, K. Kukkonen-Harjula, *et al.*, Associations of physical activity and sedentary behavior with academic skills–a follow-up study among primary school children. *PLoS One* **9**, e107031 (2014).

[81] S. Masley, R. Roetzheim, and T. Gualtieri, Aerobic exercise enhances cognitive flexibility. *J Clin Psychol Med Settings* **16**, 186–193 (2009).

[82] A. Diamond, Executive functions. *Annu Rev Psychol* **64**, 135–168 (2013).

[83] J. M. Fuster, Frontal lobe and cognitive development. *J Neurocytol* **31**, 373–385 (2002).

[84] J. M. Halperin, and D. M. Healey, The influences of environmental enrichment, cognitive enhancement, and physical exercise on brain development:

can we alter the developmental trajectory of ADHD? *Neurosci Biobehav Rev* **35**, 621–634 (2011).

[85] S. R. Quartz, and T. J. Sejnowski, The neural basis of cognitive development: a constructivist manifesto. *Behav Brain Sci* **20**, 537–556; discussion 556–596 (1997).

[86] M. C. Diamond, Response of the brain to enrichment. *An Acad Bras Cienc* **73**, 211–220 (2001).

[87] H. van Praag, G. Kempermann, and F. H. Gage, Running increases cell proliferation and neurogenesis in the adult mouse dentate gyrus. *Nat Neurosci* **2**, 266–270 (1999).

[88] R. S. Duman, Neurotrophic factors and regulation of mood: role of exercise, diet and metabolism. *Neurobiol Aging* **26** Suppl 1, 88–93 (2005).

[89] A. C. Pereira, D. E. Huddleston, A. M. Brickman, *et al.*, An in vivo correlate of exercise-induced neurogenesis in the adult dentate gyrus. *Proc Natl Acad Sci USA* **104**, 5638–5643 (2007).

[90] H. van Praag, Exercise and the brain: something to chew on. *Trends Neurosci* **32**, 283–290 (2009).

[91] M. V. Lourenco, R. L. Frozza, G. B. de Freitas, *et al.*, Exercise-linked FNDC5/irisin rescues synaptic plasticity and memory defects in Alzheimer's models. *Nat Med* **25**, 165–175 (2019).

[92] M. Berchicci, M. B. Pontifex, E. S. Drollette, C. Pesce, C. H. Hillman, and F. Di Russoa, From cognitive motor preparation to visual processing: the benefits of childhood fitness to brain health. *Neuroscience* **298**, 211–219 (2015).

[93] M. B. Pontifex, L. B. Raine, C. R. Johnson, *et al.*, Cardiorespiratory fitness and the flexible modulation of cognitive control in preadolescent children. *J Cogn Neurosci* **23**, 1332–1345 (2011).

[94] M. W. Voss, L. Chaddock, J. S. Kim, *et al.*, Aerobic fitness is associated with greater efficiency of the network underlying cognitive control in preadolescent children. *Neuroscience* **199**, 166–176 (2011).

[95] L. Chaddock, K. I. Erickson, R. S. Prakash, *et al.*, Basal ganglia volume is associated with aerobic fitness in preadolescent children. *Dev Neurosci* **32**, 249–256 (2010).

[96] C. H. Hillman, K. I. Erickson, and A. F. Kramer, Be smart, exercise your heart: exercise effects on brain and cognition. *Nat Rev Neurosci* **9**, 58–65 (2008).

[97] I. M. van der Fels, S. C. Te Wierike, E. Hartman, M. T. Elferink-Gemser, J. Smith, and C. Visscher, The relationship between motor skills and cognitive skills in 4–16 year old typically developing children: a systematic review. *J Sci Med Sport* **18**, 697–703 (2015).

[98] C. H. Hillman, M. B. Pontifex, L. B. Raine, D. M. Castelli, E. E. Hall, and A. F. Kramer, The effect of acute treadmill walking on cognitive control and academic achievement in preadolescent children. *Neuroscience* **159**, 1044–1054 (2009).

[99] K. Byun, K. Hyodo, K. Suwabe, *et al.*, Positive effect of acute mild exercise on executive function via arousal-related prefrontal activations: an fNIRS study. *NeuroImage* **98**, 336–345 (2014).

[100] M. B. Pontifex, A. L. McGowan, M. C. Chandler, *et al.*, A primer on investigating the after effects of acute bouts of physical activity on cognition. *Psychol Sport Exerc* **40**, 1–22 (2019).

[101] M.-F. F. Mavilidi, A. Okely, P. Chandler, S. Louise Domazet, and F. Paas, Immediate and delayed effects of integrating physical activity into preschool children's learning of numeracy skills. *J Exp Child Psychol* **166**, 502–519 (2018).

[102] J. K. Ma, L. L. Mare, and B. J. Gurd, Classroom-based high-intensity interval activity improves off-task behaviour in primary school students. *Appl Physiol Nutr Metab* **39**, 1332–1337 (2014).

[103] B. Fenesi, Lucibello, K., Kim, J. A., and Heisz, J. J., Sweat So You don't forget: exercise breaks during a university lecture increase on-task attention and learning. *J Appl Res Mem Cogn* **7**, 261–269 (2018).

[104] T. M. DiLorenzo, E. P. Bargman, R. Stucky-Ropp, *et al.*, Long-term effects of aerobic exercise on psychological outcomes. *Prev Med* **28**, 75–85 (1999).

[105] J. A. Blumenthal, C. F. Emery, D. J. Madden, *et al.*, Long-term effects of exercise on psychological functioning in older men and women. *J Gerontol* **46**, P352–P361 (1991).

[106] J. R. Best, B. K. Chiu, C. L. Hsu, L. S. Nagamatsu, and T. Liu-Ambrose, Long-term effects of resistance exercise training on cognition and brain volume in older women: results from a randomized controlled trial. *J Int Neuropsychol Soc* **21**, 745–756 (2015).

[107] C. Vivar, and H. van Praag, Running changes the brain: the long and the short of it. *Physiology (Bethesda)* **32**, 410–424 (2017).

[108] M. K. Jedrziewski, D. C. Ewbank, H. Wang, and J. Q. Trojanowski, Exercise and cognition: results from the National Long Term Care Survey. *Alzheimers Dement* **6**, 448–455 (2010).

[109] A. L. Fedewa, and S. Ahn, The effects of physical activity and physical fitness on children's achievement and cognitive outcomes: a meta-analysis. *Res Q Exerc Sport* **82**, 521–535 (2011).

[110] C. Lees, and J. Hopkins, Effect of aerobic exercise on cognition, academic achievement, and psychosocial function in children: a systematic review of randomized control trials. *Prev Chronic Dis* **10**, E174 (2013).

[111] D. N. Ardoy, J. M. Fernández-Rodríguez, D. Jiménez-Pavón, R. Castillo, J. R. Ruiz, F. B. Ortega, A physical education trial improves adolescents' cognitive performance and academic achievement: the EDUFIT study. *Scand J Med Sci Sports* **24**, e52–61 (2014).

[112] C. L. Davis, P. D. Tomporowski, J. E. McDowell, *et al.*, Exercise improves executive function and achievement and alters brain activation in overweight children: a randomized, controlled trial. *Health Psychol* **30**, 91–98 (2011).

[113] E. B. Fontes, H. Bortolotti, K. Grandjean da Costa, *et al.*, Modulation of cortical and subcortical brain areas at low and high exercise intensities. *Br J Sports Med*, 1–7 (2019).

[114] M. C. Gallotta, G. P. Emerenziani, S. Iazzoni, M. Meucci, C. Baldari, and L. Guidetti, Impacts of coordinative training on normal weight and overweight/obese children's attentional performance. *Front Hum Neurosci* **9**, (2015).

[115] H. Budde, C. Voelcker-Rehage, S. Pietrabyk-Kendziorra, P. Ribeiro, and G. Tidow, Acute coordinative exercise improves attentional performance in adolescents. *Neurosci Lett* **441**, 219–223 (2008).

[116] C. L. Tsai, The effectiveness of exercise intervention on inhibitory control in children with developmental coordination disorder: using a visuospatial attention paradigm as a model. *Res Dev Disabil* **30**, 1268–1280 (2009).

[117] M. D. Chen, H. Y. Tsai, C. C. Wang, and Y. P. Wuang, The effectiveness of racket-sport intervention on visual perception and executive functions in children with mild intellectual disabilities and borderline intellectual functioning. *Neuropsychiatr Dis Treat* **11**, 2287–2297 (2015).

[118] D. Grissmer, K. J. Grimm, S. M. Aiyer, W. M. Murrah, and J. S. Steele, Fine motor skills and early comprehension of the world: two new school readiness indicators. *Dev Psychol* **46**, 1008–1017 (2010).

[119] V. R. Fernandes, M. L. Scipião Ribeiro, T. Melo, *et al.*, Motor coordination correlates with academic achievement and cognitive function in children. *Front Psychol* **7**, 318 (2016).

[120] J. E. Donnelly, C. H. Hillman, D. Castelli, *et al.*, Physical activity, fitness, cognitive function, and academic achievement in children: a systematic review. *Med Sci Sports Exerc* **48**, 1197–1222 (2016).

[121] C. Crova, I. Struzzolino, R. Marchetti, *et al.*, Cognitively challenging physical activity benefits executive function in overweight children. *J Sports Sci* **32**, 201–211 (2014).

[122] M. Schmidt, K. Jäger, F. Egger, C. M. Roebers, and A. Conzelmann, Cognitively engaging chronic physical activity, but not and aerobic exercise, affects executive functions in primary school children: a group-randomized controlled trial. *J Sport Exerc Psychol* **37**, 575–591 (2015).

[123] C. Pesce, I. Masci, R. Marchetti, S. Vazou, A. Sääkslahti, and P. D. Tomporowski, Deliberate play and preparation jointly benefit motor and cognitive development: mediated and moderated effects. *Front Psychol* **7**, (2016).

[124] A. S. Singh, E. Saliasi, V. van den Berg, *et al.*, Effects of physical activity interventions on cognitive and academic performance in children and adolescents: a novel combination of a systematic review and recommendations from an expert panel. *Br J Sports Med* **53**, 640–647 (2019).

[125] C. Álvarez-Bueno, C. Pesce, I. Cavero-Redondo, M. Sánchez-López, J. A. Martínez-Hortelano, and V. Martínez-Vizcaíno, The effect of physical

activity interventions on children's cognition and metacognition: a systematic review and meta-analysis. *J Am Acad Child Adolesc Psychiatry* **56**, 729–738 (2017).

[126] S. Vazou, C. Pesce, K. Lakes, and A. Smiley-Oyen, More than one road leads to Rome: a narrative review and meta-analysis of physical activity intervention effects on cognition in youth. *Int J Sport Exerc Psychol* **17**, 153–178 (2019).

[127] C. Nanthakumar, The benefits of yoga in children. *J Integr Med* **16**, 14–19 (2018).

[128] Y. Du, B. Liu, Y. Sun, L. G. Snetselaar, R. B. Wallace, and W. Bao, Trends in adherence to the physical activity guidelines for Americans for aerobic activity and time spent on sedentary behavior among US adults, 2007 to 2016. *JAMA Netw Open* **2**, e197597 (2019).

[129] H. W. Kohl, and H. D. Cook, *Educating the Student Body: Taking Physical Activity and Physical Education to School*. H. W. K. H. D. Cook, Ed., (National Academies Press, Washington, 2013).

[130] K. L. Piercy, R. P. Troiano, and R. M. Ballard, The physical activity guidelines for Americans. *JAMA* **320**, 2020–2028 (2018).

[131] L. M. Fadel, V. R. Ulbricht, C. R. Batista, and T. Vanzin, *Gamificação na Educação*. São Paulo (Pimenta Cultural, 2014).

[132] S. Deterding, Gamification: designing for motivation. *Interactions* **19**, 14–17 (2012).

[133] G. Zichermann, and C. Cunningham, Gamification by design : implementing game mechanics in web and mobile apps. (O'Reilly Media, Sebastopol, Calif., ed. 1st., 2011), xix, 182 p.

[134] S. Sepehr, and M. Head, in Proceedings of the First International Conference on Gameful Design, Research, and Applications. (2013), pp. 2–9.

[135] M. Csikszentmihalyi, *Flow: The Psychology of Optimal Experience*. (Harper & Row, New York, ed. 1st, 1990), xii, 303 p.

[136] J. Henrich, S. J. Heine, and A. Norenzayan, The weirdest people in the world? *Behav Brain Sci* **33**, 61–83; discussion 83–135 (2010).

[137] J. Henrich, S. J. Heine, and A. Norenzayan, Most people are not WEIRD. *Nature* **466**, 29 (2010).

[138] D. A. Wagner, and N. M. Castillo, Learning at the bottom of the pyramid: constraints, *Comparability and Policy in Developing Countries. Journal Articles (Literacy.org)*. **29**, (2014).

[139] A. P. Goldin, M. J. Hermida, D. E. Shalom, *et al.*, Far transfer to language and math of a short software-based gaming intervention. *Proc Natl Acad Sci USA* **111**, 6443–6448 (2014).

[140] D. Lomas, K. Patel, J. L. Forlizzi, and K. R. Koedinger, in SIGCHI Conference on Human Factors in Computing Systems. (2013), vol. CHI '13, pp. 89–98.

[141] M. Lopez-Rosenfeld, F. Carrillo, G. Garbulsky, D. Fernandez Slezak, and M. Sigman, Quantitative pedagogy: a digital two player game to examine communicative competence. *PLoS One* **10**, e0142579 (2015).

[142] A. Méndez, A. Martén, A. C. Pires, *et al.*, Temporal perception and delay aversion: a videogame screening tool for the early detection of ADHD. *Rev Argent Cienc Comport* **7**, 90–101 (2015).

[143] N. B. Mota, M. Copelli, and S. Ribeiro, Computational tracking of mental health in youth: Latin American contributions to a low-cost and effective solution for early psychiatric diagnosis. *New Dir Child Adolesc Dev* **152**, 59–69 (2016).

[144] D. Odic, J. V. Lisboa, R. Eisinger, M. G. Olivera, A. Maiche, and J. Halberda, Approximate number and approximate time discrimination each correlate with school math abilities in young children. *Acta Psychol (Amst)* **163**, 17–26 (2016).

[145] I. Magnisalis, S. Demetriadis, and A. Karakostas, Adaptive and intelligent systems for collaborative learning support: a review of the field. *IEEE Transactions on Learning Technologies* **4**, 5–20 (2011).

[146] E. Walker, N. Rummel, and K. R. Koedinger, Adaptive intelligent support to improve peer tutoring in algebra. *Int J Artif Intell Educ* **24**, 33–61 (2014).

[147] N. B. Mota, J. Weissheimer, B. Madruga, *et al.*, A naturalistic assessment of the organization of children's memories predicts cognitive functioning and reading ability. *Mind Brain Educ* **10**, 184–195 (2016).

[148] N. B. Mota, N. A. P. Vasconcelos, N. Lemos, *et al.*, Speech graphs provide a quantitative measure of thought disorder in psychosis. *PLoS One* **7**, e34928 (2012).

[149] N. B. Mota, R. Furtado, P. P. Maia, M. Copelli, and S. Ribeiro, Graph analysis of dream reports is especially informative about psychosis. *Sci Rep* **4**, 3691 (2014).

[150] N. B. Mota, M. Copelli, and S. Ribeiro, Thought disorder measured as random speech structure classifies negative symptoms and schizophrenia diagnosis 6 months in advance. *NPJ Schizophr* **3**, 1–10 (2017).

[151] N. B. Mota, M. Sigman, G. Cecchi, M. Copelli, and S. Ribeiro, The maturation of speech structure in psychosis is resistant to formal education. *NPJ Schizophr* **4**, 25 (2018).

[152] G. Bedi, F. Carrillo, G. Cecchi, *et al.*, Automated analysis of free speech predicts psychosis onset in high-risk youths. *NPJ Schizophr* **1**, 15030 (2015).

[153] G. Calogiuri, H. Nordtug, and A. Weydahl, The potential of using exercise in nature as an intervention to enhance exercise behavior: results from a pilot study. *Percept Mot Skills* **121**, 350–370 (2015).

[154] I. L. J. Lemmers-Jansen, A. J. Fett, S. S. Shergill, M. T. R. van Kesteren, and L. Krabbendam, Girls-Boys: an investigation of gender differences in the behavioral and neural mechanisms of trust and reciprocity in adolescence. *Front Hum Neurosci* **13**, 257 (2019).

[155] P. Freire, and D. P. Macedo, *Literacy: Reading the Word & the World.* (Bergin & Garvey Publishers, South Hadley, Massachusetts, 1987).

[156] N. Report, "Report of the National Reading Panel: Teaching children to read : an evidence-based assessment of the scientific research literature on

reading and its implications for reading instruction: Reports of the subgroups.," (National Institute of Child Health and Human Development, National Institutes of Health, Washington, D.C., 2000).
[157] CNESCO, paper presented at the Conférence de consensus: Lire, comprendre, apprendre, Lyon, ENS 16 et 17 mars 2016., 2016.
[158] S. Dehaene, *Les neurones de la lecture*. (Odile Jacob, Paris, 2007), 478. p.
[159] L. Cohen, S. Dehaene, L. Naccache, *et al.*, The visual word form area: spatial and temporal characterization of an initial stage of reading in normal subjects and posterior split-brain patients. *Brain* **123(Pt 2)**, 291–307 (2000).
[160] L. Cohen, S. Lehéricy, F. Chochon, C. Lemer, S. Rivaud, and S. Dehaene, Language-specific tuning of visual cortex? Functional properties of the Visual Word Form Area. *Brain* **125**, 1054–1069 (2002).
[161] S. Dehaene, H. G. Le Clec, J. B. Poline, D. Le Bihan, and L. Cohen, The visual word form area: a prelexical representation of visual words in the fusiform gyrus. *Neuroreport* **13**, 321–325 (2002).
[162] L. Cohen, and S. Dehaene, Specialization within the ventral stream: the case for the visual word form area. *NeuroImage* **22**, 466–476 (2004).
[163] R. Kolinsky, and A. Verhaeghe, How literacy affects vision: further data on the processing of mirror images by illiterate adults. *Neurociência da Linguagem* **7**, 52–65 (2011).
[164] R. Kolinsky, A. Verhaeghe, T. Fernandes, E. J. Mengarda, L. Grimm-Cabral, and J. Morais, Enantiomorphy through the looking glass: literacy effects on mirror-image discrimination. *J Exp Psychol Gen* **140**, 210–238 (2011).
[165] S. Dehaene, F. Pegado, L. W. Braga, *et al.*, How learning to read changes the cortical networks for vision and language. *Science* **330**, 1359–1364 (2010).
[166] F. Pegado, K. Nakamura, L. Cohen, and S. Dehaene, Breaking the symmetry: mirror discrimination for single letters but not for pictures in the visual word form area. *NeuroImage* **55**, 742–749 (2011).
[167] F. Pegado, K. Nakamura, L. W. Braga, *et al.*, Literacy breaks mirror invariance for visual stimuli: a behavioral study with adult illiterates. *J Exp Psychol Gen* **143**, 887–894 (2014).
[168] F. Pegado, K. Nakamura, and T. Hannagan, How does literacy break mirror invariance in the visual system? *Front Psychol* **5**, 703 (2014).
[169] S. Dehaene, K. Nakamura, A. Jobert, C. Kuroki, S. Ogawa, and L. Cohen, Why do children make mirror errors in reading? Neural correlates of mirror invariance in the visual word form area. *NeuroImage* **49**, 1837–1848 (2010).
[170] A. R. Torres, F. Pegado, and S. Ribeiro, in 4th House Symposium of the Brain Institute of the Federal University of Rio Grande do Norte (UFRN). (Brain Institute of UFRN, Natal, Brazil, 2018), pp. OR08.
[171] H. L. Roediger, and J. D. Karpicke, Test-enhanced learning: taking memory tests improves long-term retention. *Psychol Sci* **17**, 249–255 (2006).

[172] P. K. Agarwal, and H. L. Roediger 3rd, Lessons for learning: how cognitive psychology informs classroom practice. *Phi Delta Kappan* **100**, 8–12 (2018).

[173] A. C. Butler, J. D. Karpicke, and H. L. Roediger 3rd, Correcting a metacognitive error: feedback increases retention of low-confidence correct responses. *J Exp Psychol Learn Mem Cogn* **34**, 918–928 (2008).

[174] C. A. Rowland, and E. L. DeLosh, Mnemonic benefits of retrieval practice at short retention intervals. *Memory* **23**, 403–419 (2015).

[175] N. J. Cepeda, E. Vul, D. Rohrer, J. T. Wixted, and H. Pashler, Spacing effects in learning: a temporal ridgeline of optimal retention. *Psychol Sci* **19**, 1095–1102 (2008).

[176] L. K. Son, Spacing one's study: evidence for a metacognitive control strategy. *J Exp Psychol Learn Mem Cogn* **30**, 601–604 (2004).

[177] J. D. Karpicke, A. C. Butler, and H. L. Roediger, 3rd, Metacognitive strategies in student learning: do students practise retrieval when they study on their own? *Memory* **17**, 471–479 (2009).

[178] A. P. Yonelinas, The nature of recollection and familiarity: a review of 30 years of research. *J Mem Lang* **46**, 441–517 (2002).

[179] R. M. Montoya, R. S. Horton, J. L. Vevea, M. Citkowicz, and E. A. Lauber, A re-examination of the mere exposure effect: the influence of repeated exposure on recognition, familiarity, and liking. *Psychol Bull* **143**, 459–498 (2017).

[180] A. B. Blake, M. Nazarian, and A. D. Castel, The Apple of the mind's eye: everyday attention, metamemory, and reconstructive memory for the Apple logo. *Q J Exp Psychol (Hove)* **68**, 858–865 (2015).

[181] H. L. Roediger, and M. A. Pyc, Inexpensive techniques to improve education: applying cognitive psychology to enhance educational practice. *J Appl Res Mem Cogn* **1**, 242–248 (2012).

[182] J. Dunlosky, K. A. Rawson, E. J. Marsh, M. J. Nathan, and D. T. Willingham, Improving students' learning with effective learning techniques: promising directions from cognitive and educational psychology. *Psychol Sci Public Interest* **14**, 4–58 (2013).

[183] A. L. Putnam, V. W. Sungkhasettee, and H. L. Roediger, 3rd, Optimizing learning in college: tips from cognitive psychology. *Perspect Psychol Sci* **11**, 652–660 (2016).

[184] F. Ballarini, M. C. Martinez, M. Diaz Perez, D. Moncada, and H. Viola, Memory in elementary school children is improved by an unrelated novel experience. *PLoS One* **8**, e66875 (2013).

[185] T. Durgahee, Facilitating reflection: from a sage on stage to a guide on the side. *Nurse Educ Today* **18**, 158–164 (1998).

[186] E. Mazur, *Peer Instruction : A User's Manual*. Prentice Hall series in educational innovation (Prentice Hall, Upper Saddle River, N.J., 1997), xv, 253 p.

[187] J. T. Bruer, Education and the brain: a bridge too far. *Educ Res* **26**, 4–16 (1997).

[188] J. S. Bowers, The practical and principled problems with educational neuroscience. *Psychol Rev* **123**, 600–612 (2016).
[189] U. Goswami, Neuroscience and education. *Br J Educ Psychol* **74**, 1–14 (2004).
[190] U. Goswami, The foundations of psychological understanding. *Dev Sci* **9**, 545–550 (2006).
[191] U. Goswami, and P. Bryant, Childrens Social Development: Peer Interaction Classroom Learning. Primary Review Research Survey 2/1b (University of Cambridge Faculty of Education, Cambridge, 2007).
[192] J. B. Balardin, G. A. Zimeo Morais, R. A. Furucho, L. R. Trambaiolli, and J. R. Sato, Impact of communicative head movements on the quality of functional near-infrared spectroscopy signals: negligible effects for affirmative and negative gestures and consistent artifacts related to raising eyebrows. *J Biomed Opt* **22**, 046010 (2017).
[193] R. E. Vanderwert, and C. A. Nelson, The use of near-infrared spectroscopy in the study of typical and atypical development. *NeuroImage* **85** Pt 1, 264–271 (2014).
[194] Y. Minagawa-Kawai, K. Mori, J. C. Hebden, and E. Dupoux, Optical imaging of infants' neurocognitive development: recent advances and perspectives. *Dev Neurobiol* **68**, 712–728 (2008).
[195] N. Takeuchi, T. Mori, Y. Suzukamo, and S. Izumi, Activity of Prefrontal Cortex in Teachers and Students during Teaching of an Insight Problem. Mind, Brain, and Education (2019).
[196] X. Cui, D. M. Bryant, and A. L. Reiss, NIRS-based hyperscanning reveals increased interpersonal coherence in superior frontal cortex during cooperation. *NeuroImage* **59**, 2430–2437 (2012).
[197] A. M. Battro, C. I. Calero, A. P. Goldin, *et al.*, The cognitive neuroscience of the teacher–student interaction. *Mind Brain Educ* **7**, 177–181 (2013).
[198] D. C. Hyde, D. A. Boas, C. Blair, and S. Carey, Near-infrared spectroscopy shows right parietal specialization for number in pre-verbal infants. *NeuroImage* **53**, 647–652 (2010).
[199] L. A. Petitto, M. S. Berens, I. Kovelman, M. H. Dubins, K. Jasinska, and M. Shalinsky, The "Perceptual Wedge Hypothesis" as the basis for bilingual babies' phonetic processing advantage: new insights from fNIRS brain imaging. *Brain Lang* **121**, 130–143 (2012).
[200] Y. Moriguchi, and K. Hiraki, Prefrontal cortex and executive function in young children: a review of NIRS studies. *Front Hum Neurosci* **7**, 867 (2013).
[201] M. Yabe, S. Oshima, S. Eifuku, *et al.*, Effects of storytelling on the childhood brain: near-infrared spectroscopic comparison with the effects of picture-book reading. *Fukushima J Med Sci* **64**, 125–132 (2018).
[202] G. Brockington, J. B. Balardin, G. A. Zimeo Morais, *et al.*, From the laboratory to the classroom: the potential of functional near-infrared spectroscopy in educational neuroscience. *Front Psychol* **9**, 1840 (2018).

[203] L. S. Vygotsky, *The Collected Works of LS Vygotsky: Problems of the Theory and History of Psychology*. (Springer Science & Business Media, 1997).

[204] R. M. Carter, D. L. Bowling, C. Reeck, and S. A. Huettel, A distinct role of the temporal-parietal junction in predicting socially guided decisions. *Science* **337**, 109–111 (2012).

[205] M. Soltanlou, C. Artemenko, A.-C. Ehlis, *et al*., Reduction but no shift in brain activation after arithmetic learning in children: a simultaneous fNIRS-EEG study. *Sci Rep* **8**, 1707 (2018).

[206] G. J. Duncan , C. J. Dowsett, A. Claessens, *et al*., School readiness and later achievement. *Dev Psychol* **43**, 1428–1446 (2007).

[207] R. Steinmayr, M. Ziegler, and B. Trauble, Do intelligence and sustained attention interact in predicting academic achievement? *Learn Indiv Differ* **20**, 14–18 (2010).

[208] B. Bloem, R. B. Poorthuis, and H. D. Mansvelder, Cholinergic modulation of the medial prefrontal cortex: the role of nicotinic receptors in attention and regulation of neuronal activity. *Front Neural Circuits* **8**, 17 (2014).

[209] M. L. Vo, A. M. Jacobs, L. Kuchinke, *et al*., The coupling of emotion and cognition in the eye: introducing the pupil old/new effect. *Psychophysiology* **45**, 130–140 (2008).

[210] A. T. Duchowski, *Eye Tracking Methodology: Theory and Practice*. (Springer, London, 2nd ed., 2007), xxiv, 328 p.

Index